The sign-solvability of a linear system implies that the signs of the entries of the solution (or at least some of the entries) are determined solely on the basis of the signs of the coefficients of the system. That it might be worthwhile and possible to investigate such linear systems was recognized by Samuelson in his classic book *Foundations of Economic Analysis*. Sign-solvability is part of a larger study which seeks to understand the special circumstances under which an algebraic, analytic, or geometric property of a matrix can be determined from the combinatorial arrangement of the positive, negative, and zero elements of the matrix. These are thus properties shared by all members of a qualitative class of matrices. Several classes of matrices arise in this way, notably sign-nonsingular matrices, L-matrices, S-matrices, and sign-stable matrices. The essential idea of a sign-nonsingular matrix arose in a different context in the key 1963 paper "Dimer statistics and phase transitions" by P. W. Kastelyn.

The large and diffuse body of literature connected with sign-solvability is presented as a coherent whole for the first time in this book, displaying it as a beautiful interplay between combinatorics (especially graph theory) and linear algebra. Results in the literature are presented in a new and organized way with many new connections established and with many new results and proofs. One of the features of this book is that algorithms that are implicit in many of the proofs have been explicitly described and their complexity has been commented on.

T0291506

CAMBRIDGE TRACTS IN
MATHEMATICS

General Editors
B. BOLLOBAS, P. SARNAK, C. T. C. WALL

116   Matrices of sign-solvable linear systems

RICHARD A. BRUALDI
University of Wisconsin – Madison

BRYAN L. SHADER
University of Wyoming

Matrices of sign-solvable linear systems

CAMBRIDGE
UNIVERSITY PRESS

CAMBRIDGE UNIVERSITY PRESS
Cambridge, New York, Melbourne, Madrid, Cape Town, Singapore, São Paulo, Delhi

Cambridge University Press
The Edinburgh Building, Cambridge CB2 8RU, UK

Published in the United States of America by Cambridge University Press, New York

www.cambridge.org
Information on this title: www.cambridge.org/9780521105828

First published 1995
This digitally printed version 2009

A catalogue record for this publication is available from the British Library

Library of Congress Cataloguing in Publication data
Brualdi, Richard A.
Matrices of sign-solvable linear systems / Richard A. Brualdi and
Bryan L. Shader.
p.   cm. – (Cambridge tracts in mathematics; 116)
Includes bibliographical references and index.
ISBN 0-521-48296-8
1. Matrices   I. Shader, Bryan L.   II. Title.   III. Title: Sign-solvable
linear systems.   IV. Series.
QA188.B79   1995
512.9′434 – dc20   94–40931
                                                       CIP

ISBN 978-0-521-48296-7 hardback
ISBN 978-0-521-10582-8 paperback

To Mona (so unexpected and so special)
(from RAB)

To Chanyoung
(from BLS)

Contents

Preface

The possibility of writing this book occurred to us in the late fall of 1991 when we were both participating in the program on Applied Linear Algebra at the Institute for Mathematics and its Applications (IMA) in Minnesota. A few years earlier we had been attracted to the subject of sign-solvability because of the beautiful interplay it afforded among linear algebra, combinatorics, and theoretical computer science (combinatorial algorithms). The subject, begun in 1947 by the economist P. Samuelson, was developed from various perspectives in the linear algebra, combinatorics, and economics literature. We thought that it would be a worthwhile project to organize the subject and to give a unified and self-contained presentation. Because there were no previous books or even survey papers on the subject, the tasks of deciding what was fundamental and how the material should be ordered for exposition had to be thought out very carefully. Our organization of the material has resulted in new connections among various results in the literature. In addition, many new results and many new and simpler proofs of previously established results are given throughout the book. We began the book in earnest in early 1992 and completed approximately three quarters of it while in residence at the IMA. After we returned to our home institutions, with the other duties that that entails, it was difficult to find the time for completing the book.

One of the features of this book is that we have explicitly described algorithms that are implicit in many of the proofs and have commented on their complexity. Throughout we have given credit for results that have previously occurred in the literature. There is a bibliography at the end of each chapter as well as a master bibliography (including some papers not cited in the text) at the end of the book.

That it might be worthwhile to investigate systems of linear equations for which the signs of the solution could be determined knowing only the signs of its coefficients was recognized by Samuelson in his book *Foundations of Economic Analysis*. The mathematical study of sign-solvability, in particular of sign-nonsingular matrices, was begun by L. Bassett, J. Maybee and J. Quirk in their paper "Qualitative economics and the scope of the correspondence principle" in 1968. Since the appearance in 1984 of the paper "Signsolvability revisited" by V. Klee, R. Ladner, and R. Manber, there has been renewed interest in the subject. Indeed we were first attracted to sign-solvability and related topics by this paper. The essential idea of a sign-nonsingular matrix arose in a different context in the 1963 paper "Dimer statistics and phase transitions" by P.W. Kastelyn. A key paper in the development that proceeded from Kastelyn's work is the 1975 paper "A characterization of convertible (0, 1)-matrices" by C.H.C. Little. The connection between the two different points of view was made in RAB's 1988 paper "Counting permutations with restricted positions: Permanents of (0, 1)-matrices. A tale in four parts."

We wish to thank the IMA for providing a stimulating environment in which to work during 1991–1992, the financial support given to RAB and the post-doctoral fellowship awarded to BLS. We are grateful to Victor Klee for the encouragement he has given us in completing this project. During the period this book was written, RAB was also partially supported by NSF Grant No. DMS-9123318.

Richard A. Brualdi
Bryan L. Shader

1

Sign-solvability

1.1 A problem in economics

Qualitative economics is usually considered to have originated with the work of Samuelson [11, Chapter III] who discussed the possibility of determining unambiguously the qualitative behavior of solution values of a system of equations. In his pioneering paper [6] (see also [4] and [7, 8]) Lancaster put it this way: *Economists believed for a very long time, and most economists would still hope it to be so, that a considerable body of sensible economic propositions could be expressed in a qualitative way, that is, in a form in which the algebraic sign of some effect is predicted from a knowledge of the signs, only, of the relevant structural parameters of the system.*

Consider the following example, similar to one discussed in Samuelson [11], of a market for a product, say bananas, where the price and quantity are determined by the intersection of its supply and demand curves. We introduce a *shift* parameter α into the demand curve, and assume that an increase in α shifts the demand curve upward and to the right. For instance, α might represent people's taste for bananas, and as people's taste for bananas increases so does their demand for bananas. Let $S(p)$ denote the number x of bananas that farmers will produce if the price per banana is p. Simple economic principles tell us that as the price p increases farmers will supply more bananas. This gives a supply curve as indicated in Figure 1.1.

Let $D(p, \alpha)$ denote the number x of bananas that consumers will demand if the price per banana is p and people's taste for bananas is α. Again simple economic principles tell us that for α fixed, the demand for bananas decreases as the price p increases. For p fixed, as people's taste α for bananas increases so does their demand for bananas. This gives a family of demand curves as indicated in Figure 1.2.

1

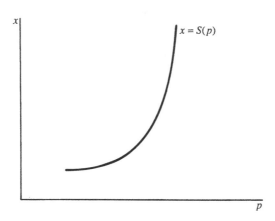

Fig. 1.1.

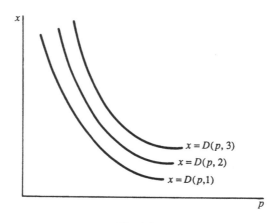

Fig. 1.2.

The equilibrium equations, where supply equals demand, are

$$
\begin{aligned}
S(p) - x &= 0 \\
D(p, \alpha) - x &= 0.
\end{aligned}
\tag{1.1}
$$

The equilibrium points (p_α, x_α) are pictured in Figure 1.3. This figure suggests that as α increases so do p_α and x_α. That this is indeed the case can be justified mathematically as follows.

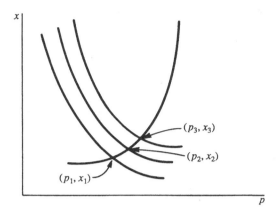

Fig. 1.3.

Taking partial derivatives with respect to α in (1.1), we obtain

$$\frac{\partial S}{\partial p}\frac{\partial p}{\partial \alpha} - \frac{\partial x}{\partial \alpha} = 0$$

$$\frac{\partial D}{\partial p}\frac{\partial p}{\partial \alpha} + \frac{\partial D}{\partial \alpha} - \frac{\partial x}{\partial \alpha} = 0.$$

(1.2)

Equivalently,

$$\begin{bmatrix} \frac{\partial S}{\partial p} & -1 \\ \frac{\partial D}{\partial p} & -1 \end{bmatrix} \begin{bmatrix} \frac{\partial p}{\partial \alpha} \\ \frac{\partial x}{\partial \alpha} \end{bmatrix} = \begin{bmatrix} 0 \\ -\frac{\partial D}{\partial \alpha} \end{bmatrix}.$$

(1.3)

The conclusions just given, which were based on simple economic principles, are equivalent to

$$\frac{\partial S}{\partial p} > 0, \quad \frac{\partial D}{\partial p} < 0, \quad \text{and} \quad \frac{\partial D}{\partial \alpha} > 0.$$

(1.4)

In (1.3) we replace by their signs those quantities whose signs are determined by (1.4) and obtain

$$\begin{bmatrix} + & - \\ - & - \end{bmatrix} \begin{bmatrix} \frac{\partial p}{\partial \alpha} \\ \frac{\partial x}{\partial \alpha} \end{bmatrix} = \begin{bmatrix} 0 \\ - \end{bmatrix}.$$

(1.5)

Every matrix with the same sign pattern as the 2 by 2 matrix in (1.5) has a negative determinant and hence is invertible. It follows by inspection (or use Cramer's rule) that

$$\frac{\partial p}{\partial \alpha} > 0 \text{ and } \frac{\partial x}{\partial \alpha} > 0$$

independent of the magnitudes of the quantities in (1.4). We conclude that p_α and x_α are increasing functions of α.

The preceding example is a special case of a more general situation. Let $x_1, x_2, \ldots, x_n$ and α be $n+1$ variables satisfying the n functional relationships

$$f_i(x_1, \ldots, x_n, \alpha) = 0 \quad (i = 1, 2, \ldots, n). \tag{1.6}$$

If the directions of the rates of change of the f_i with respect to the x_j and α are known, can we determine the direction of the rates of change of the x_j with respect to α? Taking partial derivatives of the f_i with respect to α we obtain the linear system

$$A \begin{bmatrix} u_1 \\ \vdots \\ u_n \end{bmatrix} = \begin{bmatrix} c_1 \\ \vdots \\ c_n \end{bmatrix} \tag{1.7}$$

where $A = [a_{ij}]$ is the matrix of order n with

$$a_{ij} = \frac{\partial f_i}{\partial x_j} \quad (i = 1, 2, \ldots, n; \, j = 1, 2, \ldots, n),$$

and

$$u_j = \frac{\partial x_j}{\partial \alpha} \quad (j = 1, 2, \ldots, n)$$

$$c_i = -\frac{\partial f_i}{\partial \alpha} \quad (i = 1, 2, \ldots, n).$$

Our question is equivalent to: Can we solve for the signs of the u_j knowing only the signs of the a_{ij} and the c_i? This is the origin of the study of sign-solvable linear systems that we discuss in more detail in the next section.

1.2 Sign-solvable linear systems

We define the *sign* of a real number a by

$$\operatorname{sign} a = \begin{cases} +1 & \text{if } a > 0 \\ 0 & \text{if } a = 0 \\ -1 & \text{if } a < 0. \end{cases}$$

The *sign pattern* of a real matrix A is the $(0, 1, -1)$-matrix obtained from A by replacing each entry by its sign. A real matrix A determines a *qualitative class* $\mathcal{Q}(A)$ consisting of all matrices with the same sign pattern as A. The *zero pattern* of A is the $(0,1)$-matrix obtained from A by replacing each nonzero entry by 1. A *nonnegative matrix* is a matrix whose sign pattern is a $(0, 1)$-matrix. A *positive matrix* is a matrix whose sign pattern contains only 1's. We denote a matrix each of whose entries equals 1 by J. If the matrix is m by n, then we also write $J_{m,n}$, and this is abbreviated to J_n if $m = n$.

Consider a system of m equations in n unknowns given by

$$Ax = b \tag{1.8}$$

where

$$A = \begin{bmatrix} a_{11} & a_{12} & \cdots & a_{1n} \\ a_{21} & a_{22} & \cdots & a_{2n} \\ \vdots & \vdots & \ddots & \vdots \\ a_{m1} & a_{m2} & \cdots & a_{mn} \end{bmatrix}$$

and

$$b = \begin{bmatrix} b_1 \\ b_2 \\ \vdots \\ b_m \end{bmatrix}$$

are real matrices. The linear system (1.8) is sign-solvable provided we can solve for the signs of the entries of x knowing only the signs of the entries of A and of b. More precisely, (1.8) is *sign-solvable* provided that for each matrix $\widetilde{A}$ in the qualitative class $\mathcal{Q}(A)$ and for each matrix $\tilde{b}$ in the qualitative class $\mathcal{Q}(b)$,

$$\widetilde{A}x = \tilde{b}$$

is solvable and

$$\{\tilde{x} : \text{ there exists } \widetilde{A} \in \mathcal{Q}(A) \text{ and } \tilde{b} \in \mathcal{Q}(b) \text{ with } \widetilde{A}\tilde{x} = \tilde{b}\} \tag{1.9}$$

is entirely contained in one qualitative class. If $Ax = b$ is sign-solvable, then (1.9) is called the *qualitative solution class* of $Ax = b$ and is denoted by $\mathcal{Q}(Ax = b)$. Suppose z satisfies $Az = b$ and w is in $\mathcal{Q}(z)$. There exists a nonnegative invertible diagonal matrix D such that $w = Dz$. Then w satisfies $(AD^{-1})w = b$ and AD^{-1} is in $\mathcal{Q}(A)$. It follows that if $Ax = b$ is sign-solvable, then $\mathcal{Q}(Ax = b)$ is the entire qualitative class $\mathcal{Q}(z)$. We also note that if $Ax = b$ is sign-solvable and E is an invertible diagonal matrix of order n, then $(AE)x = b$ is sign-solvable and $\mathcal{Q}((AE)x = b) = \mathcal{Q}(E^{-1}z)$. In particular, by taking $E = -I_n$ we see that if $Ax = b$ is sign-solvable, then so is $Ax = -b$. The linear system

$$\begin{bmatrix} 1 & -1 \\ -1 & -1 \end{bmatrix} \begin{bmatrix} x_1 \\ x_2 \end{bmatrix} = \begin{bmatrix} 0 \\ -1 \end{bmatrix}$$

from (1.5) is an example of a sign-solvable linear system.

Let A be an m by n matrix and let B be an m by p matrix. Then $AX = B$ is *sign-solvable* provided that for each $\widetilde{A}$ in $\mathcal{Q}(A)$ and each $\widetilde{B}$ in $\mathcal{Q}(B)$ there is an n by p matrix $\widetilde{X}$ such that $\widetilde{A}\widetilde{X} = \widetilde{B}$ and

$$\{\widetilde{X}: \text{ there exists } \widetilde{A} \in \mathcal{Q}(A) \text{ and } \widetilde{B} \in \mathcal{Q}(B) \text{ with } \widetilde{A}\widetilde{X} = \widetilde{B}\}$$

is contained in one qualitative class. Thus the sign-solvability of $AX = B$ is equivalent to the sign-solvability of $Ax = b$ for every column b of B.

Theorem 1.2.1 *The homogeneous linear system $Ax = 0$ is sign-solvable if and only if every matrix in the qualitative class $\mathcal{Q}(A)$ has linearly independent columns.*

Proof First suppose that every matrix $\widetilde{A}$ in $\mathcal{Q}(A)$ has linearly independent columns. Then the only solution to $\widetilde{A}x = 0$ is the trivial solution. Thus $Ax = 0$ is sign-solvable and $\mathcal{Q}(Ax = 0) = \{0\}$. Now suppose that $Ax = 0$ is sign-solvable. Since 0 is a solution of $Ax = 0$, we conclude that $\mathcal{Q}(Ax = 0) = \{0\}$ and hence that every matrix $\widetilde{A}$ in $\mathcal{Q}(A)$ has linearly independent columns.   □

A matrix is an *L-matrix* provided every matrix in its qualitative class has linearly independent rows. The number of rows in an L-matrix does not exceed the number of its columns. By Theorem 1.2.1, *the homogeneous linear system $Ax = 0$ is sign-solvable if and only if the transpose A^T of A is an L-matrix.*[1]

Corollary 1.2.2 *If the linear system $Ax = b$ is sign-solvable, then A^T is an L-matrix.*

Proof Suppose that $Ax = b$ is sign-solvable but A^T is not an L-matrix. By Theorem 1.2.1, there exist $\widetilde{A}$ in $\mathcal{Q}(A)$ and $z \neq 0$ such that $\widetilde{A}z = 0$. Let $\widetilde{x}$ be a solution of $\widetilde{A}x = b$. Then $\widetilde{A}(\widetilde{x} + cz) = b$ for all real numbers c. We may choose c so that the sign patterns of $\widetilde{x} + cz$ and $\widetilde{x}$ are different and thus contradict the sign-solvability of $Ax = b$.   □

As an example we show that the matrix

$$A = \begin{bmatrix} 1 & 1 & 1 & -1 \\ 1 & 1 & -1 & 1 \\ 1 & -1 & 1 & 1 \end{bmatrix} \tag{1.10}$$

[1] In the literature an L-matrix is often defined to be a matrix for which every matrix in its qualitative class has linearly independent *columns*. This is convenient for discussing the connections between L-matrices and sign-solvable systems, but we have found it more convenient in the general study of L-matrices to require, as we have, linearly independent rows.

is an L-matrix. Let $\widetilde{A}$ be a matrix in $Q(A)$. The sign pattern of $\widetilde{A}$ implies that no row of $\widetilde{A}$ is a multiple of another row. Every 3 by 1 sign pattern of 1's and -1's is the sign pattern of some column of $\widetilde{A}$ or its negative. It follows that no nontrivial linear combination of the rows of $\widetilde{A}$ equals zero, and hence the rows of $\widetilde{A}$ are linearly independent.

If A is an invertible diagonal matrix, then $Ax = b$ is a sign-solvable linear system. We now show that the sign-solvability of $Ax = b$ for all b implies that the rows of A can be permuted to obtain an invertible diagonal matrix.

Theorem 1.2.3 *The linear system $Ax = b$ is sign-solvable for all b if and only if A is a square matrix and there exists a permutation matrix P such that PA is an invertible diagonal matrix.*

Proof Suppose that $Ax = b$ is sign-solvable for all b. Then the linear system $Ax = b$ has a unique solution for all b. Hence A is an invertible square matrix and $A^{-1}b$ is the unique solution. Assume that some row of A^{-1}, say the first row, contains nonzero entries c and d in columns j and k, respectively, where $j \neq k$. Choose b so that its entry in row j has the same sign as c, its entry in row k has the same sign as $-d$, and its remaining entries are 0. Then there exist $\check{b}$ and $\hat{b}$ in $Q(b)$ such that the first entry of $A^{-1}\check{b}$ is positive and that of $A^{-1}\hat{b}$ is negative, contradicting the sign-solvability of $Ax = b$. Hence each row of A^{-1} contains a unique nonzero entry, implying that $A^{-1}Q$ is a diagonal matrix for some permutation matrix Q. Therefore $Q^{-1}A$ is an invertible diagonal matrix. The converse is obvious. $\qquad\square$

If A is an L-matrix, then A^T is an L-matrix if and only if A is square. A square L-matrix is also called a *sign-nonsingular matrix*, abbreviated SNS-matrix. In the next theorem we use the determinant in order to obtain characterizations of SNS-matrices. These characterizations were already evident in the early papers [11, 6, 1]. A square matrix A has a *signed determinant* provided the determinants of the matrices in $Q(A)$ all have the same sign. We recall that the *standard determinant expansion* of a matrix $A = [a_{ij}]$ of order n is

$$\det A = \sum_{\sigma} \operatorname{sgn}(\sigma) a_{1i_1} a_{2i_2} \cdots a_{ni_n} \qquad (1.11)$$

where the summation extends over all permutations $\sigma = (i_1, i_2, \ldots, i_n)$ of $\{1, 2, \ldots, n\}$ and $\operatorname{sgn}(\sigma)$ denotes the sign of the permutation σ.

Lemma 1.2.4 *Let $A = [a_{ij}]$ be a matrix of order n. Then A has a signed determinant if and only if one of the following holds:*

  (i) *Every term in the standard determinant expansion of A is zero.*

(ii) *There is a nonzero term in the standard determinant expansion of A and every such term has the same sign.*

Proof If either (i) or (ii) holds, then clearly A has a signed determinant. Now assume that A has a signed determinant and that (i) does not hold. Let $\sigma = (i_1, i_2, \ldots, i_n)$ be a permutation for which the corresponding term

$$t_\sigma = \mathrm{sgn}(\sigma) a_{1i_1} a_{2i_2} \cdots a_{ni_n}$$

in the determinant expansion of A is not zero. The sign of the determinant of the matrix in $Q(A)$ obtained by multiplying each entry of A not occurring in t_σ by a positive number ϵ is the same as the sign of t_σ for ϵ sufficiently small. Since A has a signed determinant, (ii) holds.   □

Theorem 1.2.5 *Let $A = [a_{ij}]$ be a matrix of order n. Then the following are equivalent:*

(i) *A is an SNS-matrix.*

(ii) *$\det A \neq 0$ and A has a signed determinant.*

(iii) *There is a nonzero term in the standard determinant expansion of A and every nonzero term has the same sign.*

Proof The equivalence of (i) and (ii) is a consequence of the facts that $Q(A)$ is a connected set and the determinant is a continuous function. The equivalence of (ii) and (iii) is a consequence of Lemma 1.2.4.   □

Let n be an integer with $n \geq 2$, and let

$$H_n = \begin{bmatrix} -1 & 1 & 0 & \cdots & 0 & 0 \\ -1 & -1 & 1 & \cdots & 0 & 0 \\ -1 & -1 & -1 & \cdots & 0 & 0 \\ \vdots & \vdots & \vdots & \ddots & \vdots & \vdots \\ -1 & -1 & -1 & \cdots & -1 & 1 \\ -1 & -1 & -1 & \cdots & -1 & -1 \end{bmatrix} \qquad (1.12)$$

be the lower Hessenberg matrix with -1's on and below the main diagonal and 1's on the superdiagonal. It follows by induction on n and the Laplace expansion of the determinant by the first row that each nonzero term in the standard determinant expansion of H_n is $(-1)^n$. Hence by Theorem 1.2.5, H_n is an SNS-matrix for all $n \geq 2$.   □

Let $A = [a_{ij}]$ be a matrix of order n. Then A has an *identically zero determinant* provided each of the $n!$ terms in the standard determinant expansion is 0.

It follows from Lemma 1.2.4 that A has an identically zero determinant if and only if the determinant of each matrix in $\mathcal{Q}(A)$ is zero. The Frobenius–König theorem [2] asserts that there is a nonzero term in the standard determinant expansion of A if and only if A does not have a p by q zero submatrix for any positive integers p and q with $p + q = n + 1$. Thus A has an identically zero determinant if and only if there exist positive integers p and q with $p + q = n + 1$ such that A has a p by q zero submatrix. Theorem 1.2.5 implies that any matrix which is obtained from an SNS-matrix by replacing some of its nonzero entries with zeros is either an SNS-matrix or has an identically zero determinant. It also implies that if A is an SNS-matrix and $a_{ij} \neq 0$, then the matrix of order $n - 1$ obtained from A by deleting row i and column j is either an SNS-matrix or has an identically zero determinant. More generally, if A is an SNS-matrix and B is a square submatrix of A which does not have an identically zero determinant, then the matrix obtained from A by deleting the rows and columns which intersect B either is an SNS-matrix or has an identically zero determinant.

If C is an m by n matrix and u is an m by 1 column vector, then $C(i \leftarrow u)$ denotes the matrix obtained from C by replacing its ith column by u ($i = 1, 2, \ldots, n$).

Let A be a nonsingular matrix of order n and let b be an n by 1 column vector. Then Cramer's rule asserts that the unique solution $x = (x_1, x_2, \ldots, x_n)^T$ of $Ax = b$ satisfies

$$x_i = \frac{\det A(i \leftarrow b)}{\det A} \quad (i = 1, 2, \ldots, n).$$

In particular, the (i, j)-entry of A^{-1} equals

$$\frac{\det A(i \leftarrow e_j)}{\det A} = (-1)^{i+j} \frac{\det A_{j,i}}{\det A} \quad (i, j = 1, 2, \ldots, n)$$

where e_j is the n by 1 column vector whose only nonzero entry is a 1 in the jth position and $A_{j,i}$ is the matrix obtained by deleting row j and column i of A. We now obtain a Cramer-type rule for sign-solvable systems.

Theorem 1.2.6 *Let A be a matrix of order n and let b be an n by 1 column vector. Then $Ax = b$ is sign-solvable if and only if A is an SNS-matrix and each of the matrices $A(i \leftarrow b)$ is either an SNS-matrix or has an identically zero determinant.*

Proof First assume that $Ax = b$ is sign-solvable. By Corollary 1.2.2, A is an SNS-matrix. Let $\widetilde{A}$ and $\widetilde{b}$ be arbitrary matrices in $\mathcal{Q}(A)$ and $\mathcal{Q}(b)$, respectively.

By Cramer's rule, $\widetilde{A}x = \tilde{b}$ has a unique solution $\tilde{x} = (\tilde{x}_1, \tilde{x}_2, \ldots, \tilde{x}_n)^T$ where

$$\tilde{x}_i = \frac{\det \widetilde{A}(i \leftarrow \tilde{b})}{\det \widetilde{A}}.$$

By Theorem 1.2.5, $\det \widetilde{A}$ and $\det A$ have the same sign. Hence, for each $i = 1, 2, \ldots, n$, the sign of $\det \widetilde{A}(i \leftarrow \tilde{b})$ is the same as the sign of $\det A(i \leftarrow b)$ for all $\widetilde{A}$ in $Q(A)$ and all $\tilde{b}$ in $Q(b)$. It follows that for each i, either $A(i \leftarrow b)$ has an identically zero determinant, or by Theorem 1.2.5 that $A(i \leftarrow b)$ is an SNS-matrix. The converse is an immediate consequence of Cramer's rule.   □

Corollary 1.2.7 *If* $Ax = b$ *is a sign-solvable linear system where* A *is a square matrix and* c *is obtained from* b *by replacing some of its nonzero entries with zeros, then* $Ax = c$ *is sign-solvable.*

Recall that a square matrix A is an SNS-matrix if and only if $\widetilde{A}^{-1}$ exists for all $\widetilde{A}$ in $Q(A)$. If A is an SNS-matrix, then the matrices $\widetilde{A}^{-1}$ need not have the same sign pattern. For example, the matrices

$$\begin{bmatrix} -1 & 1 & 0 \\ -1 & -1 & 1 \\ -1 & -1 & -1 \end{bmatrix} \text{ and } \begin{bmatrix} -1 & 1 & 0 \\ -1 & -1 & 1 \\ -2 & -1 & -1 \end{bmatrix}$$

are SNS-matrices with the same sign pattern, but the signs of the entries in the $(3,1)$ position of their inverses are different. An SNS-matrix A such that the matrices in $\{\widetilde{A}^{-1} : \widetilde{A} \in Q(A)\}$ have the same sign pattern is called a *strong SNS-matrix*, which we abbreviate to S^2NS-matrix. It follows that a matrix A of order n is an S^2NS-matrix if and only if $AX = I_n$ is sign-solvable where I_n is the identity matrix of order n. Any matrix of the form $A = PD$ where P is a permutation matrix and D is an invertible diagonal matrix is an example of an S^2NS-matrix.

The following corollary is an immediate consequence of Cramer's rule.

Corollary 1.2.8 *Let* $A = [a_{ij}]$ *be a matrix of order n. Then* A *is an* S^2NS-*matrix if and only if*

(i) *A is an SNS-matrix, and*

(ii) *for each i and j with $a_{ij} = 0$, the submatrix $A_{i,j}$ of A of order $n - 1$ obtained by deleting row i and column j is either an SNS-matrix or has an identically zero determinant.*

Let n be an integer with $n \geq 2$ and let G_n be the lower Hessenberg matrix

$$\begin{bmatrix} 1 & -1 & 0 & \cdots & 0 & 0 \\ 0 & 1 & -1 & \cdots & 0 & 0 \\ 0 & 0 & 1 & \cdots & 0 & 0 \\ \vdots & \vdots & \vdots & \ddots & \vdots & \vdots \\ 0 & 0 & 0 & \cdots & 1 & -1 \\ 1 & 0 & 0 & \cdots & 0 & 1 \end{bmatrix} \tag{1.13}$$

with 1's on the main diagonal, -1's on the superdiagonal, and a 1 in the lower left corner. Both of the nonzero terms in the standard determinant expansion of G_n are positive and hence G_n is an SNS-matrix. Each of the submatrices of G_n of order $n-1$ is an SNS-matrix since each has exactly one nonzero term in its standard determinant expansion. Hence, by Corollary 1.2.8, G_n is an S^2NS-matrix. Let F_n be the matrix of order n with 1's on and above the main diagonal and -1's below the main diagonal. Then $G_n^{-1} = (1/2)F_n$, and hence

$$\{\widetilde{G}_n^{-1} : \widetilde{G}_n \in \mathcal{Q}(G_n)\} \subseteq \mathcal{Q}(F_n). \tag{1.14}$$

If $n \geq 3$, the matrix F_n is not an SNS-matrix and, in particular, (1.14) is a proper containment.

Corollary 1.2.9 *If $AX = B$ is a sign-solvable linear system where A and B are square matrices of order n and B does not have an identically zero determinant, then A is an S^2NS-matrix.*

Proof Since B does not have an identically zero determinant, it follows from Corollary 1.2.7 that there exist an invertible diagonal matrix D and a permutation matrix P such that $AX = PD$ is sign-solvable. Thus $D^{-1}P^{-1}AX = I_n$ is sign-solvable. We conclude that $D^{-1}P^{-1}A$, and hence A, is an S^2NS-matrix. $\square$

Let A be a square matrix of order n and let b be an n by 1 column vector. It follows from Theorem 1.2.6 and its proof that $Ax = b$ is sign-solvable and the vectors in its qualitative solution class have no zero coordinates if and only if A and each of the matrices $A(i \leftarrow b)$ is an SNS-matrix. An n by $n+1$ matrix B is called an S^*-*matrix* provided that each of the $n+1$ matrices obtained by deleting a column of B is an SNS-matrix. Clearly every S^*-matrix is an L-matrix. The linear system $Ax = b$ is sign-solvable and the vectors in its qualitative solution class have no zero coordinates if and only if the matrix $\begin{bmatrix} A & -b \end{bmatrix}$ obtained from A by appending $-b$ as a last column is an S^*-matrix.

Let

$$A = \begin{bmatrix} 1 & -1 & 0 \\ 1 & 1 & -1 \\ 1 & 1 & 1 \end{bmatrix} \text{ and } b = \begin{bmatrix} 0 \\ 0 \\ 1 \end{bmatrix}.$$

Then $Ax = b$ is a sign-solvable linear system with $\mathcal{Q}(Ax = b) = \mathcal{Q}((1, 1, 1)^T)$. Thus the matrix

$$\begin{bmatrix} A & -b \end{bmatrix} = \begin{bmatrix} 1 & -1 & 0 & 0 \\ 1 & 1 & -1 & 0 \\ 1 & 1 & 1 & -1 \end{bmatrix}$$

is an S^*-matrix.

The following characterization of S^*-matrices is an immediate consequence of Cramer's rule.

Corollary 1.2.10 *Let B be an n by $n + 1$ matrix. Then B is an S^*-matrix if and only if there exists a vector w with no zero coordinates such that the right null spaces of the matrices $\widetilde{B}$ in $\mathcal{Q}(B)$ are contained in $\{0\} \cup \mathcal{Q}(w) \cup \mathcal{Q}(-w)$.*

An n by $n + 1$ matrix B is an *S-matrix* provided it is an S^*-matrix and the right null space of B contains a vector all of whose coordinates are positive. If B is an S^*-matrix and D is an invertible diagonal matrix, then BD is also an S^*-matrix. Let B be an n by $n + 1$ S^*-matrix and let $w = (w_1, w_2, \ldots, w_{n+1})$ be the sign pattern of a nonzero vector in the right null space of B. Let D be the diagonal matrix of order $n + 1$ with diagonal entries $w_1, w_2, \ldots, w_{n+1}$. Then BD is an S-matrix. Thus an S^*-matrix can be made into an S-matrix by multiplying certain columns by -1.

The next theorem shows that the sign-solvability of a general linear system $Ax = b$ can be determined from the right null spaces of the augmented matrices $\begin{bmatrix} \widetilde{A} & -\tilde{b} \end{bmatrix}$ $(\widetilde{A} \in \mathcal{Q}(A), \tilde{b} \in \mathcal{Q}(b))$.

Theorem 1.2.11 *The linear system $Ax = b$ is sign-solvable if and only if $\widetilde{A}x = \tilde{b}$ is solvable for all $\widetilde{A} \in \mathcal{Q}(A)$ and $\tilde{b} \in \mathcal{Q}(b)$ and there exists a vector w such that the right null spaces of the matrices $\begin{bmatrix} \widetilde{A} & -\tilde{b} \end{bmatrix}$ are contained in $\{0\} \cup \mathcal{Q}(w) \cup \mathcal{Q}(-w)$.*

Proof First suppose that $Ax = b$ is sign-solvable. Then $\widetilde{A}x = \tilde{b}$ is solvable for all $\widetilde{A} \in \mathcal{Q}(A)$ and $\tilde{b} \in \mathcal{Q}(b)$. By Corollary 1.2.2, A^T is an L-matrix. Suppose that $\begin{bmatrix} \widetilde{A} & -\tilde{b} \end{bmatrix} z = 0$. Let c equal the last coordinate of z and let z' be obtained from z by deleting its last coordinate. If $c = 0$, then since A^T is an

L-matrix, $z = 0$. If $c \neq 0$, then $\widetilde{A}z' = cb$, and hence $cz' \in \mathcal{Q}(Ax = b)$. The converse follows in a similar way. $\qquad\square$

We now show that the study of sign-solvable linear systems is equivalent to the study of L-matrices and S^*-matrices (Manber[9] and Klee, Ladner and Manber[5]).

Let A be an m by n matrix, and let α be a subset of $\{1, 2, \ldots, m\}$, and let β be a subset of $\{1, 2, \ldots, n\}$. Then $A[\alpha, \beta]$ denotes the submatrix of A determined by the rows whose index is in α and the columns whose index is in β. If $\alpha = \{1, 2, \ldots, m\}$ or $\beta = \{1, 2, \ldots, n\}$ we shorten this notation to $A[:, \beta]$ and $A[\alpha, :]$, respectively. The submatrix $A[\overline{\alpha}, \overline{\beta}]$ determined by the rows whose indices are complementary to those of α and the columns whose indices are complementary to those of β is also denoted by $A(\alpha, \beta)$. If A is square and $\alpha = \beta$, then we write $A[\alpha]$ instead of $A[\alpha, \alpha]$ and $A(\alpha)$ instead of $A(\alpha, \alpha)$. If z is an m by 1 column vector then we write $z[\alpha]$ instead of $z[\alpha, \{1\}]$.

Theorem 1.2.12 *Let $A = [a_{ij}]$ be an m by n matrix and let b be an m by 1 column vector. Assume that $z = (z_1, z_2, \ldots, z_n)^T$ is a solution of the linear system $Ax = b$. Let*

$$\beta = \{j : z_j \neq 0\} \text{ and } \alpha = \{i : a_{ij} \neq 0 \text{ for some } j \in \beta\}.$$

Then $Ax = b$ is sign-solvable if and only if the matrix

$$\left[\begin{array}{cc} A[\alpha, \beta] & -b[\alpha] \end{array} \right]$$

is an S^*-matrix and the matrix $A(\alpha, \beta)^T$ is an L-matrix.[2]

Proof Without loss of generality we assume that $\beta = \{1, 2, \ldots, \ell\}$ and that $\alpha = \{1, 2, \ldots, k\}$ for some nonnegative integers k and ℓ. Note that if $\ell = 0$ then $k = 0$. It follows from the definitions of α and β that

$$A = \left[\begin{array}{cc} A_1 & A_3 \\ O & A_2 \end{array} \right]$$

where A_1 is a k by ℓ matrix with no zero rows. Let $b' = b[\{1, 2, \ldots, k\}]$. Then the linear system $Ax = b$ can be written as

$$\begin{array}{rcl} A_1 x^{(1)} + A_3 x^{(2)} & = & b' \\ A_2 x^{(2)} & = & 0. \end{array} \tag{1.15}$$

---

[2] We must regard a 0 by k matrix as an (empty) L-matrix for each nonnegative integer k. In particular, a matrix of order 0 is an SNS-matrix. A 0 by 1 matrix is thus an (empty) S^*-matrix, indeed an S-matrix. But in proving theorems about L-, S-, and S^*-matrices, we implicitly assume that the matrices are nonempty.

Assume that $Ax = b$ is sign-solvable. Then every vector $[\tilde{x}^{(1)} \ \tilde{x}^{(2)}]^T$ in the qualitative solution class $\mathcal{Q}(Ax = b)$ satisfies $\tilde{x}^{(2)} = 0$ and hence $A_1 x^{(1)} = b'$ is sign-solvable with $\mathcal{Q}(A_1 x^{(1)} = b') = \mathcal{Q}((z_1, \ldots, z_\ell)^T)$. By Corollary 1.2.2, A^T is an L-matrix. If $\ell = 0$, then $A_2 = A$ and hence A_2^T is an L-matrix. Assume that $\ell > 0$. The fact that A^T is an L-matrix implies that A_1^T is an L-matrix and that $k \geq \ell$. We now show that $\ell = k$. Without loss of generality we assume that the first ℓ rows of A_1 are linearly independent. Then the linear system

$$A_1[\{1, 2, \ldots, \ell\}, :]x^{(1)} = b[\{1, 2, \ldots, \ell\}]$$

has as its unique solution the vector $x^{(1)} = z[\{1, 2, \ldots, \ell\}, :]$ with no zero entries. If $k > \ell$, then since each row of A_1 has a nonzero entry, there exists a matrix $\widetilde{A}_1$ in $\mathcal{Q}(A_1)$ such that $\widetilde{A}_1[\{1, 2, \ldots, \ell\}, :] = A_1[\{1, 2, \ldots, \ell\}, :]$ and

$$\widetilde{A}_1 z[\{1, 2, \ldots, k\}] \neq b',$$

contradicting the sign-solvability of $A_1 x^{(1)} = b$. Hence $\ell = k$ and the matrix

$$\begin{bmatrix} A_1 & -b' \end{bmatrix} \tag{1.16}$$

is an S^*-matrix.

Now we show that A_2^T is an L-matrix. Let $\widetilde{A}_2$ be any matrix in $\mathcal{Q}(A_2)$ and suppose $\widetilde{A}_2 \tilde{u} = 0$. Since A_1 is a square L-matrix, the linear system $A_1 x^{(1)} = b' - A_3 \tilde{u}$ has a solution. Let $\widetilde{A}$ be the matrix in $\mathcal{Q}(A)$ obtained from A by replacing A_2 with $\widetilde{A}_2$. Since the solution of $\widetilde{A}x = b$ belongs to the qualitative class of z, we conclude that $\tilde{u} = 0$ and thus A_2^T is an L-matrix. Therefore, if $Ax = b$ is sign-solvable, then the matrix (1.16) is an S^*-matrix and the matrix A_2^T is an L-matrix.

Conversely, assume that the matrix (1.16) is an S^*-matrix and A_2^T is an L-matrix. Let $\widetilde{A}$ be a matrix in $\mathcal{Q}(A)$ and b be a column vector in $\mathcal{Q}(b)$. Then

$$\widetilde{A} = \begin{bmatrix} \widetilde{A}_1 & \widetilde{A}_3 \\ O & \widetilde{A}_2 \end{bmatrix}$$

where $\widetilde{A}_i$ belongs to $\mathcal{Q}(A_i)$, $(i = 1, 2, 3)$. The linear system $\widetilde{A}x = \tilde{b}$ is of the form

$$\widetilde{A}_1 x^{(1)} + \widetilde{A}_3 x^{(2)} = \tilde{b}'$$
$$\widetilde{A}_2 x^{(2)} = 0.$$

Since A_2^T is an L-matrix and since (1.16) is an S^*-matrix, the above equations have a unique solution $\tilde{z} = [\tilde{z}^{(1)} \ \tilde{z}^{(2)}]^T$ where $\tilde{z}^{(2)} = 0$ and $\tilde{z}^{(1)}$ is the solution of the sign-solvable linear system $\widetilde{A}_1 x^{(1)} = \tilde{b}'$. Therefore $Ax = b$ is a sign-solvable linear system. $\qquad\square$

The following corollary is an immediate consequence of Theorem 1.2.12.

Corollary 1.2.13 *Let $Ax = b$ be a linear system such that A has no zero rows. Then $Ax = b$ is sign-solvable and the vectors in its qualitative solution class have no zero coordinates if and only if the matrix $\begin{bmatrix} A & -b \end{bmatrix}$ is an S^*-matrix (in particular, A is a square matrix).*

Corollary 1.2.13 implies that the assumption in Corollary 1.2.10 that the matrix B is n by $n + 1$ can be replaced by the assumption that B has no zero rows.

We now remove the restriction in Corollary 1.2.7 that the coefficient matrix is square.

Corollary 1.2.14 *Let $Ax = b$ be a sign-solvable linear system and let c be obtained from b by replacing some of its nonzero entries with zeros. Then $Ax = c$ is sign-solvable.*

Proof We use the notation in Theorem 1.2.12. By that theorem $A(\alpha, \beta)^T$ is an L-matrix and hence $Ax = c$ is sign-solvable if and only if $A[\alpha, \beta]x^{(1)} = c[\alpha]$ is sign-solvable. The theorem also implies that $A[\alpha, \beta]$ is a square matrix and $A[\alpha, \beta]x^{(1)} = b[\alpha]$ is sign-solvable. The corollary now follows from Corollary 1.2.7. □

It follows from Theorem 1.2.12 that if $Ax = b$ is sign-solvable, then there exist permutation matrices P and Q such that $P \begin{bmatrix} A & -b \end{bmatrix} Q$ has the form

$$\begin{bmatrix} B_1 & B_3 \\ O & B_2 \end{bmatrix} \tag{1.17}$$

where B_1 is a k by $k + 1$ S^*-matrix and B_2^T is an L-matrix. If $b = 0$, then B_1 is the vacuous 0 by 1 S^*-matrix. If $Ax = b$ has a solution with no zero coordinates and A is an m by n matrix, then the matrix B_2 is the vacuous $m - n$ by 0 matrix. The next theorem implies that the matrices B_1 and B_2 are uniquely determined up to permutation of their rows and columns, and that every matrix of the form (1.17) corresponds to some sign-solvable system $Ax = b$.

Theorem 1.2.15 *Let B be a matrix of the form (1.17) where the k by $k + 1$ matrix B_1 is an S^*-matrix and B_2^T is an L-matrix. Let b be a column of B and let A be the matrix obtained from B by deleting the column b. Then $Ax = b$ is sign-solvable if and only if b is one of the first $k + 1$ columns of B.*

Proof If b is not one of the first $k + 1$ columns of B, then the columns of A are linearly dependent and hence by Corollary 1.2.2, $Ax = b$ is not sign-solvable. Now let b be one of the first $k + 1$ columns of B. The assumptions on B and Theorem 1.2.11 imply that $Ax = -b$, and hence $Ax = b$, is sign-solvable.

$\square$

By Theorem 1.2.12, the problem of recognizing sign-solvable linear systems is equivalent to the problems of recognizing S-matrices and recognizing L-matrices. Thus the fundamental matrices of sign-solvable linear systems are S-matrices and L-matrices. Since every qualitative class of matrices equals $\mathcal{Q}(A)$ for a unique $(0, 1, -1)$-matrix A, the theory of S- and L-matrices is primarily combinatorial in nature. In the next chapter we study basic properties of L-matrices. The qualitative approach to linear systems arose in economics, and it continues to occur in the economics literature (e.g., in [3] a regional economic policy model, called the QUEBEC model, is analyzed from a qualitative point of view). However, the highly restrictive conditions needed for the sign-solvability of a linear system severely limit its applicability [10].

Bibliography

[1] L. Bassett, J. Maybee, and J. Quirk. Qualitative economics and the scope of the correspondence principle, *Econometrica*, 36:544–63, 1968.

[2] R.A. Brualdi and H.J. Ryser. *Combinatorial Matrix Theory*, Cambridge University Press, New York, 1991.

[3] P. Fontaine, M. Garbely, and M. Gilli. Qualitative solvability in economic models, *Computer Science in Economics and Management*, 4:285–301, 1991.

[4] W.M. Gorman. A wider scope for qualitative economics, *Rev. Econ. Studies*, 31:65–8, 1964.

[5] V. Klee, R. Ladner, and R. Manber. Signsolvability revisited, *Linear Alg. Appls.*, 59:131–57, 1984.

[6] K. Lancaster. The scope of qualitative economics, *Rev. Econ. Studies*, 29:99–132, 1962.

[7] K. Lancaster. The theory of qualitative linear systems, *Econometrica*, 33:395–408, 1964.

[8] K. Lancaster. Partitionable systems and qualitative economics, *Rev. Econ. Studies*, 31:69–72, 1964.

[9] R. Manber. Graph-theoretical approach to qualitative solvability of linear systems, *Linear Alg. Appls.*, 48:457–70, 1982.

[10] T. Rader. Impossibility of qualitative economics: Excessively strong correspondence principles in production-exchange economies, *Zeitschrift für Nationalökonomie*, 32:397–416, 1972.

[11] P.A. Samuelson. *Foundations of Economic Analysis,* Harvard University Press, Cambridge, 1947, Atheneum, New York, 1971.

2

L-matrices

2.1 Signings

Let A be an m by n matrix. Recall that A is an L-matrix if and only if every matrix in the qualitative class $\mathcal{Q}(A)$ has linearly independent rows. If A is an L-matrix, then every matrix obtained from A by appending column vectors is also an L-matrix. If A is an L-matrix and each of the m by $n-1$ matrices obtained from A by deleting a column is not an L-matrix, then A is called a *barely* L-*matrix* [1]. Thus a barely L-matrix is an L-matrix in which every column is essential. If A is an L-matrix, then we can obtain a barely L-matrix by deleting certain columns of A. An SNS-matrix, that is, a square L-matrix, is a barely L-matrix. But there are barely L-matrices which are not square. The 3 by 4 matrix (1.10) is an L-matrix, and it follows from Theorem 1.2.5 that each of its submatrices of order 3 is not an SNS-matrix. Hence (1.10) is a barely L-matrix.

A *signing* of order k is a nonzero $(0, 1, -1)$-diagonal matrix of order k. A *strict signing* is a signing that is invertible. Let $D = \mathrm{diag}(d_1, d_2, \ldots, d_k)$ be a signing of order k with diagonal entries $d_1, d_2, \ldots, d_k$. If $k = m$, then the matrix DA is a *row signing of the matrix* A, and if D is a strict signing, then DA is a *strict row signing* of A. If $k = n$, then the matrix AD is a *column signing of the matrix* A, and if D is strict, then AD is a *strict column signing* of A. A signing $D' = \mathrm{diag}(d'_1, d'_2, \ldots, d'_k)$ is an *extension* of the signing D provided that $D' \neq D$ and $d'_i = d_i$ whenever $d_i \neq 0$. A vector is *balanced* provided either it is a zero vector or it has both a positive entry and a negative entry. A vector v is *unsigned* provided it is not balanced. Thus v is unsigned provided $v \neq 0$, and the nonzero entries of v have the same sign. A *balanced row signing of the matrix* A is a row signing of A in which all columns are balanced. A *balanced column signing* of A is a column signing of A is which all rows are balanced.

18

According to the definition, the matrix A is an L-matrix if and only if each matrix in the infinite set $\mathcal{Q}(A)$ has linearly independent rows. If A is a square matrix, then Theorem 1.2.5 provides a combinatorial characterization of SNS-matrices, namely that there is a nonzero term in the standard determinant expansion and every such term has the same sign. There is no direct analogue of this characterization for rectangular L-matrices. However, L-matrices [4], barely L-matrices [1], S^*-matrices [1], and S-matrices [4] can be characterized combinatorially.

Theorem 2.1.1 *Let A be an m by n matrix. Then*

(i) *A is an L-matrix if and only if every row signing of A contains a unsigned column;*

(ii) *A is a barely L-matrix if and only if*

    (a) *A is an L-matrix, and*

    (b) *for each $i = 1, 2, \ldots, n$ there is a row signing of A for which column i is the only unsigned column;*

(iii) *A is an S^*-matrix if and only if $n = m + 1$ and every row signing of A contains at least two unsigned columns;*

(iv) *A is an S^*-matrix if and only if $n = m + 1$ and there exists a strict signing D such that AD and $A(-D)$ are the only balanced column signings of A;*

(v) *A is an S-matrix if and only if A is an S^*-matrix and every row of A is balanced;*

(vi) *A is an S-matrix if and only if $n = m + 1$ and $AI_m = A$ and $A(-I_m) = -A$ are the only balanced column signings of A.*

Proof First assume that there is a signing D such that every column of DA is balanced. This implies that there exists a matrix $\widetilde{B}$ in $\mathcal{Q}(A)$ such that each of the column sums of $D\widetilde{B}$ equals zero. Hence the rows of $\widetilde{B}$ are linearly dependent and A is not an L-matrix. Now assume that A is not an L-matrix. Then there is a matrix $\widetilde{A}$ whose rows are linearly dependent. Hence there exists a nonzero diagonal matrix $E = \mathrm{diag}(e_1, e_2, \ldots, e_m)$ such that $(e_1, e_2, \ldots, e_m)\widetilde{A} = 0$. Let E' be the signing obtained from E by replacing the e_i by their signs. Then each column of $E'A$ is balanced. Therefore (i) holds.

We now prove that (ii) holds. Assume that A is a barely L-matrix. Then A is an L-matrix and by (i) every row signing of A contains a unsigned column. Let i be an integer with $1 \leq i \leq n$. The matrix A_i obtained from A by deleting column i is not an L-matrix and hence by (i) there is a balanced row signing DA_i of A_i. It follows that column i is the only unsigned column of DA.

Hence (a) and (b) hold. Conversely, if (a) and (b) hold, then for each i the matrix obtained from A by deleting column i has a balanced row signing, and it follows that A is a barely L-matrix.

The fact that (iii) holds is an immediate consequence of (i) and the definition of an S^*-matrix. Let $D = \operatorname{diag}(d_1, d_2, \ldots, d_{n+1})$ be a signing, and let $u = (d_1, d_2, \ldots, d_{n+1})$. Then each row of AD is balanced if and only if there exists a matrix $\tilde{A}$ in $\mathcal{Q}(A)$ such that $\tilde{A}u = 0$. Assertion (iv) is now an immediate consequence of Corollary 1.2.10.

Next we prove that (v) holds. If A is an S-matrix, then A is an S^*-matrix with a positive vector in its right null space and hence every row of A is balanced. Conversely, if A is an S^*-matrix and every row of A is balanced, then there exists a matrix in $\mathcal{Q}(A)$ such that the vector of all 1's is in its right null space, and hence A is an S-matrix. Assertion (vi) is a consequence of (iv) and (v). $\qquad\square$

We now construct two infinite families $\{\Lambda_k : k \geq 1\}$ and $\{\Gamma_k : k \geq 1\}$ of barely L-matrices. Let $\Lambda_1 = [1]$ and for $k \geq 2$ we define Λ_k inductively by

$$\Lambda_k = \left[\begin{array}{ccc|ccc} & \Lambda_{k-1} & & & \Lambda_{k-1} & \\ \hline 1 & \cdots & 1 & -1 & \cdots & -1 \end{array} \right].$$

The matrix Λ_k is a k by 2^{k-1} $(1, -1)$-matrix such that every k by 1 $(1, -1)$-column vector or its negative is a column of Λ_k. For example,

$$\Lambda_2 = \begin{bmatrix} 1 & 1 \\ 1 & -1 \end{bmatrix} \quad \text{and} \quad \Lambda_3 = \begin{bmatrix} 1 & 1 & 1 & 1 \\ 1 & -1 & 1 & -1 \\ 1 & 1 & -1 & -1 \end{bmatrix}.$$

The matrix Γ_k is the $(0, 1)$-matrix whose columns are all the $2k + 1$ by 1 column vectors with $k + 1$ 1's and k 0's. The $\binom{2k+1}{k+1}$ columns of Γ_k are taken to be in lexicographic order. For example,

$$\Gamma_1 = \begin{bmatrix} 0 & 1 & 1 \\ 1 & 0 & 1 \\ 1 & 1 & 0 \end{bmatrix} \quad \text{and} \quad \Gamma_2 = \begin{bmatrix} 0 & 0 & 0 & 0 & 1 & 1 & 1 & 1 & 1 & 1 \\ 0 & 1 & 1 & 1 & 0 & 0 & 0 & 1 & 1 & 1 \\ 1 & 0 & 1 & 1 & 0 & 1 & 1 & 0 & 0 & 1 \\ 1 & 1 & 0 & 1 & 1 & 0 & 1 & 0 & 1 & 0 \\ 1 & 1 & 1 & 0 & 1 & 1 & 0 & 1 & 0 & 0 \end{bmatrix}.$$

Let $D = \operatorname{diag}(d_1, d_2, \ldots, d_m)$ be a signing. Let $z = (z_1, z_2, \ldots, z_m)^T$ where $z_i = d_i$ if $d_i \neq 0$ and $z_i = 1$ otherwise. Either z or $-z$ is a column of Λ_m. Hence $D\Lambda_M$ has a unsigned column. Suppose that $w = (w_1, w_2, \ldots, w_m)^T$ is the ith column of Λ_m and let $W = \operatorname{diag}(w_1, w_2, \ldots, w_m)$. Then column i is the only unsigned column of $W\Lambda_m$. It now follows from Theorem 2.1.1 that Λ_m is a barely L-matrix.

Now let $D = \mathrm{diag}(d_1, d_2, \ldots, d_{2k+1})$ be a signing of order $2k + 1$. First suppose that either D or $-D$ is nonnegative. Then $D\Gamma_k$ has a nonzero column and all such columns are unsigned. Now suppose that D contains both a 1 and a -1. Then either D contains at least $k + 1$ nonnegative diagonal entries or at least $k + 1$ nonpositive diagonal entries and it follows that $D\Gamma_k$ has a unsigned column. Suppose that $w = (w_1, w_2, \ldots, w_{2k+1})^T$ is the ith column of Γ_k. Let $W' = \mathrm{diag}(w'_1, w'_2, \ldots, w'_{2k+1})$ be the diagonal matrix where $w'_i = w_i$ if $w_i = 1$ and $w'_i = -1$ otherwise. Then column i is the only unsigned column of $W'\Gamma_k$. By Theorem 2.1.1, we conclude that Γ_k is a barely L-matrix.

In determining whether or not a matrix A is an L-matrix it does not suffice in general to consider only strict row signings in Theorem 2.1.1. For example, let

$$A = \begin{bmatrix} 1 & 1 & 0 \\ 1 & 1 & 0 \\ 1 & 1 & 1 \end{bmatrix}.$$

Then every strict row signing of A unisigns its last column, but clearly A is not an L-matrix. However, there are zero patterns for which strict row signings suffice.

Theorem 2.1.2 *Let A be an m by n $(0, 1, -1)$-matrix which does not have a p by q zero submatrix for any positive integers p and q with $p + q \geq m$. Then A is an L-matrix if and only if every strict row signing of A has a unsigned column.*

Proof If A is an L-matrix, then every row signing, in particular every strict row signing, of A has a unsigned column. Now assume that A is not an L-matrix. Let DA be a balanced row signing of A such that the signing D has the smallest number of zeros. Suppose that D is not a strict signing. Without loss of generality we assume that $D = D_1 \oplus O$ where D_1 is a strict signing of order $k < m$ and that

$$A = \begin{bmatrix} B_1 & O \\ B_3 & B_2 \end{bmatrix}$$

where B_1 is a k by ℓ matrix with no zero columns. The matrix $D_1 B_1$ is balanced and has no zero columns. Our assumptions imply that B_2 has more rows than columns, and hence the rows of B_2 are linearly dependent. By (i) of Theorem 2.1.1, there is a signing D_2 such that $D_2 B_2$ is balanced. Then $D_1 \oplus D_2$ is an extension of D which balances the columns of A. This contradicts the choice of D, and hence D is a strict signing which balances the columns of A. $\qquad\square$

An L-matrix is a matrix A for which one can assert the linear independence of its rows solely on the basis of the sign pattern of A. As we shall see throughout this book, L-matrices form a rich and difficult class of matrices. The subclass of L-matrices for which one can assert the linear independence of rows solely on the basis of the zero pattern has a simple characterization. Indeed, any m by n matrix which has an invertible triangular submatrix of order m is an L-matrix. These matrices and their permutations are the only matrices A for which one can conclude A is an L-matrix knowing only the zero pattern of A. This can be proved inductively using the observation that if each nonzero column of a matrix A has at least two nonzero entries, then some matrix with the same zero pattern as A has linearly dependent rows.

2.2 Indecomposability

Let

$$A = \left[\begin{array}{cc} B_1 & O \\ B_3 & B_2 \end{array} \right] \tag{2.1}$$

be an m by n matrix. If B_1 and B_2 are L-matrices, then A is also an L-matrix. Conversely, if A is an L-matrix, then B_1 is an L-matrix but B_2 is not necessarily an L-matrix. For example,

$$A = \left[\begin{array}{ccc|cc} 1 & 1 & 1 & 0 & 0 \\ -1 & 1 & 1 & 1 & 1 \\ 0 & -1 & 1 & 1 & 1 \end{array} \right]$$

is an L-matrix (it has a SNS-submatrix of order 3), but $A[\{2, 3\}, \{4, 5\}]$ is not an L-matrix. If A is a barely L-matrix and B_1 and B_2 are L-matrices, then B_1 and B_2 are barely L-matrices. In fact, we shall prove in Lemma 2.2.8 that if A and B_1 are barely L-matrices, then B_2 must be an L-matrix and hence a barely L-matrix.

A matrix A is L-*decomposable* provided that there exist permutation matrices P and Q such that

$$A = P \left[\begin{array}{cc} B_1 & O \\ B_3 & B_2 \end{array} \right] Q \tag{2.2}$$

where B_1 and B_2 are (nonvacuous) L-matrices. Note that an L-decomposable matrix is necessarily an L-matrix. An L-*indecomposable* matrix is an L-matrix which is not L-decomposable. An m by n L-matrix such that there does not exist a p by q zero submatrix for all positive integers p and q with $p + q \geq m$ is necessarily L-indecomposable.

It follows inductively that if A is an L-matrix, then the rows and columns of A can be permuted to obtain a matrix of the form

$$
\begin{bmatrix}
A_1 & O & \cdots & O \\
A_{21} & A_2 & \cdots & O \\
\vdots & \vdots & \ddots & \vdots \\
A_{k1} & A_{k2} & \cdots & A_k
\end{bmatrix}
\tag{2.3}
$$

where k is a positive integer and matrices $A_1, A_2, \ldots, A_k$ are L-indecomposable matrices. We call $A_1, A_2, \ldots, A_k$ a *set of L-indecomposable components* of the L-matrix A. In this section, which is largely based on [1], we study the L-indecomposable components of a barely L-matrix.

The L-indecomposable components of an L-matrix are not necessarily unique. For example, the matrix

$$
A = \left[\begin{array}{ccc|cc}
1 & -1 & 1 & 0 & 0 \\
1 & 1 & 1 & 0 & 0 \\
\hline
0 & 0 & 1 & 1 & -1 \\
0 & 0 & 0 & 1 & 1
\end{array}\right]
$$

has a decomposition with two L-indecomposable components. After permutations we obtain the decomposition

$$
\left[\begin{array}{cc|c|cc}
1 & 1 & 0 & 0 & 0 \\
\hline
1 & -1 & 1 & 0 & 0 \\
\hline
0 & 0 & 1 & -1 & 1 \\
0 & 0 & 1 & 1 & 1
\end{array}\right]
$$

of A with three L-indecomposable components. If A is a barely L-matrix, then clearly the L-matrices $A_1, A_2, \ldots, A_k$ in (2.3) are barely L-matrices.

The preceding notions of decomposability and indecomposability depend on the sign pattern of a matrix. However, the following notions of decomposability and indecomposability for square matrices depend only on the zero pattern (see [2]). Let A be a matrix of order n which does not have an identically zero determinant. The matrix A is called *partly decomposable* provided that there exist permutation matrices P and Q such that (2.2) holds where B_1 and B_2 are square (nonvacuous) matrices. If A is not partly decomposable, then A is *fully indecomposable*. Thus A is fully indecomposable if and only if A does not have a p by q zero submatrix for any positive integers p and q with $p + q \geq n$. It follows from the Frobenius–König theorem that no matrix obtained from a fully indecomposable matrix by deleting a row and a column has an identically zero determinant. Hence each matrix obtained from a fully indecomposable

SNS-matrix by deleting the row and column of a nonzero entry is an SNS-matrix. Theorem 2.1.2 implies that if A is a fully indecomposable matrix, then A is an SNS-matrix if and only if each strict row signing of A has a unisigned column [6].

Theorem 2.2.1 *Let A be an SNS-matrix of order n. Then A is L-indecomposable if and only if A is fully indecomposable.*

Proof First assume that there exist permutation matrices P and Q such that (2.2) holds where B_1 and B_2 are L-matrices. Then each of the matrices B_1 and B_2 has at least as many columns as rows. Since A is a square matrix, B_1 and B_2 are also square matrices and hence A is partly decomposable. Now assume that (2.2) holds where B_1 and B_2 are square matrices. Since $\det A = \pm \det B_1 \det B_2$, it follows from Theorem 1.2.5 that B_1 and B_2 are SNS-matrices. Hence A is L-decomposable. $\qquad\square$

Again let A be a matrix of order n which does not have an identically zero determinant. It is a standard result in combinatorial matrix theory [2] that the rows and columns of A can be permuted to obtain a matrix of the form (2.3) where k is a positive integer and the matrices $A_1, A_2, \ldots, A_k$ are fully indecomposable. Moreover, the matrices $A_1, A_2, \ldots, A_k$ that occur as diagonal blocks in (2.3) are uniquely determined up to permutations of their rows and columns, but their ordering in (2.3) is not necessarily unique. The matrices $A_1, A_2, \ldots, A_k$ are the *fully indecomposable components* of A. It follows from Theorem 2.2.1 that the L-indecomposable components of an SNS-matrix A are the fully indecomposable components of A and are uniquely determined up to row and column permutations. The main result of this section is more generally that a barely L-matrix has a unique set of L-indecomposable components (up to row and column permutations).

If A is an m by n matrix, then we denote the set of signings D such that column i is the unique unisigned column of DA by $\mathcal{A}_i$ ($i = 1, 2, \ldots, n$).

Lemma 2.2.2 *Let A be an m by n matrix. Then:*

(i) *If A is an L-matrix and for some integer i, D is a signing for which each column of DA other than column i is balanced, then D belongs to $\mathcal{A}_i$.*

(ii) *If A is an L-matrix, then A is a barely L-matrix if and only if each $\mathcal{A}_i$ is nonempty.*

(iii) *If A has the form (2.1) where B_2 is an L-matrix, then at least one $\mathcal{A}_i$ does not contain a strict signing.*

Proof Parts (i) and (ii) follow from definitions. If A has the form (2.1) where B_2 is an L-matrix, then each strict row signing of A contains a unisigned column which intersects B_2, and hence $\mathcal{A}_1$ does not contain a strict signing. □

Lemma 2.2.3 *Let A be an m by n matrix of the form*

$$\begin{bmatrix} B_1 & O \\ B_3 & B_2 \end{bmatrix} \tag{2.4}$$

where B_1 is a k by ℓ matrix with no zero columns and $k < m$. Let $D = D_1 \oplus D_2$ be a signing such that D_1 is a strict signing of order k, and $D_2 = O$. Then either B_2 is an L-matrix or there exists an extension D' of D such that every column unsigned by D' is also unsigned by D.

Proof Assume B_2 is not an L-matrix. Let D_2' be a signing such that $D_2' B_2$ is a balanced row signing of B_2 (if B_2 is vacuous then D_2' can be any signing of order $m - k$). Let $D' = D_1 \oplus D_2'$. Since B_1 has no zero columns and D_1 is a strict signing, each balanced column of $A D$ is also a balanced column of $A D'$. □

Corollary 2.2.4 *Let A be an m by n L-indecomposable matrix. Assume that there exists an integer i such that $\mathcal{A}_i$ is nonempty and let D be a signing in $\mathcal{A}_i$. Then either D is a strict signing or there exists a strict signing D' in $\mathcal{A}_i$ which is an extension of D.*

Proof Suppose that D is not a strict signing. It follows from Lemma 2.2.3 and the L-indecomposability of A that there is an extension E of D such that the only possible unsigned column of EA is column i. Since A is an L-matrix, column i of EA must be unsigned and hence E is in $\mathcal{A}_i$. If E is not a strict signing, we may repeat the argument with E in place of D and continue like this until we obtain a strict signing. □

We now characterize L-indecomposable, barely L-matrices in terms of strict signings.

Theorem 2.2.5 *Let A be an m by n matrix with no zero rows. Then A is an L-indecomposable, barely L-matrix if and only if*

(i) *every strict row signing of A contains a unisigned column, and*

(ii) *$\mathcal{A}_i$ contains a strict signing for each $i = 1, 2, \ldots, n$.*

Proof First assume that A is an L-indecomposable, barely L-matrix. Then A is an L-matrix and hence (i) holds. Since A is a barely L-matrix, each A_i is nonempty. Since A is L-indecomposable, Corollary 2.2.4 implies that (ii) holds. Now assume that (i) and (ii) hold. Suppose that A is not an L-matrix. Then there is a balanced row signing of A and by (i) no such row signing is strict. Let DA be a balanced row signing of A such that no extension of D balances each column of A. Without loss of generality we assume that A and D have the form given in the statement of Lemma 2.2.3. It follows from that lemma that B_2 is an L-matrix. Since A has no zero rows, B_1 is not vacuous, and it follows that no strict row signing can unsign only column 1 of A, contradicting (ii). Hence A is an L-matrix, and by (ii) and (iii) of Lemma 2.2.2, A is an L-indecomposable, barely L-matrix.   □

In verifying that the matrices Λ_k and Γ_k defined in section 2.1 are barely L-matrices, we actually verified that given any column there is a strict row signing for which it is the only unsigned column. Applying Theorem 2.2.5 we obtain the first assertion of the following corollary.

Corollary 2.2.6 *Each of the matrices Λ_k and Γ_k ($k = 1, 2, 3, \ldots$) is an L-indecomposable, barely L-matrix. A $(1, -1)$-matrix with k rows is a barely L-matrix if and only if it equals $\Lambda_k DP$ for some strict signing D and some permutation matrix P.*

Proof If a $(1, -1)$-matrix A with k rows is an L-matrix, then since each strict row signing unsigns some column, each k by 1 $(1, -1)$-column vector or its negative is a column of A. This implies the second assertion of the corollary.   □

Let $A = [a_{ij}]$ be an m by n $(0, 1)$-matrix with no zero rows. Then A is the vertex-edge incidence matrix of a hypergraph $H = H(A)$ with m vertices $1, 2, \ldots, m$ and n edges $\{i : a_{ij} = 1\}$ ($j = 1, 2, \ldots, n$). A strict row signing of A corresponds to a 2-coloring of the vertices of H with the "colors" $+$ and $-$. The hypergraph H is 2-colorable if and only if there is a 2-coloring of the vertices of H such that each edge contains vertices of both colors, that is, if and only if there is a balanced strict row signing of A. Seymour [5] calls a hypergraph a *condenser*, provided H is not 2-colorable, but each hypergraph obtained from H by removing an edge is 2-colorable. It follows from Theorem 2.2.5 that a $(0, 1)$-matrix A is an L-indecomposable, barely L-matrix if and only if $H(A)$ is a condenser. The following theorem [5] imposes a severe restriction on $(0,1)$-matrices which are L-indecomposable, barely L-matrices.

Theorem 2.2.7 *Let $A = [a_{ij}]$ be an m by n L-indecomposable, barely L matrix of 0's and 1's. Let α_j be the positions of the 1's in column j of A. Then there exists a subset β_j of $\{1, 2, \ldots, n\}$ such that $A[\alpha_j, \beta_j]$ is a permutation matrix.*

Proof If $m = 1$, then $n = 1$ and the theorem holds. Now assume that $m > 1$. Since A is L-indecomposable, α_j contains at least two elements. By (ii) of Theorem 2.2.5 there exists a strict signing D in $\mathcal{A}_j$. Suppose that k is in α_j. Let D_k be the strict signing obtained from D by negating its kth diagonal entry. Since A is an L-matrix, some column of $D_k A$, say column ℓ, is unsigned. Since α_j has at least two elements, $\ell \neq k$. Thus column ℓ of DA is balanced. Since A is a $(0,1)$-matrix, we see that $a_{k\ell} = 1$ and $a_{i\ell} = 0$ for i in $\alpha \setminus \{k\}$. The theorem now follows. $\square$

We now obtain some structural properties of barely L-matrices.

Lemma 2.2.8 *Let A be an L-matrix of the form*

$$\begin{bmatrix} B_1 & O \\ B_3 & B_2 \end{bmatrix} \tag{2.5}$$

where B_1 is a barely L-matrix. Then B_2 is an L-matrix.

Proof We first prove the lemma under the assumption that B_1 is L-indecomposable. Suppose to the contrary that B_2 is not an L-matrix. Then there is balanced row signing $D_2 B_2$ of B_2. Since A is an L-matrix, there is an i such that column i of $(O \oplus D_2)A$ and hence column i of $D_2 B_3$ are unsigned. Since B_1 is a barely L-matrix it follows from Lemma 2.2.2 that there is a strict row signing $D_1 B_1$ of B_1 for which only column i is unsigned. Since B_1 can have no zero columns and D_1 is a strict signing, each column of $D_1 B_1$ other than column i is a balanced nonzero vector. But then one of $(\pm D_1 \oplus D_2)A$ is a balanced row signing of A, contradicting the assumption that A is an L-matrix. Hence B_2 is an L-matrix.

If B_1 is L-decomposable, then induction on the number of rows of A completes the proof. $\square$

Corollary 2.2.9 *Let A be a matrix of the form (2.5) where B_1 and B_2 are barely L-matrices. Then A is a barely L-matrix.*

Proof Since B_1 and B_2 are L-matrices, A is an L-matrix. We prove that A is a barely L-matrix by showing that each matrix A' obtained from A by deleting a column is not an *L*-matrix. Since B_1 is a barely *L*-matrix, if the deleted column intersects B_1, then A' is not an *L*-matrix. Since B_2 is also a barely *L*-matrix, it follows from Lemma 2.2.8 that if the deleted column intersects B_2, then A' is not an *L*-matrix. Hence A is a barely *L*-matrix.   □

By definition a barely *L*-matrix is a matrix whose columns form a minimal set of columns which determine an *L*-matrix. We next show that a barely *L*-matrix could also be defined as a matrix whose rows form a maximal set of rows which determine an *L*-matrix.

Theorem 2.2.10 *Let A be an L-matrix. Then A is a barely L-matrix if and only if there does not exist a row vector u such that*

$$\left[\frac{A}{u} \right] \tag{2.6}$$

is an L-matrix.

Proof First assume that A is not a barely *L*-matrix. Then there is a column of A whose deletion results in an *L*-matrix. If u is the row vector with a 1 in the position corresponding to this column and 0's elsewhere, then (2.6) is an *L*-matrix. Now assume that A is a barely *L*-matrix and let u be any row vector. If (2.6) is an *L*-matrix, then Lemma 2.2.8 applied to the matrix obtained from (2.6) by appending a column of 0's implies that the 1 by 1 zero matrix is an *L*-matrix. Hence, for all u, (2.6) is not an *L*-matrix.   □

Finally we show that up to row and column permutations, a barely *L*-matrix has a unique set of *L*-indecomposable components.

Theorem 2.2.11 *Let A be an m by n barely L-matrix. Then there is an ordered partition $(\alpha_1, \ldots, \alpha_k)$ of $\{1, 2, \ldots, m\}$ and an ordered partition $(\beta_1, \ldots, \beta_k)$ of $\{1, 2, \ldots, n\}$ such that $A[\alpha_i, \beta_i]$ is an L-indecomposable matrix for each $i = 1, 2, \ldots, k$ and $A[\alpha_i, \beta_j] = O$ for all $i < j$. The unordered partitions $\{\alpha_1, \ldots, \alpha_k\}$ and $\{\beta_1, \ldots, \beta_k\}$ are uniquely determined by A. Moreover, the matrices $A[\alpha_1, \beta_1], \ldots, A[\alpha_k, \beta_k]$ are barely L-matrices.*

Proof The existence of such ordered partitions $(\alpha_1, \ldots, \alpha_k)$ and $(\beta_1, \ldots, \beta_k)$ is a consequence of definition and has already been pointed out for general *L*-matrices. Suppose $(\gamma_1, \ldots, \gamma_\ell)$ and $(\delta_1, \ldots, \delta_\ell)$ are ordered partitions of $\{1, 2, \ldots, m\}$ and $\{1, 2, \ldots, n\}$, respectively, such that $A[\gamma_i, \delta_i]$ is *L*-indecomposable for each $i = 1, 2, \ldots, k$ and $A[\gamma_i, \delta_j] = O$ for all $i < j$.

Let $A_i = A[\alpha_i, \beta_i]$ $(i = 1, 2, \ldots, k)$ and let $B_j = A[\gamma_j, \delta_j]$ $(j = 1, 2, \ldots, \ell)$. Since A is a barely L-matrix, each of the matrices A_i and B_j is a barely L-matrix. Let p be the smallest integer such that $\alpha_p \cap \gamma_\ell \neq \emptyset$ and let q be the smallest integer such that $\beta_q \cap \delta_\ell \neq \emptyset$. If $p < q$, then B_ℓ has a zero row, contradicting the assumption that B_ℓ is an L-matrix. If $p > q$, then A_q has a zero column, contradicting the assumption that A_q is a barely L-matrix. Therefore $p = q$ and the matrix $A[\alpha_p \cap \gamma_\ell, \beta_p \cap \delta_\ell]$ is a nonvacuous matrix. The minimality of p now implies that $A[\alpha_p \cap \gamma_\ell, \delta_\ell \setminus \beta_p] = O$. Since B_ℓ is an L-matrix, the matrix $A[\alpha_p \cap \gamma_\ell, \beta_p \cap \delta_\ell]$ is also an L-matrix. Suppose that $\alpha_p \cap \gamma_\ell \neq \alpha_p$ and that $\beta_p \cap \delta_\ell \neq \beta_p$. Then $A[\alpha_p \setminus \gamma_\ell, \beta_p \cap \delta_\ell] = O$. Since A_p is an L-matrix, $A[\alpha_p \setminus \gamma_\ell, \beta_p \setminus \delta_\ell]$ is an L-matrix. We now have an L-decomposition of A_p, contradicting its L-indecomposability. Thus either (i) $\alpha_p \subseteq \gamma_\ell$ or (ii) $\beta_p \subseteq \delta_\ell$ holds.

If (ii) but not (i) holds, then A_p has a zero row, contradicting the fact that A_p is an L-matrix. If (i) but not (ii) holds, then $A[\alpha_p, \beta_p \cap \delta_\ell]$ is an L-matrix, and we contradict the fact that A_p is a barely L-matrix. Therefore (i) and (ii) both hold, and hence

$$B_\ell = \begin{bmatrix} A_p & O \\ X & Y \end{bmatrix}$$

for some matrices X and Y. If Y is not vacuous, then since A_p is a barely L-matrix, Lemma 2.2.8 implies that B_ℓ is L-decomposable. Hence Y is vacuous. Since B_ℓ is a barely L-matrix, Y cannot have columns. Since A_p is a barely L-matrix, it follows from Theorem 2.2.10 that Y cannot have rows. Therefore $\alpha_p = \gamma_\ell$ and $\beta_p = \delta_\ell$. We now delete the rows of A whose index is in γ_ℓ and the columns whose index is in δ_ℓ, and the theorem follows by induction on k. □

Let A be an SNS-matrix of order n. Then A is a barely L-matrix and, as already pointed out in the discussion following Theorem 2.2.1, the L-indecomposable components of A are the fully indecomposable components of A and hence are determined by the zero pattern of A. Thus there exist ordered partitions $(\alpha_1, \ldots, \alpha_k)$ and $(\beta_1, \ldots, \beta_k)$ of $\{1, 2, \ldots, n\}$ such that if B is any SNS-matrix with the same zero pattern as A, then the L-indecomposable components of B are the matrices $B[\alpha_1, \beta_1], \ldots, B[\alpha_k, \beta_k]$. While the L-indecomposable components of a rectangular barely L-matrix are uniquely determined by its sign pattern, they are not in general uniquely determined by its zero pattern. For example, let A be the matrix

$$\left[\begin{array}{cccccccccc|ccccc|c|cccc}
1 & 0 & 0 & 0 & 0 & 1 & 1 & 1 & 1 & 1 & 0 & 0 & 0 & 0 & 0 & 0 & 0 & 0 & 0 & 0 \\
0 & 1 & 1 & 1 & 0 & 1 & 1 & 1 & 0 & 0 & 0 & 0 & 0 & 0 & 0 & 0 & 0 & 0 & 0 & 0 \\
0 & 1 & 1 & 0 & 1 & 1 & 0 & 0 & 1 & 1 & 0 & 0 & 0 & 0 & 0 & 0 & 0 & 0 & 0 & 0 \\
1 & 1 & 0 & 1 & 1 & 0 & 1 & 0 & 1 & 0 & 0 & 0 & 0 & 0 & 0 & 0 & 0 & 0 & 0 & 0 \\
1 & 0 & 1 & 1 & 1 & 0 & 0 & 1 & 0 & 1 & 0 & 0 & 0 & 0 & 0 & 0 & 0 & 0 & 0 & 0 \\ \hline
0 & 0 & 0 & 0 & 0 & 1 & 0 & 0 & 0 & 0 & 1 & 1 & 0 & 0 & 0 & 0 & 0 & 0 & 0 & 0 \\
0 & 0 & 0 & 0 & 0 & 0 & 1 & 0 & 0 & 0 & 0 & 1 & 1 & 0 & 0 & 0 & 0 & 0 & 0 & 0 \\
0 & 0 & 0 & 0 & 0 & 0 & 0 & 1 & 0 & 0 & 0 & 0 & 1 & 1 & 0 & 0 & 0 & 0 & 0 & 0 \\
0 & 0 & 0 & 0 & 0 & 0 & 0 & 0 & 1 & 0 & 0 & 0 & 0 & 1 & 1 & 0 & 0 & 0 & 0 & 0 \\
0 & 0 & 0 & 0 & 0 & 0 & 0 & 0 & 0 & 1 & 1 & 0 & 0 & 0 & 1 & 0 & 0 & 0 & 0 & 0 \\ \hline
0 & 0 & 0 & 0 & 0 & 0 & 0 & 0 & 0 & 0 & 1 & 1 & 1 & 1 & 1 & 1 & 0 & 0 & 0 & 0 \\ \hline
0 & 0 & 0 & 0 & 0 & 0 & 0 & 0 & 0 & 0 & 1 & 1 & 1 & 0 & 0 & 0 & 1 & 1 & 1 & 0 \\
0 & 0 & 0 & 0 & 0 & 0 & 0 & 0 & 0 & 0 & 1 & 0 & 0 & 1 & 1 & 0 & -1 & 1 & 0 & -1 \\
0 & 0 & 0 & 0 & 0 & 0 & 0 & 0 & 0 & 0 & 0 & 1 & 0 & 1 & 0 & 1 & -1 & 0 & 1 & 1 \\
0 & 0 & 0 & 0 & 0 & 0 & 0 & 0 & 0 & 0 & 0 & 0 & 1 & 0 & 1 & 1 & 0 & 1 & -1 & 1
\end{array}\right]$$

Let $(\alpha_1, \alpha_2, \alpha_3, \alpha_4)$ be the ordered partition of $\{1, 2, \ldots, 15\}$ where $\alpha_1 = \{1, 2, 3, 4, 5, \}$, $\alpha_2 = \{6, 7, 8, 9, 10\}$, $\alpha_3 = \{11\}$ and $\alpha_4 = \{12, 13, 14, 15\}$, and let $(\beta_1, \beta_2, \beta_3, \beta_4)$ be the ordered partition of $\{1, 2, \ldots, 20\}$ where $\beta_1 = \{1, 2, 3, 4, 5, 6, 7, 8, 9, 10\}$, $\beta_2 = \{11, 12, 13, 14, 15\}$, $\beta_3 = \{16\}$, and $\beta_4 = \{17, 18, 19, 20\}$. The matrix $A[\alpha_1, \beta_1]$ can be obtained from Γ_2 by column permutations and hence by Corollary 2.2.6 is an L-indecomposable, barely L-matrix. It is easy to verify that each of the matrices $A[\alpha_i, \beta_i]$ $(i = 2, 3, 4)$ is a fully indecomposable SNS-matrix, and hence each is an L-indecomposable, barely L-matrix. By Corollary 2.2.9, A is a barely L-matrix, and the L-indecomposable components of A are the matrices $A[\alpha_i, \beta_i]$ with $i = 1, 2, 3, 4$.

Let B be the matrix

$$\left[\begin{array}{c|cccc|c|c|c|c|c|cccccccccc}
1 & 0 & 0 & 0 & 0 & 1 & 1 & 1 & 1 & 1 & 0 & 0 & 0 & 0 & 0 & 0 & 0 & 0 & 0 & 0 \\ \hline
0 & 1 & 1 & 1 & 0 & 1 & 1 & 1 & 0 & 0 & 0 & 0 & 0 & 0 & 0 & 0 & 0 & 0 & 0 & 0 \\
0 & -1 & 1 & 0 & -1 & 1 & 0 & 0 & 1 & 1 & 0 & 0 & 0 & 0 & 0 & 0 & 0 & 0 & 0 & 0 \\
1 & -1 & 0 & 1 & 1 & 0 & 1 & 0 & 1 & 0 & 0 & 0 & 0 & 0 & 0 & 0 & 0 & 0 & 0 & 0 \\
1 & 0 & 1 & -1 & 1 & 0 & 0 & 1 & 0 & 1 & 0 & 0 & 0 & 0 & 0 & 0 & 0 & 0 & 0 & 0 \\ \hline
0 & 0 & 0 & 0 & 0 & 1 & 0 & 0 & 0 & 0 & 1 & 1 & 0 & 0 & 0 & 0 & 0 & 0 & 0 & 0 \\ \hline
0 & 0 & 0 & 0 & 0 & 0 & 1 & 0 & 0 & 0 & 0 & 1 & 1 & 0 & 0 & 0 & 0 & 0 & 0 & 0 \\ \hline
0 & 0 & 0 & 0 & 0 & 0 & 0 & 1 & 0 & 0 & 0 & 0 & 1 & 1 & 0 & 0 & 0 & 0 & 0 & 0 \\ \hline
0 & 0 & 0 & 0 & 0 & 0 & 0 & 0 & 1 & 0 & 0 & 0 & 0 & 1 & 1 & 0 & 0 & 0 & 0 & 0 \\ \hline
0 & 0 & 0 & 0 & 0 & 0 & 0 & 0 & 0 & 1 & 1 & 0 & 0 & 0 & 1 & 0 & 0 & 0 & 0 & 0 \\ \hline
0 & 0 & 0 & 0 & 0 & 0 & 0 & 0 & 0 & 0 & 1 & 1 & 1 & 1 & 1 & 1 & 0 & 0 & 0 & 0 \\
0 & 0 & 0 & 0 & 0 & 0 & 0 & 0 & 0 & 0 & 1 & 1 & 1 & 0 & 0 & 0 & 1 & 1 & 1 & 0 \\
0 & 0 & 0 & 0 & 0 & 0 & 0 & 0 & 0 & 0 & 1 & 0 & 0 & 1 & 1 & 0 & 1 & 1 & 0 & 1 \\
0 & 0 & 0 & 0 & 0 & 0 & 0 & 0 & 0 & 0 & 0 & 1 & 0 & 1 & 0 & 1 & 1 & 0 & 1 & 1 \\
0 & 0 & 0 & 0 & 0 & 0 & 0 & 0 & 0 & 0 & 0 & 0 & 1 & 0 & 1 & 1 & 0 & 1 & 1 & 1
\end{array}\right]$$

The matrix B has the same zero pattern as A but not the same sign pattern. Let $(\gamma_1, \gamma_2, \ldots, \gamma_8)$ be the ordered partition of $\{1, 2, \ldots, 15\}$ where $\gamma_1 = \{1\}$,

$\gamma_2 = \{2, 3, 4, 5\}$, $\gamma_3 = \{6\}$, $\gamma_4 = \{7\}$, $\gamma_5 = \{8\}$, $\gamma_6 = \{9\}$, $\gamma_7 = \{10\}$, and $\gamma_8 = \{11, 12, 13, 14, 15\}$, and let $(\delta_1, \delta_2, \ldots, \delta_8)$ be the ordered partition of $\{1, 2, \ldots, 20\}$ where $\delta_1 = \{1\}$, $\delta_2 = \{2, 3, 4, 5\}$, $\delta_3 = \{6\}$, $\delta_4 = \{7\}$, $\delta_5 = \{8\}$, $\delta_6 = \{9\}$, $\delta_7 = \{10\}$, and $\delta_8 = \{11, 12, 13, 14, 15, 16, 17, 18, 19, 20\}$. The matrices $B[\gamma_i, \delta_i]$ $(i = 1, 2, 3, 4, 5, 6, 7, 8)$ are L-indecomposable, barely L-matrices and hence are the L-indecomposable components of the barely L-matrix B. Thus, while the barely L-matrices A and B have the same zero pattern, the unordered partitions of the rows and of the columns determined by their L-indecomposable components are different.

2.3 *L*-canonical form

Let A be an m by n barely L-matrix. By Theorem 2.2.11, there exist permutation matrices P and Q such that PAQ has the form

$$
\begin{bmatrix}
A[\alpha_1, \beta_1] & O & \cdots & O \\
A[\alpha_2, \beta_1] & A[\alpha_2, \beta_2] & \cdots & O \\
\vdots & \vdots & \ddots & \vdots \\
A[\alpha_k, \beta_1] & A[\alpha_k, \beta_2] & \cdots & A[\alpha_k, \beta_k]
\end{bmatrix}
\tag{2.7}
$$

where $A[\alpha_1, \beta_1]$, $A[\alpha_2, \beta_2]$, $\ldots$, $A[\alpha_k, \beta_k]$ are L-indecomposable, barely L-matrices. The matrices $A[\alpha_1, \beta_1]$, $A[\alpha_2, \beta_2]$, $\ldots$, $A[\alpha_k, \beta_k]$ are the L-indecomposable components of A and are uniquely determined up to row and column permutations. We call (2.7) an *L-canonical form* of the barely L-matrix A. The order of the L-indecomposable components of A in its L-canonical forms is not necessarily unique. We use the zero pattern to determine which of the orderings are possible.

Let Π_A be the digraph with vertices $1, 2, \ldots, k$ and an arc from vertex i to vertex j if and only if $i \neq j$ and $A[\alpha_j, \beta_i] \neq O$. The digraph Π_A has no directed cycles. Defining $j \preceq i$ if and only if $j = i$, or $j \neq i$ and there exists a path from i to j in Π_A determines a partially ordered set $\mathcal{P}_A$ with elements $1, 2, \ldots, k$. Let $r_1, r_2, \ldots, r_k$ be a permutation of $\{1, 2, \ldots, k\}$. Then there exist permutation matrices P' and Q' such that $P'AQ'$ has the form

$$
\begin{bmatrix}
A[\alpha_{r_1}, \beta_{r_1}] & O & \cdots & O \\
A[\alpha_{r_2}, \beta_{r_1}] & A[\alpha_{r_2}, \beta_{r_2}] & \cdots & O \\
\vdots & \vdots & \ddots & \vdots \\
A[\alpha_{r_k}, \beta_{r_1}] & A[\alpha_{r_k}, \beta_{r_2}] & \cdots & A[\alpha_{r_k}, \beta_{r_k}]
\end{bmatrix}
\tag{2.8}
$$

and hence is an L-canonical form of A, if and only if $r_1, r_2, \ldots, r_k$ is a linear extension of the partially ordered set $\mathcal{P}_A$.[1]

We now show how signings can be used in order to compute the L-canonical forms of A. Let $D = \mathrm{diag}(d_1, d_2, \ldots, d_m)$ be a signing. The *support* of D is defined to be the set

$$\mathrm{supp}\, D = \{i : d_i \neq 0, 1 \le i \le m\}.$$

Let

$$\mathrm{supp}\, \mathcal{A}_i = \cup_{D \in \mathcal{A}_i} \mathrm{supp}\, D$$

be the union of the supports of the signings D for which only column i of DA is unisigned $(i = 1, 2, \ldots, n)$.

Lemma 2.3.1 *Assume that ℓ is an element of β_j. Then*

$$\mathit{supp}\, \mathcal{A}_\ell = \cup_{i \preceq j} \alpha_i.$$

Moreover, there exists a signing D such that D is in $\mathcal{A}_\ell$ and $supp\, D = supp\, \mathcal{A}_\ell$.

Proof   There exist permutation matrices U and V such that

$$UAV = \begin{bmatrix} C_1 & O \\ C_3 & C_2 \end{bmatrix}$$

where C_1 and C_2 are barely L-matrices in L-canonical form such that the L-indecomposable components of C_1 are the matrices $A[\alpha_i, \beta_i]$ with $i \preceq j$. Since C_2 is an L-matrix, we have $\mathrm{supp}\, \mathcal{A}_\ell \subseteq \cup_{i \preceq j} \alpha_i$. For each i with $i \prec j$ there is a nonzero entry below $A[\alpha_i, \beta_i]$ in C_1. Since each $A[\alpha_i, \beta_i]$ is a barely L-matrix, given any column of $A[\alpha_i, \beta_i]$ there is a strict row signing of $A[\alpha_i, \beta_i]$ such that it, and hence its negative, unisigns only that column. There is strict row signing of $A[\alpha_j, \beta_j]$ whose only unsigned column is the column which is contained in column ℓ of A. Starting with this strict row signing of $A[\alpha_j, \beta_j]$, we may recursively construct a strict row signing of C_1 which unsigns only the column of C_1 contained in column ℓ of A. This strict row signing of C_1 determines a row signing of A with support equal to $\cup_{i \preceq j} \alpha_i$ for which column ℓ of A is the only unsigned column.   □

It follows from Lemma 2.3.1 that the relation on $\{1, 2, \ldots, n\}$ defined by $i \sim j$ if and only $\mathrm{supp}\, \mathcal{A}_i = \mathrm{supp}\, \mathcal{A}_j$ is an equivalence relation whose equivalence classes are the sets $\beta_1, \beta_2, \ldots, \beta_k$. We consider the k different sets among the sets $\mathrm{supp}\, \mathcal{A}_i$ $(i = 1, 2, \ldots, n)$ and partially order them by inclusion. The

[1]That is, if $r_i \preceq r_j$, then $i \le j$.

resulting partially ordered set is denoted by $\mathcal{S}_A$. Lemma 2.3.1 implies that $\mathcal{S}_A$ is isomorphic to the partially ordered set $\mathcal{P}_A$.

Combining the various parts of this section we obtain the following theorem.

Theorem 2.3.2 *Let A be an m by n barely L-matrix. Let*

$$supp\, \mathcal{A}_{i_1}, supp\, \mathcal{A}_{i_2}, \ldots, supp\, \mathcal{A}_{i_k} \tag{2.9}$$

be the k elements of $\mathcal{S}_A$. Let

$$\alpha_j = supp\, \mathcal{A}_{i_j} \setminus \cup \{supp\, \mathcal{A}_{i_\ell} : supp\, \mathcal{A}_{i_\ell} \subset supp\, \mathcal{A}_{i_j}\, j \neq \ell\} \quad (j = 1, 2, \ldots, k),$$

and let

$$\beta_j = \{\ell : supp\, \mathcal{A}_\ell = supp\, \mathcal{A}_{i_j}\} \quad (j = 1, 2, \ldots, k).$$

Then the L-indecomposable components of A are the matrices $A[\alpha_j, \beta_j]$ ($j = 1, 2, \ldots, k$). Moreover, the matrix (2.7) is the L-canonical form of A if and only if (2.9) is a linear extension of $\mathcal{S}_A$.

We conclude this section with the following remarks concerning computational issues. Let A be an m by n matrix. A row signing of A is balanced if and only if its negative is balanced. Thus one can determine whether or not A is an L-matrix by checking $(3^m - 1)/2$ row signings of A. By Theorem 2.1.2, if A does not have a p by q zero submatrix with $p + q \geq m$, it suffices to check only strict row signings and hence only $(2^m - 1)/2$ row signings. In this section we have seen how signings can be used to compute the L-canonical form of a barely L-matrix. The L-matrix recognition problem *Is the matrix A not an L-matrix?* belongs to the class of NP-problems. It has been shown in [4] that the L-matrix recognition problem is in the subclass of NP-complete problems, and as a result it is highly unlikely that there is a polynomial-time recognition algorithm.[2] It follows from Theorem 1.2.12 that since the L-matrix recognition problem is NP-complete, so is the sign-solvability recognition problem. The L-matrix recognition problem remains NP-complete even if $n = m + \lfloor m^{1/k} \rfloor$ where k is a fixed positive integer. However, the complexity of the L-matrix recognition problem for square matrices has not been resolved.

Bibliography

[1] R.A. Brualdi, K.L. Chavey, and B.L. Shader. Rectangular L-matrices, *Linear Alg. Appls.*, 196:37–61, 1994.

---

[2]We refer the reader to the book [3] for a discussion of the theory of NP-completeness.

[2] R.A. Brualdi and H.J. Ryser. *Combinatorial Matrix Theory*, Cambridge University Press, New York, 1991.

[3] M.R. Garey and D.S. Johnson. *Computers and Intractability, A guide to the theory of NP-completeness*, W.H. Freeman and Company, San Francisco, 1979.

[4] V. Klee, R. Ladner, and R. Manber. Signsolvability revisited, *Linear Alg. Appls.*, 59:131–57, 1984.

[5] P.D. Seymour. On the two-colouring of hypergraphs, *Quart. J. Math. Oxford*, 25:303–12, 1974.

[6] B.L. Shader. Maximal convertible matrices, *Congressus Numerantium*, 81:161–72, 1991.

3

Sign-solvability and digraphs

3.1 Partial sign-solvability

Let $Ax = b$ be a linear system with m equations in n variables $x_1, x_2, \ldots, x_n$. Then the variable x_j is *sign-determined by* $Ax = b$ provided $\widetilde{A}x = \tilde{b}$ is solvable for each $\widetilde{A}$ in $\mathcal{Q}(A)$ and for each $\tilde{b}$ in $\mathcal{Q}(b)$, and all the numbers in the set

$$\{\tilde{x}_j : \text{ there exists } \widetilde{A} \in \mathcal{Q}(A) \text{ and } \tilde{b} \in \mathcal{Q}(b) \text{ with } \widetilde{A}\tilde{x} = \tilde{b}\}$$

have the same sign. If x_j is sign-determined by $Ax = b$, then we denote the common sign by

$$\text{sign}\,(x_j : Ax = b).$$

Clearly, $Ax = b$ is sign-solvable if and only if each x_j is sign-determined by $Ax = b$. The linear system $Ax = b$ is *partially sign-solvable* provided at least one x_j is sign-determined. If $Ax = b$ is sign-solvable, then by Corollary 1.2.2 and Theorem 3.2.1, A^T is an L-matrix. This conclusion need not hold if $Ax = b$ is only partially sign-solvable. For example, x_3 is sign-determined by the linear system

$$\begin{bmatrix} 1 & 1 & 0 \\ 0 & 0 & 1 \\ 0 & 0 & 1 \end{bmatrix} \begin{bmatrix} x_1 \\ x_2 \\ x_3 \end{bmatrix} = \begin{bmatrix} 1 \\ 0 \\ 0 \end{bmatrix},$$

but the transpose of the coefficient matrix is not an L-matrix. In this section we show that a partially sign-solvable linear system has a special block structure.

We introduce in this section a canonical form for matrices which is useful in the study of partial sign-solvability. First we recall the following basic fact.

Lemma 3.1.1 *The linear system $Ax = b$ is solvable if and only if $z^T b = 0$ for all vectors z in the left null space of A, that is, $z^T A = 0$ implies $z^T b = 0$.*

We use this lemma in order to obtain some information about linear systems $Ax = b$ such that $\widetilde{A}x = \widetilde{b}$ is solvable (but not necessarily sign-solvable) for all $\widetilde{A}$ in $\mathcal{Q}(A)$ and $\widetilde{b}$ in $\mathcal{Q}(b)$. First we prove the following.

Lemma 3.1.2 *Let* $b = (b_1, b_2, \ldots, b_m)^T$. *Then the linear system* $Ax = \widetilde{b}$ *is solvable for all* $\widetilde{b}$ *in* $\mathcal{Q}(b)$ *if and only if for every vector* $z = (z_1, z_2, \ldots, z_m)^T$ *in the left null space of* A *we have*

$$z_i b_i = 0 \quad (i = 1, 2, \ldots, m). \tag{3.1}$$

Proof Clearly $z^T \widetilde{b} = 0$ for all $\widetilde{b}$ in $\mathcal{Q}(b)$ if and only if (3.1) holds. The result now follows from Lemma 3.1.1. $\qquad\qquad\square$

The *support of a vector* is the set of integers j for which its jth entry is nonzero. Thus the support of a signing as defined in Chapter 1 is the support of its diagonal vector. The *left null-support of a matrix* A is defined to be the union of the supports of the vectors in the left null space of A and is denoted by $\mathcal{N}_\ell(A)$. The *left null-support of the qualitative class* $\mathcal{Q}(A)$ is the union of the left null-supports of the matrices in $\mathcal{Q}(A)$ and is denoted by $\mathcal{N}_\ell(\mathcal{Q}(A))$. Note that A is an L-matrix if and only if $\mathcal{N}_\ell(\mathcal{Q}(A)) = \emptyset$, and that $\mathcal{N}_\ell(\mathcal{Q}(A))$ is the union of the supports of the signings D such that each column of DA is balanced. We define the *right null-support* of A and of $\mathcal{Q}(A)$ in a similar way and use the notations $\mathcal{N}_r(A)$ and $\mathcal{N}_r(\mathcal{Q}(A))$, respectively.

Applying Lemma 3.1.2 to each matrix in the qualitative class of A, we obtain the following.

Corollary 3.1.3 *Let* A *be an* m *by* n *matrix and let* $b = (b_1, b_2, \ldots, b_m)^T$ *be a column vector. Then* $\widetilde{A}x = \widetilde{b}$ *is solvable for all* $\widetilde{A}$ *in* $\mathcal{Q}(A)$ *and all* $\widetilde{b}$ *in* $\mathcal{Q}(b)$ *if and only if* $b_i = 0$ *for all* i *in* $\mathcal{N}_\ell(\mathcal{Q}(A))$.

Theorem 3.1.4 *Let* A *be an* m *by* n *matrix. Then there exists a unique subset* α *of* $\{1, 2, \ldots, m\}$ *and a unique subset* β *of* $\{1, 2, \ldots, n\}$ *such that*

 (i) $A[\alpha, \beta]$ *has no zero columns,*
 (ii) $A[\alpha, \overline{\beta}] = O$,
 (iii) $A[\overline{\alpha}, \overline{\beta}]$ *is an* L-*matrix, and*
 (iv) *there exists a signing* D *with support* α *such that each column of* DA *is balanced.*

In fact,

$$\alpha = \mathcal{N}_\ell(A) \text{ and } \beta = \{j : a_{ij} \neq 0 \text{ for some } i \in \alpha\}. \tag{3.2}$$

Proof Let $D = \operatorname{diag}(d_1, d_2, \ldots, d_m)$ be a signing such that each column of DA is balanced and D has maximal support among all such signings. Let $\alpha = \operatorname{supp} D$ and β be the set of all j such that $a_{ij} \neq 0$ for some i in α. Without loss of generality we assume that $\alpha = \{1, 2, \ldots, k\}$ and $\beta = \{1, 2, \ldots, \ell\}$. Thus

$$A = \begin{bmatrix} B_1 & O \\ B_3 & B_2 \end{bmatrix}$$

where B_1 is a k by ℓ matrix with no zero columns. By Lemma 2.2.3 and our choice of D, B_2 is an L-matrix. This proves the existence of an α and β satisfying (i), (ii), (iii), and (iv). Since B_2 is an L-matrix, if E is a signing such that each column of EA is balanced, then $\operatorname{supp} E \subseteq \operatorname{supp} D$. Hence (3.2) holds.

Now assume that α and β are such that (i), (ii), (iii), and (iv) hold. Since $A[\overline{\alpha}, \overline{\beta}]$ is an L-matrix it follows that $\mathcal{N}_\ell(\mathcal{Q}(A)) \subseteq \alpha$. Since (iv) holds, $\alpha \subseteq \mathcal{N}_\ell(\mathcal{Q}(A))$, and the theorem follows. $\qquad\square$

Let $A = [a_{ij}]$ be an m by n matrix. Let α and β be as in (3.2), and define

$$\gamma = \mathcal{N}_r(\mathcal{Q}(A)) \text{ and } \delta = \{i : a_{ij} \neq 0 \text{ for some } j \in \gamma\}. \tag{3.3}$$

It follows from Theorem 3.1.4 applied to A^T that $A[\overline{\delta}, \overline{\gamma}]^T$ is an L-matrix and there exists a signing E with support γ such that each row of AE is balanced. There exist permutation matrices P and Q such that PAQ equals

$$\begin{bmatrix} A[\alpha \cap \overline{\delta}, \beta \cap \overline{\gamma}] & O & O & O \\ A[\alpha \cap \delta, \beta \cap \overline{\gamma}] & A[\alpha \cap \delta, \beta \cap \gamma] & O & O \\ A[\overline{\alpha} \cap \overline{\delta}, \beta \cap \overline{\gamma}] & O & A[\overline{\alpha} \cap \delta, \overline{\beta} \cap \overline{\gamma}] & O \\ A[\overline{\alpha} \cap \delta, \beta \cap \overline{\gamma}] & A[\overline{\alpha} \cap \delta, \beta \cap \gamma] & A[\overline{\alpha} \cap \delta, \overline{\beta} \cap \overline{\gamma}] & A[\overline{\alpha} \cap \delta, \overline{\beta} \cap \gamma] \end{bmatrix}.$$

$$\tag{3.4}$$

Since $A[\overline{\alpha}, \overline{\beta}]$ is an L-matrix, $A[\overline{\alpha} \cap \overline{\delta}, \overline{\beta} \cap \overline{\gamma}]$ is an L-matrix. Since $A[\overline{\delta}, \overline{\gamma}]^T$ is an L-matrix, $A[\overline{\alpha} \cap \overline{\delta}, \overline{\beta} \cap \overline{\gamma}]^T$ is an L-matrix. Thus $A[\overline{\alpha} \cap \overline{\delta}, \overline{\beta} \cap \overline{\gamma}]$ is a square matrix and hence an SNS-matrix. It now follows from Lemma 2.2.8 that $A[\overline{\alpha} \cap \delta, \overline{\beta} \cap \gamma]$ and $A[\alpha \cap \overline{\delta}, \beta \cap \overline{\gamma}]^T$ are L-matrices.

In general, let U and V be permutation matrices such that UAV has the form

$$
\begin{bmatrix}
C_1 & O & O & O \\
C_{21} & C_2 & O & O \\
C_{31} & O & C_3 & O \\
C_{41} & C_{42} & C_{43} & C_4
\end{bmatrix}
\tag{3.5}
$$

where

   (i) each row and column of C_2 contains a nonzero entry,

  (ii) C_1^T and C_4 are L-matrices and C_3 is an SNS-matrix,

 (iii) there exists a strict signing D' such that each column of

$$
D' \begin{bmatrix} C_1 & O \\ C_{21} & C_2 \end{bmatrix}
$$

      is balanced, and

 (iv) there exists a strict signing E' such that each row of

$$
\begin{bmatrix} C_2 & O \\ C_{42} & C_4 \end{bmatrix} E'
$$

      is balanced.

Then (3.5) is called the *N-canonical form*[1] of A. Note that whether or not the matrix (3.5) is in N-canonical form does not depend on the matrices C_{31}, C_{41}, and C_{43}. The ordered partition $(\pi_1, \pi_2, \pi_3, \pi_4)$ of $\{1, 2, \ldots, m\}$ determined by the rows of C_1, C_2, C_3, and C_4 is called the *ordered row partition of the N-canonical form* (3.5). The *ordered column partition* of (3.5) is the ordered partition $(\theta_1, \theta_2, \theta_3, \theta_4)$ of $\{1, 2, \ldots, n\}$ determined by the columns of C_1, C_2, C_3, and C_4. It follows from Theorem 3.1.4 that (3.4) is an N-canonical form of A.

Theorem 3.1.5 *Let A be an m by n matrix. Then A has an N-canonical form, and its ordered row and column partitions are uniquely determined by A.*

Proof Let α and β be the unique sets described in Theorem 3.1.4 and let γ and δ be the corresponding sets for A^T described in (3.3). Let $(\pi_1, \pi_2, \pi_3, \pi_4)$

---

[1] Here the N stands for null space.

and $(\theta_1, \theta_2, \theta_3, \theta_4)$ be the ordered row and column partitions for an N-canonical form of A. It follows from Theorem 3.1.4 and the definition of N-canonical form that

$$\pi_1 \cup \pi_2 = \alpha, \ \theta_1 \cup \theta_2 = \beta, \ \theta_2 \cup \theta_4 = \delta, \text{ and } \pi_2 \cup \pi_4 = \gamma.$$

Since the π_i and θ_i are uniquely determined by these equations, and since by Theorem 3.1.4 the sets α, β, γ, and δ are uniquely determined by A, the theorem follows. $\qquad\square$

The N-canonical form of an S^*-matrix A is easily determined as follows. Since A is an L-matrix, $\mathcal{N}_\ell(\mathcal{Q}(A)) = \emptyset$. From Corollary 1.2.10 we obtain $\mathcal{N}_r(\mathcal{Q}(A)) = \{1, 2, \ldots, m + 1\}$. Hence the ordered row and column partitions of the N-canonical form of A are, respectively,

$$(\emptyset, \emptyset, \emptyset, \{1, 2, \ldots, m\}) \text{ and } (\emptyset, \emptyset, \emptyset, \{1, 2, \ldots, m + 1\}).$$

We now use the N-canonical form in order to study partially sign-solvable linear systems. First we obtain the following necessary condition for a variable x_j to be sign-determined by the linear system $Ax = b$.

Lemma 3.1.6 *If the variable x_j is sign-determined by the linear system $Ax = b$, then j is not in $\mathcal{N}_r(\mathcal{Q}(A))$.*

Proof Suppose that j is in $\mathcal{N}_r(\mathcal{Q}(A))$. Then there exist a matrix $\widetilde{A}$ in $\mathcal{Q}(A)$ and a vector $z = (z_1, z_2, \ldots, z_n)^T$ such that $z_j \neq 0$ and $\widetilde{A}z = 0$. If $\widetilde{A}x = b$ is not solvable, then $Ax = b$ is not partially sign-solvable, and hence x_j is not sign-determined. Suppose that u is a vector satisfying $\widetilde{A}u = b$. Then $\widetilde{A}(u + \lambda z) = b$ for all real numbers λ, and hence x_j is not sign-determined. $\qquad\square$

Although partial sign-solvability of $Ax = b$ does not in general imply that A^T is an L-matrix, we do have the following.

Corollary 3.1.7 *Let A be an m by n matrix and assume that A does not have a p by q zero submatrix for any integers p and q with $p + q \geq n$. If $Ax = b$ is partially sign-solvable, then A^T is an L-matrix.*

Proof Suppose that $Ax = b$ is partially sign-solvable. By Lemma 3.1.6 $\mathcal{N}_r(\mathcal{Q}(A))$ is a proper subset of $\{1, 2, \ldots, n\}$. Hence, if D is a strict signing of order n, then some row of AD is unsigned. Applying Theorem 2.1.2 to A^T, we conclude that A^T is an L-matrix. □

Let $Ax = b$ be a linear system. By reordering equations and relabelling variables we may assume that A is in N-canonical form (3.5). Hence the linear system $Ax = b$ has the partitioned form

$$
\begin{bmatrix}
C_1 & O & O & O \\
C_{21} & C_2 & O & O \\
C_{31} & O & C_3 & O \\
C_{41} & C_{42} & C_{43} & C_4
\end{bmatrix}
\begin{bmatrix}
x^{(1)} \\
x^{(2)} \\
x^{(3)} \\
x^{(4)}
\end{bmatrix}
=
\begin{bmatrix}
b^{(1)} \\
b^{(2)} \\
b^{(3)} \\
b^{(4)}
\end{bmatrix}.
\tag{3.6}
$$

Theorem 3.1.8 *Consider the linear system $Ax = b$ of the form (3.6) whose coefficient matrix A is in N-canonical form. Then*

(i) *no variable in $x^{(2)}$ or $x^{(4)}$ is sign-determined by $Ax = b$;*

(ii) *$\tilde{A}x = \tilde{b}$ is solvable for all $\tilde{A}$ in $\mathcal{Q}(A)$ and all $\tilde{b}$ in $\mathcal{Q}(b)$ if and only if $b^{(1)} = 0$ and $b^{(2)} = 0$;*

(iii) *if $b^{(1)} = 0$ and $b^{(2)} = 0$, then each variable x_j in $x^{(1)}$ is sign-determined and satisfies $sign(x_j : Ax = b) = 0$, and a variable in $x^{(3)}$ is sign-determined by $Ax = b$ if and only if it is sign-determined by $C_3x^{(3)} = b^{(3)}$.*

Proof The assertion (i) is a consequence of the definition of the N-canonical form and Lemma 3.1.6. The assertion (ii) is a consequence of the definition of the N-canonical form and Corollary 3.1.3. Now suppose that $b^{(1)} = 0$ and $b^{(2)} = 0$. By (ii) $\tilde{A}x = \tilde{b}$ is always solvable. Since C_1^T is an L-matrix, each variable x_j in $x^{(1)}$ is sign-determined and $sign(x_j : Ax = b) = 0$. Since each variable x_j in $x^{(1)}$ satisfies $sign(x_j : Ax = b) = 0$, it follows that a variable in $x^{(3)}$ is sign-determined by $Ax = b$ if and only if it is signed-determined by $A_3x^{(3)} = b^{(3)}$. □

Corollary 3.1.9 *Consider the linear system $Ax = b$ of the form (3.6) whose coefficient matrix A is in N-canonical form, and where $b^{(1)} = 0$ and $b^{(2)} = 0$. Then $Ax = b$ is partially sign-solvable if and only if either C_1 has at least one column or the linear system $C_3x^{(3)} = b^{(3)}$, whose coefficient matrix C_3 is an SNS-matrix, is partially sign-solvable.*

We conclude this section with an example. Let A be a matrix of the form

$$\begin{bmatrix} 1 & 1 & 1 & 0 & 0 & 0 & 0 & 0 & 0 & 0 & 0 & 0 \\ 0 & 1 & 1 & 0 & 0 & 0 & 0 & 0 & 0 & 0 & 0 & 0 \\ 0 & -1 & 1 & 0 & 0 & 0 & 0 & 0 & 0 & 0 & 0 & 0 \\ 0 & 0 & -1 & 0 & 0 & 0 & 0 & 0 & 0 & 0 & 0 & 0 \\ -1 & 0 & 0 & -1 & 1 & 0 & 0 & 0 & 0 & 0 & 0 & 0 \\ 0 & 0 & 0 & 1 & -1 & 0 & 0 & 0 & 0 & 0 & 0 & 0 \\ * & * & * & 0 & 0 & 1 & -1 & 0 & 0 & 0 & 0 & 0 \\ * & * & * & 0 & 0 & 1 & 1 & -1 & 0 & 0 & 0 & 0 \\ * & * & * & 0 & 0 & 1 & 1 & 1 & 0 & 0 & 0 & 0 \\ * & * & * & -1 & 0 & * & * & * & 1 & 0 & 0 & 0 \\ * & * & * & 0 & 0 & * & * & * & 1 & 1 & -1 & 0 \\ * & * & * & 0 & 0 & * & * & * & 1 & 1 & 1 & -1 \end{bmatrix}$$

The last diagonal block and the transpose of the first diagonal block are L-matrices, and the third diagonal block is an SNS-matrix. Since each of the columns of $A[\{1, 2, 3, 4, 5, 6\}, :]$ is balanced, it follows that $\mathcal{N}_\ell(\mathcal{Q}(A)) = \{1, 2, 3, 4, 5, 6\}$. Since each of the rows of $A[:, \{4, 5, 9, 10, 11, 12\}]$ is balanced, it also follows that $\mathcal{N}_r(\mathcal{Q}(A)) = \{4, 5, 9, 10, 11, 12\}$. Therefore the matrix A is in N-canonical form. Now consider the linear system $Ax = b$ where $b = (b_1, b_2, \ldots, b_{12})^T$. By Theorem 3.1.8, the variables $x_4, x_5, x_9, x_{10}, x_{11}$, and x_{12} are not sign-determined by $Ax = b$. If $Ax = b$ is partially sign-solvable, then $b_1 = b_2 = b_3 = b_4 = b_5 = b_6 = 0$. Suppose that $b_1 = b_2 = b_3 = b_4 = b_5 = b_6 = 0$. Then x_1, x_2, and x_3 are sign-determined with sign 0. The variables x_6, x_7, and x_8 are sign-determined by $Ax = b$ if and only if they are sign-determined by the linear system

$$\begin{bmatrix} 1 & -1 & 0 \\ 1 & 1 & -1 \\ 1 & 1 & 1 \end{bmatrix} \begin{bmatrix} x_6 \\ x_7 \\ x_8 \end{bmatrix} = \begin{bmatrix} b_7 \\ b_8 \\ b_9 \end{bmatrix}.$$

If, for instance, $b_7 = 1$, $b_8 = b_9 = 0$, then x_6 is sign-determined with sign $+1$, x_7 is sign-determined with sign -1, and x_8 is not sign-determined. If $b_7 = b_8 = 0$ and $b_9 = 1$, then each of x_6, x_7, and x_8 is sign-determined with sign $+1$. If $b_7 = b_9 = 1$ and $b_8 = -1$, then none of x_6, x_7, and x_8 is sign-determined.

In the next section we give a digraph characterization of SNS-matrices. In section 3.3 we use digraphs to further investigate partially sign-solvable linear systems with a sign-nonsingular coefficient matrix.

3.2 Digraph characterizations

Let $\mathcal{D}$ be a digraph of order n with vertices $1, 2, \ldots, n$. A *path* α of $\mathcal{D}$ from vertex i to vertex j of length $k \geq 0$ is a sequence $i = i_0, i_1, \ldots, i_{k-1}, i_k = j$ of distinct vertices such that (i_{p-1}, i_p) is an arc of $\mathcal{D}$ for $p = 1, 2, \ldots, k$.[2] We denote the path α by

$$i_0 \to i_1 \to \cdots \to i_{k-1} \to i_k.$$

A *directed cycle* γ of $\mathcal{D}$ of length $k \geq 1$ is a sequence $j_1, j_2, \ldots, j_k, j_{k+1} = j_1$ of vertices such that $j_1, j_2, \ldots, j_k$ are distinct and (j_q, j_{q+1}) is an arc of $\mathcal{D}$ for $q = 1, 2, \ldots, k$. The directed cycle γ is denoted by

$$j_1 \to j_2 \to \cdots \to j_k \to j_1.$$

The digraph $\mathcal{D}$ is *strongly connected* provided that for each pair of distinct vertices u and v there is a path in $\mathcal{D}$ from u to v and a path from v to u. The *strong components* of $\mathcal{D}$ are the maximal strongly connected subdigraphs of $\mathcal{D}$. The strong components determine a partition of the vertices of $\mathcal{D}$ and can be listed in an order $C_1, C_2, \ldots, C_k$ where there is an arc from a vertex in C_i to a vertex in C_j only if $i \geq j$.

Now let $A = [a_{ij}]$ be a square matrix of order n. Let $\mathcal{D} = \mathcal{D}(A)$ be the digraph of order n with vertices $1, 2, \ldots, n$ and an arc (i, j) from vertex i to vertex j if and only if $i \neq j$ and $a_{ij} \neq 0$. We call $\mathcal{D}(A)$ the *digraph of the matrix A*. The digraph $\mathcal{D}(A)$ is independent of the entries on the main diagonal of A. The matrix A is *reducible* provided that there exists a permutation matrix P such that

$$PAP^T = \begin{bmatrix} B_1 & O \\ B_{21} & B_2 \end{bmatrix}$$

where B_1 and B_2 are (nonvacuous) square matrices. The matrix A is *irreducible* provided it is not reducible. Each matrix of order 1 is irreducible. It is a standard fact that A is irreducible if and only if the digraph $\mathcal{D}(A)$ is strongly connected. In general there exists a permutation matrix P such that PAP^T has the form

$$\begin{bmatrix} A_1 & O & \cdots & O \\ A_{21} & A_2 & \cdots & O \\ \vdots & \vdots & \ddots & \vdots \\ A_{k1} & A_{k2} & \cdots & A_k \end{bmatrix} \tag{3.7}$$

where $A_1, A_2, \ldots, A_k$ are irreducible matrices. The matrices $A_1, A_2, \ldots, A_k$ correspond to the strong components of $\mathcal{D}(A)$ and are called the *irreducible*

---

[2] Note that we allow $k = 0$ and thus a path of one vertex with no arcs. Such a path is called an *empty path*.

components of A. If A does not have any zeros on its main diagonal, then A is irreducible if and only if A is fully indecomposable, and more generally, the irreducible components of A are the fully indecomposable components of A. For more information on these and related matters we refer to [2].

To each arc (i, j) of the digraph $\mathcal{D}(A)$ we may assign the number sign a_{ij} and then speak of the *signed digraph* of A. The *sign of the arc* (i, j) is defined to be sign a_{ij}. Note that no arc has sign equal to 0. Let α be the path

$$i = i_0 \to i_1 \to \cdots \to i_{k-1} \to i_k = j$$

from i to j of length k. The *sign of the path* α is

$$\text{sign}\,\alpha = \text{sign}\,a_{i_0 i_1} a_{i_1 i_2} \cdots a_{i_{k-1} i_k}$$

and equals the product of the signs of its arcs. The sign of a path of length 0, being an empty product, is defined to be $+1$. Let γ be the directed cycle

$$j_1 \to j_2 \to \cdots \to j_k \to j_1$$

of length k. The *sign of the directed cycle* γ is

$$\text{sign}\,\gamma = \text{sign}\,a_{j_1 j_2} a_{j_2 j_3} \cdots a_{j_{k-1} j_k} a_{j_k j_1}$$

and equals the product of the signs of its arcs. Arcs, paths, and directed cycles with sign equal to $+1$ are called *positive* whereas those with sign equal to -1 are called *negative*.

A necessary condition for the matrix A to be a sign-nonsingular matrix is that there is a nonzero term in the standard determinant expansion of A. Let P be a permutation matrix of order n and let D be a strict signing of order n. Then A is an SNS-matrix if and only if DPA is an SNS-matrix. Thus, in investigating SNS-matrices one may assume without loss of generality that A has a *negative main diagonal*, that is, each of the entries on its main diagonal is negative, or one may also assume that A has a *positive main diagonal*. In general, it turns out that theorems about SNS-matrices have a simpler form under the assumption of a negative main diagonal.

The following characterizations of SNS- and S^2NS-matrices in terms of signed digraphs are due to Bassett, Maybee, and Quirk [1].

Theorem 3.2.1 *Let $A = [a_{ij}]$ be a matrix of order n with a negative main diagonal. Then A is an SNS-matrix if and only if every directed cycle of the signed digraph $\mathcal{D}(A)$ is negative.*

Proof First assume that A is an SNS-matrix. Let γ be the directed cycle

$$j_1 \to j_2 \to \cdots \to j_k \to j_1$$

of $\mathcal{D}(A)$. Then

$$(-1)^{k-1}a_{j_1 j_2}a_{j_2 j_3}\cdots a_{j_{k-1} j_k}a_{j_k j_1}\prod_{i\neq j_1, j_2, \ldots, j_k}a_{ii}$$

is a nonzero term in the standard determinant expansion of A and has sign equal to

$$(-1)^{n-1}\operatorname{sign}\gamma.$$

Because sign $a_{11}a_{22}\cdots a_{nn} = (-1)^n$, it then follows from Theorem 1.2.5 that sign $\gamma = -1$.

Now assume that every directed cycle γ of $\mathcal{D}(A)$ is negative. Let $\sigma = (i_1, i_2, \ldots, i_n)$ be a permutation of $\{1, 2, \ldots, n\}$ such that

$$t_\sigma = \operatorname{sgn}(\sigma)a_{1i_1}a_{2i_2}\cdots a_{ni_n}$$

is a nonzero term in the standard determinant expansion of A. We now let $\gamma_1, \gamma_2, \ldots, \gamma_\ell$ be the (permutation) cycles of σ of length greater than one and let p denote the number of cycles of length equal to one. Each γ_i corresponds to a directed cycle γ_i' of $\mathcal{D}(A)$. Since $\operatorname{sgn}(\sigma) = (-1)^{n-(\ell+p)}$ and each of the diagonal entries of A is negative, we have

$$\operatorname{sign} t_\sigma = (-1)^{n-\ell-p}(-1)^p\prod_{i=1}^{\ell}\operatorname{sign}\gamma_i = (-1)^n.$$

Hence the sign of each nonzero term in the standard determinant expansion of A is $(-1)^n$, and by Theorem 1.2.5, A is an SNS-matrix. $\qquad\square$

Let $A = [a_{ij}]$ be a matrix of order n whose determinant is not identically zero. There are, in general, many permutation matrices P and strict signings E such that the entries on the main diagonal of EPA are negative. The resulting signed digraphs $\mathcal{D}(EPA)$ may be quite different from one another. Yet since A is an SNS-matrix if and only if EPA is, either all of the signed digraphs have only negative directed cycles or none of them do. Suppose that A has a negative main diagonal and instead of assigning the numbers sign a_{ij} to the arcs (i, j) of $\mathcal{D}(A)$, we assign weights as follows:

$$\operatorname{wt}(i, j) = \begin{cases} 0 & \text{if } a_{ij} > 0 \\ 1 & \text{if } a_{ij} < 0. \end{cases}$$

Define the *weight of a directed cycle* of $\mathcal{D}(A)$ to be the sum of the weights of its arcs. Then it follows from Theorem 3.2.1 that A is an SNS-matrix if and only if the weight of every directed cycle of $\mathcal{D}(A)$ is odd.

The characterization of SNS-matrices that there is a nonzero term in the standard determinant expansion and all such terms have the same sign, can also be formulated in terms of bipartite graphs [5].

Let H_n be the $(0, 1, -1)$-matrix defined in (1.12). For example, we have

$$H_5 = \begin{bmatrix} -1 & 1 & 0 & 0 & 0 \\ -1 & -1 & 1 & 0 & 0 \\ -1 & -1 & -1 & 1 & 0 \\ -1 & -1 & -1 & -1 & 1 \\ -1 & -1 & -1 & -1 & -1 \end{bmatrix}. \tag{3.8}$$

We verified in section 1.2 that H_n is an SNS-matrix. We give another verification by applying Theorem 3.2.1. The signed digraph $\mathcal{D}(H_n)$ has positive arcs $(1, 2), (2, 3), \ldots, (n - 1, n)$ and a negative arc (i, j) for all $i > j$. Each directed cycle is of the form $p \to p + 1 \to p + 2 \to \cdots \to q \to p$ for some integers p and q with $1 \le p < q \le n - 1$, and hence has sign equal to -1. It follows from Theorem 3.2.1 that H_n is an SNS-matrix.

Corollary 3.2.2 *Let $A = [a_{ij}]$ be an SNS-matrix of order n with a negative main diagonal. Assume that $a_{rs} = 0$, and let E_{rs} be the $(0, 1)$-matrix of order n whose only nonzero entry is in the (r, s)-position. Then one of the matrices $A \pm E_{rs}$ is an SNS-matrix if and only if all paths from s to r in the signed digraph $\mathcal{D}(A)$ have the same sign. In fact, if all paths from s to r are positive, then $A - E_{rs}$ is an SNS-matrix; if all paths from s to r are negative, then $A + E_{rs}$ is an SNS-matrix; and if there are no paths from s to r, then both $A + E_{rs}$ and $A - E_{rs}$ are SNS-matrices.*

Proof Since A is an SNS-matrix, it follows from Theorem 3.2.1 that $A - E_{rs}$ is an SNS-matrix if and only if every directed cycle of the signed digraph $\mathcal{D}(A - E_{rs})$ containing the arc (r, s) is negative. Hence $A - E_{rs}$ is an SNS-matrix if and only if all paths from s to r in the signed digraph $\mathcal{D}(A)$ are positive. Similarly, $A + E_{rs}$ is an SNS-matrix if and only if all paths from s to r in the signed digraph $\mathcal{D}(A)$ are negative. The corollary now follows. $\square$

Corollary 3.2.3 *Let $A = [a_{ij}]$ be a matrix of order n with a negative main diagonal. Then A is an S^2NS-matrix if and only if*

  (i) *every directed cycle of the signed digraph $\mathcal{D}(A)$ is negative, and*
  (ii) *sign α = sign β for every pair of paths α and β with the same initial vertex and the same terminal vertex.*

Proof Let E_{ij} be the $(0,1)$-matrix of order n whose only nonzero entry is in the (i, j)-position. It follows from Corollary 1.2.8 that A is an S^2NS-matrix if and only if A is an SNS-matrix and, for each i and j with $a_{ij} = 0$, either $A + E_{ij}$ or $A - E_{ij}$ is an SNS-matrix. The corollary is now a consequence of Theorem 3.2.1 and Corollary 3.2.2.   □

An S^2NS-matrix is an SNS-matrix A for which the sign of each entry in the inverses of the matrices in $\mathcal{Q}(A)$ is determined by the sign pattern of A. We now determine when the sign of a prescribed entry in the inverses of the matrices in $\mathcal{Q}(A)$ has the same sign. First we prove the following lemma [9].

Lemma 3.2.4 *Let* $A = [a_{ij}]$ *be an SNS-matrix of order n with a negative main diagonal and let r and s be distinct integers with* $1 \leq r, s \leq n$. *Then there exists a matrix* $\widetilde{A}$ *in* $\mathcal{Q}(A)$ *such that the* (r, s)-*entry of* $\widetilde{A}^{-1}$ *is positive, respectively negative, if and only if there exists a negative, respectively positive, path from r to s in the signed digraph* $\mathcal{D}(A)$.

Proof Without loss of generality we take A to be a $(0, 1, -1)$-matrix. First assume that there exists a negative path

$$r \to i_1 \to \cdots \to i_{k-2} \to s$$

from r to s in $\mathcal{D}(A)$. Let e_s be the n by 1 column vector whose only nonzero entry is a 1 in position s. Then there exists a nonzero term in the standard determinant expansion of $\det A(r \leftarrow e_s)$ of the form

$$(-1)^{k-1}(1)a_{ri_1} \cdots a_{i_{k-2},s}(-1)^{n-k} = (-1)^{n-1}a_{ri_1} \cdots a_{i_{k-2},s}.$$

Since $a_{ri_1} \cdots a_{i_{k-2},s}$ is negative, it follows that there exists a matrix $\widetilde{A}$ in $\mathcal{Q}(A)$ such that the sign of $\det \widetilde{A}(r \leftarrow e_s)$ is $(-1)^n$. By Cramer's rule, the (r, s)-entry of $\widetilde{A}^{-1}$ is positive.

Now assume that there exists a matrix $\widetilde{A}$ in $\mathcal{Q}(A)$ such that the (r, s)-entry of $\widetilde{A}^{-1}$ is positive. It follows from Cramer's rule that there exists a term t_σ in the standard determinant expansion of $\det \widetilde{A}(r \leftarrow e_s)$ whose sign is $(-1)^n$. There exist integers $i_1, \ldots, i_{k-2}$ such that

$$t_\sigma = (-1)^{k-1}(1)a_{ri_1} \cdots a_{i_{k-2},s}t'$$

where t' is a term in the standard determinant expansion of $A(\{r, i_1, \ldots, i_{k-2}, s\})$. Then

$$t_{\sigma'} = (-1)^k t'$$

is a nonzero term in the standard determinant expansion of A. Since A is an SNS-matrix, sign $t_{\sigma'} = (-1)^n$. Thus t_σ and $t_{\sigma'}$ have the same sign. Hence sign $a_{ri_1} \cdots a_{i_{k-2},s} = -1$ and

$$r \to i_1 \to \cdots \to i_{k-2} \to s$$

is a negative path from r to s in $\mathcal{D}(A)$. Similarly, if there exists a matrix $\widetilde{A}$ in $\mathcal{Q}(A)$ such that the (r, s)-entry of $\widetilde{A}^{-1}$ is negative, then there is a positive path from r to s. $\qquad\square$

Let A be an SNS-matrix of order n with a negative main diagonal. Since sign det $A = (-1)^n$, and since sign det $A(\{i\}) = (-1)^{n-1}$ for each i, the inverse of each matrix in $\mathcal{Q}(A)$ has a negative main diagonal. For entries off the main diagonal we have the following.

Theorem 3.2.5 *Let $A = [a_{ij}]$ be an SNS-matrix of order n with a negative main diagonal, and let r and s be distinct integers with $1 \le r, s \le n$.*

(i) *The entries in the (s, r)-position of the matrices in $\{\widetilde{A}^{-1} : \widetilde{A} \in \mathcal{Q}(A)\}$ are of the same sign if and only if all paths from s to r in the signed digraph $\mathcal{D}(A)$ have the same sign.*

(ii) *If each path from s to r in $\mathcal{D}(A)$ has the same sign ϵ, then the sign of the entry in the (s, r)-position of each of the matrices in $\{\widetilde{A}^{-1} : \widetilde{A} \in \mathcal{Q}(A)\}$ is $-\epsilon$. (Here ϵ is defined to be zero if there are no paths from s to r.)*

(iii) *If A is fully indecomposable and $a_{rs} \ne 0$, then the sign of the entry in the (s, r)-position of each of the matrices in $\{\widetilde{A}^{-1} : \widetilde{A} \in \mathcal{Q}(A)\}$ is sign a_{rs}.*

Proof Lemma 3.2.4 immediately implies (i) and (ii). Assume that A is fully indecomposable and $a_{rs} \ne 0$. Since $\mathcal{D}(A)$ is strongly connected, there is a path from s to r in $\mathcal{D}(A)$. Adjoining the arc (r, s) to any such path we obtain a directed cycle which by Theorem 3.2.1 is negative. Hence the sign of each path from s to r in $\mathcal{D}(A)$ is $-$sign a_{rs}. Assertion (iii) now follows from (ii). $\qquad\square$

For example, the sign pattern of the inverse of every matrix in $\mathcal{Q}(H_5)$, where H_5 is given in (3.8), has the form

$$\begin{bmatrix} -1 & -1 & -1 & -1 & -1 \\ 1 & -1 & -1 & -1 & -1 \\ * & 1 & -1 & -1 & -1 \\ * & * & 1 & -1 & -1 \\ * & * & * & 1 & -1 \end{bmatrix}.$$

Let A be a matrix of order n of the form

$$\begin{bmatrix} A_1 & A_2 \\ A_3 & A_4 \end{bmatrix} \tag{3.9}$$

where A_4 is an SNS-matrix of order $n - k$ with a negative main diagonal. Since each principal submatrix of A_4 is an SNS-matrix, we may pivot on the last $n - k$ positions of the main diagonal and obtain

$$\begin{bmatrix} I_k & -A_2 A_4^{-1} \\ O & A_4^{-1} \end{bmatrix} \begin{bmatrix} A_1 & A_2 \\ A_3 & A_4 \end{bmatrix} = \begin{bmatrix} A_1 - A_2 A_4^{-1} A_3 & O \\ A_4^{-1} A_3 & I_{n-k} \end{bmatrix}.$$

The matrix $A_1 - A_2 A_4^{-1} A_3$ is the *Schur complement* of A_4 in A. The next corollary [7] characterizes those entries of the Schur complement whose signs are determined by the sign pattern of A.

Corollary 3.2.6 *Let $A = [a_{ij}]$ be a matrix of order n of the form* (3.9) *such that A_4 is an SNS-matrix of order $n - k$ with a negative main diagonal, and let p and q be integers with $1 \le p, q \le k$. Then the entries in the (p, q)-position of the Schur complements*

$$\widetilde{A}_1 - \widetilde{A}_2 \widetilde{A}_4^{-1} \widetilde{A}_2 \quad (\widetilde{A} \in \mathcal{Q}(A))$$

are of the same sign if and only if either $p \ne q$ and all paths in $\mathcal{D}(A)$ from p to q whose internal vertices are contained in $\{k + 1, \ldots, n\}$ have the same sign and this sign equals $-\text{sign}\, a_{pq}$ if $a_{pq} \ne 0$, or $p = q$ and all directed cycles in $\mathcal{D}(A)$ containing p whose other vertices are contained in $\{k + 1, \ldots, n\}$ have the same sign and this sign equals $-\text{sign}\, a_{pp}$ if $a_{pp} \ne 0$.

Proof Let

$$\widetilde{A} = [\tilde{a}_{ij}] = \begin{bmatrix} \widetilde{A}_1 & \widetilde{A}_2 \\ \widetilde{A}_3 & \widetilde{A}_4 \end{bmatrix}$$

be a matrix in $\mathcal{Q}(A)$. Let $\widetilde{A}_4^{-1} = [z_{ij}]$ where $k + 1 \le i, j \le n$. Since the (p, q)-entry of $\widetilde{A}_2 \widetilde{A}_4^{-1} \widetilde{A}_3$ equals

$$\sum_{i=k+1}^{n} \sum_{j=k+1}^{n} \tilde{a}_{pi} z_{ij} \tilde{a}_{jq}, \tag{3.10}$$

the (p, q)-entries of the matrices $\widetilde{A}_2 \widetilde{A}_4^{-1} \widetilde{A}_3$ $(A_i \in \mathcal{Q}(A_i), i = 1, 2, 3)$ are of the same sign if and only if the nonzero terms occurring in the sums of the form (3.10) have the same sign. The corollary now follows from Lemma 3.2.4. $\quad\square$

We next characterize in terms of signed digraphs the N-canonical form of a square matrix with a negative main diagonal.

Lemma 3.2.7 *Let $A = [a_{ij}]$ be a matrix of order n with $a_{ii} \neq 0$ for all $i = 1, 2, \ldots, n$. Let $A[\lambda_1], A[\lambda_2], \ldots, A[\lambda_k]$ be the irreducible components of A. Then p is in $\mathcal{N}_\ell(\mathcal{Q}(A))$ if and only if there exist a λ_i for which $A[\lambda_i]$ is not an SNS-matrix and a path from a vertex in λ_i to p.*

Proof Let R be the set of all vertices which are reachable by a path from a vertex in some λ_i for which $A[\lambda_i]$ is not an SNS-matrix. Clearly, R is the union of some of the sets $\lambda_1, \lambda_2, \ldots, \lambda_k$, and if λ_j is not a subset of R, then $A[\lambda_j]$ is an SNS-matrix. Hence $A[\overline{R}]$ is an SNS-matrix, and since $A[R, \overline{R}] = O$ we have $\mathcal{N}_\ell(\mathcal{Q}(A)) \subseteq R$.

Now let β be the set of j such that $a_{ij} \neq 0$ for some $i \in \mathcal{N}_\ell(\mathcal{Q}(A))$. By Theorem 3.1.4, $A[\overline{\mathcal{N}_\ell(\mathcal{Q}(A))}, \overline{\beta}]$ is an L-matrix and hence has at least as many columns as rows. Since each of the diagonal entries of A is nonzero, it follows that $A[\overline{\mathcal{N}_\ell(\mathcal{Q}(A))}, \overline{\beta}]$ is a square matrix and hence an SNS-matrix, and that $\beta = \mathcal{N}_\ell(\mathcal{Q}(A))$. No vertex in $\overline{\mathcal{N}_\ell(\mathcal{Q}(A))}$ can be reached by a path from a vertex in a λ_j where $A[\lambda_j]$ is not an SNS-matrix, and hence $R \subseteq \mathcal{N}_\ell(\mathcal{Q}(A))$. □

Theorem 3.2.8 *Let A be a square matrix with signed digraph $\mathcal{D}(A)$, and assume that A has a negative main diagonal. Also, let $(\pi_1, \pi_2, \pi_3, \pi_4)$ and $(\theta_1, \theta_2, \theta_3, \theta_4)$ be the ordered row and column partitions of the N-canonical form of A. Then*

(i) *$(\pi_1, \pi_2, \pi_3, \pi_4) = (\theta_1, \theta_2, \theta_3, \theta_4)$,*

(ii) *π_1 is the set of vertices which can be reached by a path from a vertex of a positive directed cycle but which cannot reach a vertex of a positive directed cycle,*

(iii) *π_2 is the set of vertices which can be reached by a path from a vertex of a positive directed cycle and which can reach a vertex of a positive directed cycle,*

(iv) *π_3 is the set of vertices which cannot be reached by a path from a vertex of a positive directed cycle and which cannot reach a vertex of a positive directed cycle, and*

(v) *π_4 is the set of vertices which cannot be reached by a path from a vertex of a positive directed cycle but which can reach a vertex of a positive directed cycle.*

Proof Let $\alpha = \mathcal{N}_\ell(\mathcal{Q}(A))$ and let β be the set defined in (3.2). Let $\gamma = \mathcal{N}_r(\mathcal{Q}(A))$ and let δ be the set defined in (3.3). It follows from Lemma 3.2.7 and Theorem 3.2.1 that α equals the set of vertices which can be reached by a path from a positive directed cycle of $\mathcal{D}(A)$. Since A has a negative main diagonal $\beta = \alpha$. A similar argument applied to A^T shows that γ equals the set of vertices which can reach a vertex of a positive directed cycle of $\mathcal{D}(A)$ and that $\gamma = \delta$. Since A has N-canonical form (3.5), the theorem now follows from Theorem 3.1.4. $\square$

We conclude this section by considering nonnegative SNS-matrices and SNS-matrices whose off-diagonal entries are nonnegative. First we note that by applying Theorem 3.2.1 to $-A$ we obtain the following signed digraph characterization of SNS-matrices A with a positive main diagonal.

Corollary 3.2.9 *Let A be a matrix of order n with a positive main diagonal. Then A is an SNS-matrix if and only if for each $k = 2, 3, \ldots, n$ and each directed cycle γ of length k we have sign $\gamma = (-1)^{k-1}$.*

Corollary 3.2.9 immediately implies the following.

Corollary 3.2.10 *Let A be a nonnegative matrix of order n with a positive main diagonal. Then A is an SNS-matrix if and only if $\mathcal{D}(A)$ does not have a directed cycle of even length.*

Let $\mathcal{D}$ be a digraph of order n with vertices $1, 2, \ldots, n$. The *adjacency matrix* of $\mathcal{D}$ is the $(0, 1)$-matrix $B = [b_{ij}]$ of order n where $b_{ij} = 1$ if and only if (i, j) is an arc of $\mathcal{D}$. Suppose that $\mathcal{D}$ does not have any loops. Then $b_{ii} = 0$ for all i and hence the matrix $A = I_n + B$ is a $(0,1)$-matrix. It follows from Corollary 3.2.10 that A is an SNS-matrix if and only if $\mathcal{D}$ does not have any directed cycles of even length.

As observed in [13] the problem of recognizing if a signed digraph has a positive directed cycle can be reduced to the problem of recognizing if a digraph has a directed cycle of even length. To see this, we divide each arc (i, j) whose sign is $+1$ into two arcs (i, k_{ij}) and (k_{ij}, j) by inserting a new vertex k_{ij}. The resulting digraph has a directed cycle of even length if and only if the signed digraph has a positive directed cycle. Thus the SNS-recognition problem *Is the square matrix A not an SNS-matrix?* and the even cycle recognition problem *Does the digraph $\mathcal{D}$ have a directed cycle of even length?* are equivalent. The complexity of the even cycle recognition problem for digraphs has not been resolved [8]. It is perhaps worth pointing out that there is a polynomial-time algorithm for recognizing if a digraph has a directed cycle of odd length. This follows from the observations that a digraph $\mathcal{D}$ of order n has a directed cycle of odd length if and only if it has a closed directed walk of odd length $k \leq n$, and that if A is the adjacency matrix of $\mathcal{D}$, then the number of closed directed walks of length k equals the trace of A^k.

We now use digraphs to construct SNS-matrices of 0's and 1's. Let the vertices $2, 3, \ldots, n$ be partitioned into two nonempty sets X and Y. Let $K(X, Y)$ be the complete bipartite graph with bipartition $\{X, Y\}$. Let $\overrightarrow{K(X, Y)}$ denote a digraph obtained by assigning a direction to each edge of $K(X, Y)$ in such a way that $\overrightarrow{K(X, Y)}$ has no directed cycles.[3] Let $\mathcal{D}$ be the digraph obtained from $\overrightarrow{K(X, Y)}$ by adjoining an arc from each vertex of X to vertex 1 and an arc from vertex 1 to each vertex of Y. Then each directed cycle of $\mathcal{D}$ contains vertex 1 and contains an equal number of vertices of X and Y. Hence $\mathcal{D}$ does not have any directed cycles of even length. By Corollary 3.2.10, the matrix $I_n + B$ is an SNS-matrix where B is the adjacency matrix of $\mathcal{D}$. Note that the total number of 1's which are contained in row 1 or column 1 of $I_n + B$ equals n. For example, suppose that X has k vertices and Y has ℓ vertices and suppose that each arc of $\overrightarrow{K(X, Y)}$ is oriented from a vertex in Y to a vertex in X. Then we obtain the SNS-matrix

$$A = \left[\begin{array}{c|ccc|ccc} 1 & 1 & \cdots & 1 & 0 & \cdots & 0 \\ \hline 0 & & & & & & \\ \vdots & & I_\ell & & & J_{\ell,k} & \\ 0 & & & & & & \\ \hline 1 & & & & & & \\ \vdots & & O & & & I_\ell & \\ 1 & & & & & & \end{array} \right].$$

Now let $\mathcal{D}_n$ be the digraph of order n with arcs $(1, 2), (2, 3), \ldots, (n - 1, n)$ and (i, j) for all $i > j$ for which $i - j$ is even [4]. Each directed cycle of $\mathcal{D}_n$ is of the form $p \to p + 1 \to p + 2 \to \cdots \to q \to p$ for some integers p and q with $1 \leq p < q \leq n - 1$ and $q - p$ even. Hence each directed cycle of $\mathcal{D}_n$ has odd length. By Corollary 3.2.10, the matrix $M_n = I_n + A_n$ is an SNS-matrix of 0's and 1's where A_n is the adjacency matrix of $\mathcal{D}_n$. Let $E = \text{diag}(e_1, e_2, \ldots, e_n)$ where $e_i = (-1)^{i+1}$ $(i = 1, 2, \ldots, n)$. Then M_n is

---

[3]Thus $\overrightarrow{K(X, Y)}$ is an *acyclic orientation* of $K(X, Y)$.

the matrix obtained from $-E H_n E$ by replacing each -1 by 0.[4] For example,

$$
M_5 = \begin{bmatrix} 1 & 1 & 0 & 0 & 0 \\ 0 & 1 & 1 & 0 & 0 \\ 1 & 0 & 1 & 1 & 0 \\ 0 & 1 & 0 & 1 & 1 \\ 1 & 0 & 1 & 0 & 1 \end{bmatrix} \quad \text{and} \quad M_6 = \begin{bmatrix} 1 & 1 & 0 & 0 & 0 & 0 \\ 0 & 1 & 1 & 0 & 0 & 0 \\ 1 & 0 & 1 & 1 & 0 & 0 \\ 0 & 1 & 0 & 1 & 1 & 0 \\ 1 & 0 & 1 & 0 & 1 & 1 \\ 0 & 1 & 0 & 1 & 0 & 1 \end{bmatrix}.
$$

Since for each i the total number of 1's which are contained in row i or column i is less than n, $\mathcal{D}_n$ cannot be constructed as given by using an acyclic orientation of the edges of a complete bipartite graph $K(X, Y)$. However, as we now show, a reordering of the rows of M_n gives a digraph which can. First suppose n is odd. Let P_n be the permutation matrix of order n with 1's in positions $(2, 1)$, $(3, 2)$, ..., $(n, n-1)$ and $(1, n)$. Then $P_n^T \leq M_n$ (entrywise) and hence $P_n M_n$ has only 1's on its main diagonal. Let $X = \{3, 5, \ldots, n\}$ and $Y = \{2, 4, \ldots, n-1\}$. Then $(1, y)$ and $(x, 1)$ are arcs of $\mathcal{D}(P_n M_n)$ for each $y \in Y$ and each $x \in X$. The remaining arcs form the acyclic orientation of $K(X, Y)$ where for each x in X and y in Y, (x, y) is an arc provided $x > y$ and (y, x) is an arc provided $y > x$. Now suppose that n is even, and let P_n be the permutation matrix with 1's in positions $(1, 1)$, $(3, 2)$, $(4, 3)$, ..., $(n, n-1)$, $(2, n)$. Then $P_n M_n$ has only 1's on its main diagonal. Let $X = \{1, 3, \ldots, n-1\}$ and $Y = \{4, 6, \ldots, n\}$. Then $(2, y)$ and $(x, 2)$ are arcs of $\mathcal{D}(P_n M_n)$ for each $y \in Y$ and each $x \in X$. As before, the remaining arcs form the acyclic orientation of $K(X, Y)$ where for each x in X and y in Y, (x, y) is an arc provided $x > y$ and (y, x) is an arc provided $y > x$.

We now discuss another way of constructing SNS-matrices. A digraph D is *intercyclic* provided it does not contain two directed cycles which have no common vertex. Intercyclic digraphs have been extensively studied [11, 12, 14, 15]. In [11], intercyclic digraphs are completely characterized and a polynomial-time algorithm for their recognition is given. The following shows that each intercyclic graph gives rise to an SNS-matrix.

Theorem 3.2.11 *Let D be an intercyclic digraph of order n such that there is a directed cycle containing vertex 1. Let A be the $(0, 1, -1)$-matrix such that $\mathcal{D}(A)$ is D where $a_{ij} \geq 0$ if $i \neq j$, $a_{11} = 0$ and $a_{ii} = -1$ if $i \neq 1$. Then A is an SNS-matrix.*

[4]It has been shown in [3] that det M_n equals the nth Fibonacci number and that no $(0, 1)$ lower Hessenberg matrix has a determinant with a larger absolute value.

Proof Consider a nonzero term

$$t_\sigma = \text{sgn}(\sigma) a_{1,\sigma(1)} a_{2,\sigma(2)} \cdots a_{n,\sigma(n)}$$

in the standard determinant expansion of A. Since D is intercyclic and $a_{11} = 0$, the permutation σ has exactly one cycle of length greater than 1 and $\sigma(1) \neq 1$. Let k be the length of the nontrivial cycle of σ. Without loss of generality we may assume that $\sigma(i) = i + 1$ for $i = 1, 2, \ldots, k - 1$, $\sigma(k) = 1$, and $\sigma(i) = i$ for $i > k$. Thus

$$\text{sign } t_\sigma = \text{sgn}(\sigma) a_{12} a_{23} \ldots a_{k-1,k} a_{k,1} a_{k+1,k+1} a_{k+2,k+2} \cdots a_{n,n} = (-1)^{n-1}.$$

Hence each nonzero term in the standard determinant expansion of A has the same sign, and since vertex 1 is contained in a directed cycle of D, there is at least one nonzero term in the standard determinant expansion of A. Therefore by Theorem 1.2.5, A is an SNS-matrix. $\square$

One way of constructing intercyclic digraphs is the following. Let G be a plane graph on $n + 2$ vertices whose boundary consists of the edges $\{n - 1, n\}$, $\{n, n + 1\}$, $\{n + 1, n + 2\}$, and $\{n + 2, n - 1\}$. Let H be the graph obtained from G by removing its four boundary edges, and let $\overrightarrow{H}$ be an acyclic orientation of H such that the vertices $n + 1$ and $n + 2$ have indegree 0 and vertices $n - 1$ and n have outdegree 0. Let D be the digraph obtained from $\overrightarrow{H}$ by identifying vertex $n + 1$ with vertex $n - 1$ and vertex $n + 2$ with vertex n. Clearly any directed cycle of D contains either vertex $n - 1$ or vertex n. The assumptions on G imply that every path from $n - 1$ to $n + 1$ in $\overrightarrow{H}$ and every path from n to $n + 2$ in $\overrightarrow{H}$ intersect. It now follows that D is intercyclic.

3.3 Sign-solvability and digraphs

In section 1.2 we reduced the study of the sign-solvability of a linear system to the study of L-matrices and the study of the sign-solvability of a linear system which has a positive solution and whose coefficient matrix is an SNS-matrix (that is, S-matrices). In section 3.1 we reduced the study of the partial sign-solvability of a linear system to the study of the left and right null-supports of the qualitative class of its coefficient matrix and the study of the partial sign-solvability of a linear system whose coefficient matrix is an SNS-matrix. In this section we characterize in terms of signed digraphs the partial sign-solvability and sign-solvability of a linear system with a sign-nonsingular coefficient matrix.

Throughout this section $Ax = b$ denotes a linear system with variables $x_1, x_2, \ldots, x_n$, where $A = [a_{ij}]$ is a square matrix of order n whose determinant is not identically zero and where $b = (b_1, b_2, \ldots, b_n)^T$ is a column vector. The assumption that A does not have an identically zero determinant enables us to transform the system $Ax = b$ into a form which facilitates its investigation. By permuting rows of $Ax = b$ we may assume that each diagonal entry a_{ii} of A is not zero. By multiplying $Ax = b$ on the left by a strict signing we may also assume that $b_i \le 0$ for each i. By replacing A by AD and x by Dx for some strict signing D, we may further assume that each a_{ii} is negative.

Lemma 3.3.1 *Let A be a square matrix of order n with a negative main diagonal. Let e_i be the n by 1 column vector with a 1 in the ith position and 0's elsewhere. Then the matrix $A(j \leftarrow e_i)$ has an identically zero determinant if and only if there does not exist a path from j to i in $\mathcal{D}(A)$.*

Proof In $\mathcal{D}(A(j \leftarrow e_i))$ the only arc with terminal vertex j is the arc (i, j) from i to j. The kth diagonal entry of $A(j \leftarrow e_i)$ is nonzero for each $k \ne j$. It follows that there is a nonzero term in the standard determinant expansion of $A(j \leftarrow e_i)$ if and only if there is a directed cycle in $\mathcal{D}(A(j \leftarrow e_i))$ containing the arc (i, j). The latter holds if and only if there is a path from j to i in $\mathcal{D}(A)$. $\square$

In Theorem 1.2.6 we have given necessary and sufficient conditions for the sign-solvability of $Ax = b$. It follows as in the proof of Theorem 1.2.6 that if A is an SNS-matrix, then x_j is sign-determined by $Ax = b$ if and only if the matrix $A(j \leftarrow b)$ is either an SNS-matrix or has an identically zero determinant.

The following theorem is essentially contained in [1].

Theorem 3.3.2 *Let $Ax = b$ be a linear system where A is an SNS-matrix of order n with a negative main diagonal. Assume that each entry of b is nonpositive. Let $\mathcal{P}_j$ be the set of all paths in $\mathcal{D}(A)$ with initial vertex j and with terminal vertex in $\{i : b_i < 0\}$. Then x_j is sign-determined by $Ax = b$ if and only if all paths in $\mathcal{P}_j$ have the same sign. When x_j is sign-determined, $\mathrm{sign}(x_j : Ax = b)$ equals the common sign of the paths in $\mathcal{P}_j$ if $\mathcal{P}_j \ne \emptyset$ and equals zero otherwise.*

Proof As just noted, x_j is sign-determined if and only if $A(j \leftarrow b)$ has an identically zero determinant or is an SNS-matrix. The matrix $A(j \leftarrow b)$ has an identically zero determinant if and only if $A(j \leftarrow e_i)$ has an identically zero determinant for each i with $b_i < 0$. It follows from Lemma 3.3.1 that $A(j \leftarrow b)$ has an identically zero determinant if and only if $\mathcal{P}_j = \emptyset$.[5]

Now assume that $\mathcal{P}_j \neq \emptyset$. Since $a_{jj} \neq 0$, the matrix $A(j \leftarrow -e_j)$ is an SNS-matrix. First suppose that $b_j < 0$. Then $A(j \leftarrow b)$ does not have an identically zero determinant. Since $\mathcal{D}(A(j \leftarrow b))$ has an arc (i, j) of sign -1 for each $i \neq j$ with $b_i < 0$, the result follows from Theorem 3.2.1. Now suppose that $b_j = 0$. It follows that $A(j \leftarrow b)$ is an SNS-matrix if and only if one of $A(j \leftarrow b - e_j)$ and $A(j \leftarrow -b - e_j)$ is an SNS-matrix. Again the result follows from Theorem 3.2.1.   □

Corollary 3.3.3 *Let $Ax = b$ be a sign-solvable linear system where A is a matrix of order n with a negative main diagonal. Assume that each entry of b is nonpositive. Let $\mathcal{Q}(Ax = b) = \mathcal{Q}(u)$ where $u = (u_1, u_2, \ldots, u_n)^T$. Then $u_j = 0$ if and only if there does not exist a path in $\mathcal{D}(A)$ from j to a vertex in $\{i : b_i < 0\}$.*

Given an SNS-matrix A of order n, in the next corollary we determine for each column vector b and each subset W of $\{1, 2, \ldots, n\}$ whether each of the variables x_j with $j \in W$ is sign-determined by $Ax = b$. As already noted it suffices to assume that each of the diagonal entries of A is negative and each entry of b is nonpositive.

If j is a vertex of a signed digraph $\mathcal{D}$, then α_j denotes the set of vertices of $\mathcal{D}$ which are not reachable from j by a negative path, β_j denotes the set of vertices which are not reachable from j by a positive path, and γ_j denotes the set of vertices which can be reached from j by both a positive path and a negative path. Note that $\alpha_j \cap \beta_j$ is the set of vertices of $\mathcal{D}$ not reachable from j by any path.

Corollary 3.3.4 *Let A be an SNS-matrix of order n with a negative main diagonal, and let $b = (b_1, b_2, \ldots, b_n)^T$ be a column vector each of whose entries is nonpositive. Let W be a subset of the set $\{1, 2, \ldots, n\}$ of vertices of the signed digraph $\mathcal{D}(A)$. Then all of the variables x_j with j in W are sign-determined by $Ax = b$ if and only if for some choice of δ_j in $\{\alpha_j, \beta_j\}$ $(j \in W)$ we have*

$$\{i : b_i < 0\} \subseteq \cap_{j \in W} \delta_j.$$

Proof By Theorem 3.3.2, x_j is sign-determined by $Ax = b$ if and only if $\{i : b_i < 0\}$ is contained in either α_j or β_j. The corollary now follows.   □

---

[5] Since we allow paths of length 0, $\mathcal{P}_j$ can be empty only if $b_j = 0$.

We next consider the sign-solvability of linear systems. Suppose that $Ax = b$ is sign-solvable where A is a square matrix of order n, each diagonal entry of A is negative, and each entry of b is nonpositive. By Corollary 1.2.2, A is an SNS-matrix. Let $u = (u_1, u_2, \ldots, u_n)^T$ be a column vector with $Au = b$. Consider an integer j such that $b_j < 0$. Since the path of length 0 terminating at j is positive, Theorem 3.3.2 implies that $u_j > 0$. Let $D = \mathrm{diag}(d_1, d_2, \ldots, d_m)$ be the strict signing where $d_i = -1$ if and only if u_i is negative (and hence $b_i = 0$). We may replace the linear system $Ax = b$ by $(DAD)y = Db$. Therefore, in investigating sign-solvable linear systems $Ax = b$ in which A is a square matrix, we may assume that $Ax = b$ is in the *standard form* [1, 10] defined by

  (i) $a_{ii} < 0$ for all i,
 (ii) $b_i \le 0$ for all i, and
(iii) $Ax = b$ has a solution $u = (u_1, u_2, \ldots, u_n)^T$ where $u_j \ge 0$ for all j.

Since the linear system $Ax = b$ is sign-solvable if and only if each x_j is sign-determined, Theorems 3.2.1 and 3.3.2 immediately imply the following [1, 10].

Corollary 3.3.5 *Let $Ax = b$ be a linear system in standard form where A is a matrix of order n. Then $Ax = b$ is sign-solvable if and only if*

  (i) *every directed cycle of the signed digraph $\mathcal{D}(A)$ is negative, and*
 (ii) *every path with terminal vertex in $\{i : b_i < 0\}$ is positive.*

We now examine in more detail the conditions (i) and (ii) in the characterization of sign-solvability given in Corollary 3.3.5. A vertex u of a signed digraph $\mathcal{D}$ is a *positive terminus* provided the sign of each path of $\mathcal{D}$ which terminates at u is positive. Thus u is a positive terminus if and only if no path of $\mathcal{D}$ which terminates at u contains a negative arc. Corollary 3.3.5 implies the following.

Corollary 3.3.6 *Let $Ax = b$ be a linear system in standard form where A is an SNS-matrix. Then $Ax = b$ is sign-solvable if and only if $\{i : b_i < 0\}$ is contained in the set of positive termini of $\mathcal{D}(A)$.*

Consider the matrix $A = H_n$ of order n defined in (1.12). Since the signed digraph of A has positive arcs $(1, 2), (2, 3), \ldots, (n-1, n)$ and a negative arc (i, j) for all $i > j$, the vertex n is the only positive terminus. Hence if $Ax = b$ is in standard form, then $Ax = b$ is sign-solvable if and only if $b = (0, 0, \ldots, c)^T$ where $c \le 0$.

If $\mathcal{D}$ is a signed digraph, then the digraph whose vertices are the vertices of $\mathcal{D}$ and whose arcs are the positive arcs of $\mathcal{D}$ is denoted by $\mathcal{D}^+$. Note that if each directed cycle of $\mathcal{D}$ is negative, then $\mathcal{D}^+$ is an acyclic digraph and hence has at least one *source* (a vertex with indegree equal to 0) and at least one *sink* (a vertex of outdegree equal to 0).

We now investigate the positive termini of a signed digraph. The second assertion in the next lemma is stated in [10].

Lemma 3.3.7 *Let $\mathcal{D}$ be a strongly connected signed digraph.*

   (i) *If $\mathcal{D}$ has a positive terminus, then each directed cycle of $\mathcal{D}$ contains at most one negative arc.*

  (ii) *If each directed cycle of $\mathcal{D}$ is negative, then $\mathcal{D}$ has at most one positive terminus.*

 (iii) *If each directed cycle of $\mathcal{D}$ is negative and $\mathcal{D}$ has a positive terminus u, then u is a sink of $\mathcal{D}^+$ and $\mathcal{D}^+$ has no other sink.*

Proof If $\mathcal{D}$ has a directed cycle with at least two negative arcs, then for each vertex v there exists a negative path terminating at v. Hence (i) holds. Now assume that each directed cycle of $\mathcal{D}$ is negative. Suppose u is a positive terminus of $\mathcal{D}$ and v is any other vertex. Let α and β be paths from u to v and v to u, respectively. There is a directed cycle each of whose arcs is an arc of α or of β. Since each arc of β is positive and each directed cycle is negative, α contains a negative arc and hence v is not a positive terminus. Each path from a sink of $\mathcal{D}^+$ to any other vertex must contain a negative arc of $\mathcal{D}$. Since $\mathcal{D}$ is strongly connected, u is a sink of $\mathcal{D}^+$ and there can be no other sink. Hence (ii) and (iii) hold. $\qquad\qquad\square$

Lemma 3.3.8 *Let $A = [a_{ij}]$ be a fully indecomposable SNS-matrix with a negative main diagonal. Then vertex k is a positive terminus of the signed digraph $\mathcal{D}(A)$ if and only if the matrix B obtained from A by deleting row k is an S-matrix.*

Proof First assume that k is a positive terminus of $\mathcal{D}(A)$. Then for each vertex i there is a path from i to k and every such path is positive. If $a_{kj} \neq 0$, then $A(\{k\}, \{j\})$ is an SNS-matrix. If $a_{kj} = 0$, then it follows from Corollary 3.2.2 that $A - E_{kj}$ is an SNS-matrix and hence $A(\{k\}, \{j\})$ is an SNS-matrix. Therefore B is an S^*-matrix. It follows from (iii) of Lemma 3.3.7 that each row of B is balanced and hence B is an S-matrix.

Now assume that B is an S-matrix. Then each of the matrices $A(\{k\}, \{j\})$ is an SNS-matrix. Hence each zero entry in row k of A can be replaced by 1 or -1 in such a way that the resulting matrix A' is an SNS-matrix. Since each row of B is balanced and since A' is an SNS-matrix, each entry in row k of A' is negative. Since each directed cycle of $\mathcal{D}(A')$ is negative, k is a positive terminus of $\mathcal{D}(A')$ and hence of $\mathcal{D}(A)$. $\square$

The second assertion in the following theorem was first noted in [6] (see also [10]).

Theorem 3.3.9 *Let $Ax = b$ be a linear system where A is a fully indecomposable matrix of order n and $b = (b_1, b_2, \ldots, b_n)^T$ is a column vector.*

(i) *If b has more than one nonzero entry, then $Ax = b$ is not sign-solvable.*

(ii) *Suppose that b has exactly one nonzero entry, say $b_j \neq 0$, and $Ax = b$ is in standard form. Then $Ax = b$ is sign-solvable if and only if no directed cycle of $\mathcal{D}(A)$ contains only positive arcs and j is a positive terminus of $\mathcal{D}(A)$. If $Ax = b$ is sign-solvable, then the vectors in the qualitative solution class $\mathcal{Q}(Ax = b)$ have no zero entries.*

(iii) *Suppose that b is the zero vector. Then $Ax = b$ is sign-solvable if and only if A is an SNS-matrix. If $Ax = b$ is sign-solvable, then $\mathcal{Q}(Ax = b) = \{0\}$.*

Proof Since A is fully indecomposable, for any permutation matrices P and Q of order n the digraph $\mathcal{D}(PAQ)$ is strongly connected. If $Ax = b$ is sign-solvable, then $Ax = b$ can be brought to standard form, and hence (i) follows from Corollary 3.3.6 and Lemma 3.3.7.[6]

Now suppose that b_j is the only nonzero entry of b and that $Ax = b$ is in standard form. First assume that $Ax = b$ is sign-solvable. Then A is an SNS-matrix and by Theorem 3.2.1, $\mathcal{D}(A)^+$ is acyclic. By Corollary 3.3.6, j is a positive terminus of $\mathcal{D}(A)$. Since $\mathcal{D}(A)$ is strongly connected, for each vertex i there exists a path from i to j, and hence by Corollary 3.3.3, the vectors in $\mathcal{Q}(Ax = b)$ have no zero entries. Now assume that $\mathcal{D}(A)^+$ is acyclic and that j is a positive terminus of $\mathcal{D}(A)$. By (i) of Lemma 3.3.7, each directed cycle of $\mathcal{D}(A)$ contains exactly one negative arc and hence is negative. By Corollary 3.3.6, $Ax = b$ is sign-solvable. Hence (ii) holds. By Theorem 1.2.1, $Ax = 0$ is sign-solvable if and only if A is an SNS-matrix, and hence (iii) follows. $\square$

---

[6]Assertion (i) can also be proven as follows. Assume that A is a fully indecomposable SNS-matrix of order n. Suppose that the ith entry of b is nonzero. By Theorem 2.2.5 (applied to A^T) there exists a strict signing $D = \text{diag}(d_1, d_2, \ldots, d_n)$ such that only row i of AD is unsigned. By possibly replacing D by $-D$, we may assume that row i of AD is nonnegative. It follows that there exists a matrix $\widetilde{A} \in \mathcal{Q}(A)$ such that the ith column of $\widetilde{A}^{-1}$ is $[d_1, d_2, \ldots, d_n]^T$. Thus, for some choice of $\bar{b} \in \mathcal{Q}(b)$, the solution to $\widetilde{A}x = \bar{b}$ is in $\mathcal{Q}([d_1, d_2, \ldots, d_n]^T)$. It now follows that if b has two or more nonzero entries, then $Ax = b$ is not sign-solvable.

Let $\mathcal{D}$ be a signed digraph each of whose directed cycles is negative. Let $C_1, C_2, \ldots, C_k$ be the strong components of $\mathcal{D}$ ordered so that there is an arc from a vertex in C_i to a vertex in C_j only if $i \geq j$. Each directed cycle of $\mathcal{D}$ is a directed cycle of some $\mathcal{D}_i$ and each positive terminus of $\mathcal{D}$ is a positive terminus of some C_i. It follows from Lemma 3.3.7 that each C_i contains at most one positive terminus of $\mathcal{D}$. If there is an arc (v, w) from a vertex v in C_i to a vertex w in C_j with $i \neq j$ and if C_j contains a positive terminus of $\mathcal{D}$, then v is a positive terminus of $\mathcal{D}$ and the arc (v, w) is positive. Hence we have the following [10].

Theorem 3.3.10 *Let $\mathcal{D}$ be a signed digraph each of whose directed cycles is negative. Then the strong components can be listed in the order $\mathcal{D}_1, \ldots, \mathcal{D}_\ell$, $\mathcal{D}_{\ell+1}, \ldots, \mathcal{D}_k$ such that*

(i) *each of $\mathcal{D}_{\ell+1}, \ldots, \mathcal{D}_k$ contains a unique positive terminus of $\mathcal{D}$, and none of $\mathcal{D}_1, \ldots, \mathcal{D}_\ell$ contain a positive terminus of $\mathcal{D}$;*
(ii) *each directed cycle of $\mathcal{D}_{\ell+1}, \ldots, \mathcal{D}_k$ contains exactly one negative arc;*
(iii) *there is an arc from a vertex in $\mathcal{D}_i$ to a vertex in $\mathcal{D}_j$ only if $i \geq j$;*
(iv) *if $\ell + 1 \leq j < i \leq k$, then each arc from a vertex in $\mathcal{D}_i$ to a vertex in $\mathcal{D}_j$ is a positive arc and its initial vertex is the positive terminus of $\mathcal{D}$ in $\mathcal{D}_i$.*

We now summarize our conclusions for determining whether or not a linear system $Ax = b$ with square coefficient matrix A is sign-solvable.

(i) Permute the rows of A and multiply on the left by a strict signing so that each entry on the main diagonal of A is negative. (If this is not possible, then $Ax = b$ is not sign-solvable.)
(ii) Find a solution of $Ax = b$ and put $Ax = b$ into standard form. (If this is not possible, $Ax = b$ is not sign-solvable.)
(iii) Check that A is an SNS-matrix by verifying that every directed cycle of the signed digraph of A is negative. (If there is a positive directed cycle, then by Corollary 1.2.2 $Ax = b$ is not sign-solvable.)
(iv) Simultaneously permute rows and columns to bring A to the form (3.7) where $A_{\ell+1}, \ldots, A_k$ are the irreducible components of A corresponding to the strong components of $\mathcal{D}(A)$ which contain a positive terminus of $\mathcal{D}(A)$, and $A_1, \ldots, A_\ell$ are the irreducible components of A corresponding to the strong components of $\mathcal{D}(A)$ which do not contain a positive terminus. (By Theorem 3.3.10 this can always be done.) Since the digraphs $\mathcal{D}_{\ell+1}^+, \ldots, \mathcal{D}_k^+$ are acyclic and each of their directed cycles contains exactly one negative arc, we may simultaneously permute

rows and columns so that the entries of the matrices $A_{\ell+1}, \ldots, A_k$ are nonpositive above the diagonal, and nonnegative below the diagonal. By (iii) of Lemma 3.3.7, the first row and first column of each of the matrices $A_{\ell+1}, \ldots, A_k$ corresponds to a positive terminus of $\mathcal{D}(A)$.

(v) For each i and j with $\ell + 1 \leq j < i \leq k$ check that all of the entries of the matrix A_{ij} are nonnegative and each positive entry is contained in its first row. (If not, then by (iii) and (iv) of Theorem 3.3.10 $Ax = b$ is not sign-solvable.)

(vi) If the negative entries of b are in positions corresponding to a subset of the first rows of $A_{\ell+1}, \ldots, A_k$, then $Ax = b$ is sign-solvable. (If not, then $Ax = b$ is not sign-solvable by Corollary 3.3.6.)

Note that if $Ax = b$ is sign-solvable, then b has at most $k - \ell$ nonzero entries, and the entries of the vectors in $\mathcal{Q}(Ax = b)$ corresponding to the columns of $A_1, \ldots, A_\ell$ are zero. It follows from Corollary 3.3.3 that $Ax = b$ is sign-solvable and the vectors in $\mathcal{Q}(Ax = b)$ have no zero entries if and only if $\ell = 0$ and each entry of b corresponding to a sink of $\mathcal{D}(A)^+$ is negative.

Let $A_1, A_2, \ldots, A_k$ be fully indecomposable SNS-matrices such that each entry on the main diagonal is negative, each entry in row 1 is nonpositive and the matrices obtained by deleting row 1 are S-matrices. Let A be a matrix of the form (3.7) where for each i and j with $1 \leq j < i \leq k$ all of the entries of the matrix A_{ij} are nonnegative and each positive entry is contained in its first row. Choose a column vector b so that each entry of b is nonpositive and each negative entry is contained in a row corresponding to the first row of some A_i. Further, suppose that b has been chosen so that each row of $[A \ -b]$ contains a positive entry. Then the preceding summary and Lemma 3.3.8 imply that $Ax = b$ is sign-solvable and the entries of the vectors in $\mathcal{Q}(Ax = b)$ are all positive (that is, $[A \ -b]$ is an S-matrix). It also follows from the summary that every such sign-solvable linear system arises in this way.

Bibliography

[1]   L. Bassett, J. Maybee, and J. Quirk. Qualitative economics and the scope of the correspondence principle, *Econometrica*, 36:544–63, 1968.

[2]   R.A. Brualdi and H.J. Ryser. *Combinatorial Matrix Theory*, Cambridge University Press, New York, 1991.

[3]   L. Ching. The maximum determinant of an $n \times n$ lower Hessenberg $(0, 1)$-matrix, *Linear Alg. Appls.*, 183:147–53, 1993.

[4]   F.R.K. Chung, W. Goddard, and D.J. Kleitman. Even cycles in directed graphs, *SIAM J. Disc. Math.*, 7:474–83, 1994.

[5] J.H. Drew, B.C.J. Green, C.R. Johnson, D.D. Olesky, and P. van den Driessche. Bipartite characterization of sign-nonsingularity, preprint.

[6] P. Hansen. Recognizing sign-solvable graphs, *Discrete Appl. Math.*, 6:237–41, 1983.

[7] C.R. Johnson and J.S. Maybee. *Qualitative analysis of Schur complements*, in *Applied Geometry and Discrete Mathematics* (P. Gritzmann and B. Sturmfels, eds.), Amer. Math. Soc., Providence, 359–65, 1991.

[8] V. Klee, R. Ladner, and R. Manber. Signsolvability revisited, *Linear Alg. Appls.*, 59:131–57, 1984.

[9] G. Lady and J. Maybee. Qualitatively invertible matrices, *J. Math. Social Sciences*, 6:397–407, 1983.

[10] R. Manber. Graph-theoretical approach to qualitative solvability of linear systems, *Linear Alg. Appls.*, 48:457–70, 1982.

[11] W. McCuaig. Intercyclic digraphs, *Graph Structure Theory, Contemporary Mathematics* 147, Amer. Math. Soc., Providence, 203–45, 1993.

[12] C. Thomassen. The 2-linkage problem for acyclic digraphs, *Discrete Math.*, 55:73–87, 1985.

[13] C. Thomassen. Sign-nonsingular matrices and even cycles in directed graphs, *Linear Alg. Appls.*, 75:27–41, 1986.

[14] C. Thomassen. On digraphs with no two disjoint cycles, *Combinatorics*, 7:145–50, 1987.

[15] C. Thomassen. Disjoint cycles in digraphs, *Combinatorics* 3:393–96, 1989.

4

S^*-matrices

4.1 S^*-matrices and SNS*-matrices

We first recall some basic facts from sections 1.2 and 2.1. Let A be an m by $m + 1$ matrix. By definition, A is an S^*-matrix if and only if each submatrix of A of order m is an SNS-matrix. By (iv) of Theorem 2.1.1, A is an S^*-matrix if and only if there exists a strict signing D such that AD and $A(-D)$ are the only column signings of A each of whose rows are balanced. The matrix A is an S-matrix if and only if A is an S^*-matrix and $AI_{m+1} = A$ and $A(-I_{m+1}) = -A$ are the only column signings of A each of whose rows are balanced.

Let $v_1, v_2, \ldots, v_{m+1}$ be the column vectors of A. Then $v_1, v_2, \ldots, v_{m+1}$ are the vertices of an m-simplex whose interior contains the origin if and only if the right null space of A is spanned by a vector w each of whose entries is positive. This implies a geometric description of S-matrices [6]:

> The matrix A is an S-matrix if and only if for each matrix $\widetilde{A}$ in $\mathcal{Q}(A)$, the column vectors of $\widetilde{A}$ are the vertices of an m-simplex containing the origin in its interior.

Clearly, each row of an S^*-matrix must contain at least two nonzero entries. The following theorem [7] shows that an m by $m + 1$ matrix with at least three nonzero entries in each row is not an S^*-matrix.

Theorem 4.1.1 *Let A be an m by $m + 1$ S^*-matrix. Then there is a row of A which contains exactly two nonzero entries.*

Proof Without loss of generality we assume that A is an S-matrix. Thus each row of A is balanced, and the submatrix A_j of A obtained by deleting column j is an SNS-matrix for $j = 1, 2, \ldots, m + 1$. It follows that each of the $m + 1$ submatrices $A_1, A_2, \ldots, A_{m+1}$ of A has a unsigned row. By the pigeonhole principle there exist integers i, j, and k with $i \neq j$ such that row k of A_i and row k of A_j are unsigned. Since row k of A is balanced, this implies that row k of A has exactly two nonzero entries. □

The next theorem implies that an m by $m + 1$ S^*-matrix with $m > 1$ has at most one fully indecomposable submatrix of order m [2].

Theorem 4.1.2 *Let*

$$A = \left[\ B \mid b \ \right]$$

be an S^*-matrix such that B is fully indecomposable. Then b contains exactly one nonzero entry.

Proof Since A is an S^*-matrix, b has at least one nonzero entry and $Bx = -b$ is sign-solvable. By (i) of Theorem 3.3.9 the vector b contains at most one nonzero entry. □

If

$$\left[\ A_1 \mid b_1 \ \right] \text{ and } \left[\ A_2 \mid b_2 \ \right]$$

are S^*-matrices, it is easy to verify that

$$\left[\begin{array}{c|c|c} A_1 & O & b_1 \\ \hline O & A_2 & b_2 \end{array} \right]$$

is also an S^*-matrix. Another technique for constructing S^*-matrices is contained in the following theorem.

Theorem 4.1.3 *Let B be a matrix whose last column is the vector v, and let w be a column vector obtained from v by replacing some of its nonzero entries by zero. Let*

$$A = \left[\begin{array}{ccccc|c} & & B & & & w \\ \hline 0 & \cdots & 0 & 1 & & -1 \end{array} \right].$$

Then

(i) *A is an SNS-matrix if and only if B is;*

(ii) *A is an S^*-matrix if and only if B is;*

(iii) *A is an S-matrix if and only if B is.*

Proof Let $w = (w_1, w_2, \ldots, w_n)^T$. To prove (i) we assume that B is a square matrix of order n. First suppose that B has an identically zero determinant. By the Frobenius–König theorem there exist integers p and q with $p + q = n + 1$ such that B has a p by q zero submatrix. It follows from the definition of A that A has either a $p + 1$ by q or a p by $q + 1$ zero submatrix. Since $(p + 1) + q = p + (q + 1) = n + 2$, A has an identically zero determinant. Hence neither A nor B is an SNS-matrix and (i) holds. Now suppose that B does not have an identically zero determinant. Without loss of generality we assume that each diagonal entry of B is negative. The signed digraph of A is obtained from the signed digraph of B by adjoining a new vertex $n + 1$, a positive arc from vertex $n + 1$ to vertex n, and an arc from vertex i of $\mathcal{D}(B)$ to vertex $n + 1$ with sign equal to the sign of w_i if and only if $w_i \neq 0$. Hence there is an arc in $\mathcal{D}(A)$ from i to $n + 1$ only if there is an arc in $\mathcal{D}(B)$ of the same sign from i to n. Let γ be a directed cycle of $\mathcal{D}(A)$ which contains the vertex $n + 1$. Then γ is of the form

$$n + 1 \to n \to i_1 \to \cdots \to i_k \to n + 1,$$

and

$$n \to i_1 \to \cdots \to i_k \to n$$

is a directed cycle of $\mathcal{D}(B)$ with the same sign as γ. Thus any directed cycle of $\mathcal{D}(A)$ has the same sign as some directed cycle of $\mathcal{D}(B)$. It follows from Theorem 3.2.1 that (i) holds.

To prove (ii) we assume that B is an m by $m + 1$ matrix. Since row $m + 1$ of A contains nonzero entries only in columns $m + 1$ and $m + 2$, the matrices obtained from A by deleting column $m + 1$ or $m + 2$ are SNS-matrices if and only if the matrix obtained from B by deleting column $m + 1$ is an SNS-matrix. Let j be an integer with $1 \leq j \leq m$. By (i) the matrix obtained from A by deleting column j is an SNS-matrix if and only if the matrix obtained from B by deleting column j is an SNS-matrix. Hence (ii) holds. Since each row of A is balanced if and only if each row of B is, (iii) now follows from (ii), and from (v) of Theorem 2.1.1. □

We may apply Theorem 4.1.3 by choosing v to be any column vector of B and forming A by inserting a new column equal to w, and then inserting a new row with a 1 in the column corresponding to v, a -1 in the column corresponding to w, and 0's elsewhere. Consider the S-matrix

$$B = \begin{bmatrix} -1 & 1 \end{bmatrix}. \tag{4.1}$$

Then by repeatedly choosing v to be the last column, w to be a zero column vector, and inserting w on the left and the new row on the top, we obtain the S-matrices

$$\left[\begin{array}{c|c} & 1 \\ -I_m & \vdots \\ & 1 \end{array}\right] \quad (m \geq 1).$$

If instead we repeatedly choose v to be the first column, w equal to v, and insert w on the left and the new row on the top, we obtain the m by $m + 1$ S-matrices

$$H'_{m+1} = \left[\begin{array}{ccccccc} -1 & 1 & 0 & \cdots & 0 & 0 & 0 \\ -1 & -1 & 1 & \cdots & 0 & 0 & 0 \\ -1 & -1 & -1 & \cdots & 0 & 0 & 0 \\ \vdots & \vdots & \vdots & \ddots & \vdots & \vdots & \vdots \\ -1 & -1 & -1 & \cdots & -1 & 1 & 0 \\ -1 & -1 & -1 & \cdots & -1 & -1 & 1 \end{array}\right] \quad (m \geq 1). \qquad (4.2)$$

The matrix obtained from H'_{m+1} by inserting on the bottom a row of all -1's equals the lower Hessenberg SNS-matrix H_{m+1} defined in (1.12).

In the next section we show that every $(0, 1, -1)$ S-matrix can be obtained from the 1 by 2 S-matrix in (4.1) or its negative by a construction which is slightly more general than that given in Theorem 4.1.3.[1]

We conclude this section by showing that the $S*$-matrices are precisely the matrices obtained by deleting the last row of fully indecomposable SNS-matrices which have no zeros in their last row.

Theorem 4.1.4 *Let A be an m by $m + 1$ $S*$-matrix and let a signing $D = \mathrm{diag}\,(d_1, d_2, \ldots, d_{m+1})$ be chosen so that each row of AD is balanced. Then*

$$B = \left[\begin{array}{c} A \\ \hline d_1 \quad d_2 \quad \cdots \quad d_{m+1} \end{array}\right]$$

is a fully indecomposable SNS-matrix whose last row has no zero entries. Conversely, let B' be a fully indecomposable SNS-matrix of order $m + 1$ whose last row $(e_1, e_2, \ldots, e_{m+1})$ has no zero entries. Let A' be the m by $m + 1$ matrix obtained by deleting the last row of B' and let $D' = \mathrm{diag}(\mathrm{sign}\,e_1, \mathrm{sign}\,e_2, \ldots, \mathrm{sign}\,e_{m+1})$. Then A' is an $S*$-matrix and each row of $A'D'$ is balanced.

---

[1] Since the matrix (4.1) can be obtained from the empty 0 by 1 S-matrix by the construction given in Theorem 4.1.3, we will in fact begin with the empty S-matrix.

Proof Since A is an S^*-matrix, D is a strict signing and the only column signings of A which do not have a unisigned row are AD and $A(-D)$. This implies that every column signing of B has a unisigned row. It follows from (i) of Theorem 2.1.1 that B is an SNS-matrix. Since each submatrix of A of order m is an SNS-matrix, A does not have a p by q zero submatrix with $p+q = m+1$ for any positive integers p and q. Thus, since each d_i is nonzero, B is fully indecomposable.

Conversely, since B' is a fully indecomposable SNS-matrix, the Frobenius–König theorem implies that $B(\{m + 1\}, \{j\})$ does not have an identically zero determinant, and hence $B(\{m + 1\}, \{j\})$ is an SNS-matrix for all $j = 1, 2, \ldots,$ $m + 1$. Thus A' is an S^*-matrix, and hence there exists a strict signing D'' such that each row of $A'D''$ is balanced. Since B' is an SNS-matrix, the last row of $B'D''$ must be unisigned, and hence $D'' = \pm D'$. We conclude that each row of $A'D'$ is balanced.   □

Motivated by Theorem 4.1.4 we now define a special class of SNS-matrices. An SNS-matrix is called an SNS*-*matrix*[2] provided it is fully indecomposable and has a row each of whose entries is nonzero. It follows from Theorem 4.1.4 that the study of S^*-matrices is essentially equivalent to the study of SNS*-matrices. Corresponding to any theorem about S^*-matrices there is a theorem about SNS*-matrices, and vice versa. In the next section we describe a recursive structure for these classes of matrices and then obtain simple algorithms to recognize matrices in these classes.

4.2 Recursive structure

Two vectors with the same number of entries are *conformal* provided that each pair of corresponding entries has a nonnegative product. Let $A = [a_{ij}]$ be an m by n matrix. Assume that row p of A contains exactly two nonzero entries a_{pr} and a_{ps}. Assume also that $a_{pr}a_{ps} < 0$ and that the two $m - 1$ by 1 vectors obtained by deleting row p of columns r and s are conformal. Let B be the $m - 1$ by $n - 1$ matrix obtained from A by adding column r to column s and then deleting row p and column r. We say that B is obtained from A by the *conformal contraction of columns r and s on row p*. Note that the sign pattern of B equals the sign pattern of the matrix A' obtained from A by pivoting on a_{pr} (in order to make all the entries in column r zero other than a_{pr}) and then deleting row p and column r.[3]

---

[2]Note that this is an SNS*-matrix, not an SNS^*-matrix.

[3]Alternatively, A' can be obtained by row pivoting on a_{pr} to make the entry a_{ps} zero and then deleting row p and column r.

Lemma 4.2.1 *Let $A = [a_{ij}]$ be an m by n matrix and let $B = [b_{ij}]$ be the matrix obtained from A by the conformal contraction of columns r and s on row p. Then*

(i) *if $n = m$, then A is an SNS-matrix if and only if B is;*

(ii) *if $n = m$ and columns r and s of A each have at least two nonzero entries, then A is fully indecomposable if and only if B is;*

(iii) *if $n = m + 1$, then A is an S^*-matrix if and only if B is.*

Proof Without loss of generality we assume that $p = 1$, $r = 1$, and $s = 2$. Assume that $n = m$. First suppose that A is not an SNS-matrix. Let $\widetilde{A}$ be a matrix in $\mathcal{Q}(A)$ with $\det \widetilde{A} = 0$. The matrix $\widetilde{B}$ obtained from $\widetilde{A}$ by pivoting on the entry in row 1 and column 1 satisfies $\det \widetilde{B} = 0$ and $\widetilde{B} \in \mathcal{Q}(B)$. Hence B is not an SNS-matrix. Now suppose that A is an SNS-matrix. Let $\widetilde{B} = [\tilde{b}_{ij}]$ be a matrix in $\mathcal{Q}(B)$. Let $\widetilde{A}$ be the matrix of order n defined by

$$
\begin{bmatrix}
1 & -1 & 0 & \cdots & 0 \\
u_2 & v_2 & \tilde{b}_{12} & \cdots & \tilde{b}_{1,n-1} \\
\vdots & \vdots & \vdots & \cdots & \vdots \\
u_n & v_n & \tilde{b}_{n-1,2} & \cdots & \tilde{b}_{n-1,n-1}
\end{bmatrix},
$$

where

$$
(u_i, v_i) = \begin{cases}
(0, \tilde{b}_{i-1,1}) & \text{if } a_{i1} = 0, \\
(\tilde{b}_{i-1,1}, 0) & \text{if } a_{i2} = 0, \\
\left(\dfrac{\tilde{b}_{i-1,1}}{2}, \dfrac{\tilde{b}_{i-1,1}}{2}\right) & \text{otherwise.}
\end{cases}
$$

Then $\widetilde{A}$ is in $\mathcal{Q}(A)$ and hence $\det \widetilde{A} \neq 0$. Moreover, the matrix obtained from $\widetilde{A}$ by pivoting on the entry in row 1 and column 1 equals $\widetilde{B}$, and hence $\det \widetilde{B} \neq 0$. Thus B is an SNS-matrix and (i) holds.[4]

---

[4] One could define a more general notion of conformal contraction as follows. Suppose that $a_{pr} \neq 0$ and $a_{ps_1}, \ldots, a_{ps_k}$ are the other nonzero entries in row p. Suppose also that $a_{pr} a_{ps_i} < 0$ for all i and that after one deletes row p, each of columns $s_1, \ldots, s_k$ is conformal with column r. Let B be the matrix obtained from A by adding column r to each of columns $s_1, \ldots, s_k$ and deleting row p and column r. The sign pattern of B is the same as the sign pattern of the matrix obtained from A by pivoting on a_{pr} and deleting row p and column r. If B is an SNS-matrix, then so is A. But there exist SNS-matrices A for which the matrix B is not an SNS-matrix. For example, let $p = r = 1$ and let A be the SNS-matrix

$$
\begin{bmatrix}
1 & -1 & -1 \\
-1 & -1 & 0 \\
-1 & 0 & -1
\end{bmatrix}.
$$

Then B is a matrix of order 2 each of whose entries is negative, and hence B is not an SNS-matrix.

Assume that the first two columns of A each have at least two nonzero entries. First suppose that there are positive integers k and ℓ with $k + \ell \geq n - 1$ and a k by ℓ zero submatrix of B. Then, depending on whether or not this zero submatrix intersects the first column of B, A has either a k by $\ell + 1$ or a $k + 1$ by ℓ zero submatrix with $k + \ell + 1 \geq n$. Now suppose that there are positive integers c and d with $c + d \geq n$ and a c by d zero submatrix Z of A. If Z intersects row 1 of A, then $c \geq 2$ and B has a $c - 1$ by d zero submatrix with $c - 1 + d \geq n - 1$. If Z intersects either column 1 or column 2 of A, then $d \geq 2$ and B has a c by $d - 1$ zero submatrix with $c + d - 1 \geq n - 1$. Otherwise, B has a c by d zero submatrix with $c + d \geq n$. Hence A is partly decomposable if and only if B is, and (ii) holds.

Assertion (iii) is a consequence of (i) and Theorem 4.1.4. $\square$

The next lemma implies that if a fully indecomposable SNS-matrix A has a row with only two nonzero entries and these nonzero entries have opposite signs, then a conformal contraction is possible.

Lemma 4.2.2 *Let $A = [a_{ij}]$ be a fully indecomposable SNS-matrix of order $n \geq 2$ such that $a_{11}a_{12} < 0$ and $a_{13} = \cdots = a_{1n} = 0$. Then the vectors obtained from the first two columns of A by deleting row 1 are conformal.*

Proof Let $j > 1$ be an integer such that $a_{j1}a_{j2} \neq 0$. Since A is fully indecomposable, it follows from the Frobenius–König theorem that $A(\{1, j\}, \{1, 2\})$ does not have an identically zero determinant. Thus, since A is an SNS-matrix, $A[\{1, j\}, \{1, 2\}]$ is an SNS-matrix. Hence $a_{j1}a_{j2} > 0$. $\square$

Theorem 4.1.4, Lemma 4.2.2, and Lemma 4.2.1 immediately imply the following.

Corollary 4.2.3 *Let $A = [a_{ij}]$ be an m by $m + 1$ S-matrix such that $a_{11}a_{12} \neq 0$ and $a_{13} = \cdots = a_{1,m+1} = 0$. Then the vectors obtained from the first two columns of A by deleting row 1 are conformal and the matrix obtained from A by the conformal contraction of columns 1 and 2 on row 1 is an S-matrix.*

We now characterize SNS*-matrices.

Theorem 4.2.4 *Let A be a matrix of order $n \geq 2$ containing a row each of whose entries is positive. Then A is an SNS*-matrix if and only if $\pm J_1$ can be obtained from A by successive conformal contractions on rows.*

Proof First assume that $\pm J_1$ can be obtained from A by successive conformal contractions on rows. By (i) of Lemma 4.2.1, A is an SNS-matrix, and by (ii) of the same lemma and the fact that A has a row with all positive entries, A is a fully indecomposable matrix. Hence A is an SNS*-matrix. Now assume that $A = [a_{ij}]$ is an SNS*-matrix. We prove by induction on n that $\pm J_1$ can be obtained from A by successive conformal contraction on rows. If $n = 1$, there is nothing to prove. Suppose that $n > 1$. Let row k be the row of A each of whose entries is positive. By Theorem 4.1.4, the matrix obtained from A by deleting row k is an S-matrix. Hence, by Theorem 4.1.1, A has a row, say row p, with exactly two nonzero entries, say $a_{pr} \neq 0$ and $a_{ps} \neq 0$. Since every row of an S-matrix is balanced, $a_{pr}a_{ps} < 0$. By Lemmas 4.2.2 and 4.2.1, the matrix obtained from A by the conformal contraction of columns r and s on row p is a fully indecomposable SNS-matrix, and it has a row all of whose entries are positive. The result now follows by induction. $\quad\square$

We now use Theorem 4.2.4 in order to obtain the following characterization of S-matrices [5]. As already noted, the 0 by 1 matrix is considered to be an S-matrix, and we denote it by $S_\emptyset$.

Theorem 4.2.5 *Let B be an m by $m+1$ matrix each of whose rows is balanced. Then B is an S-matrix if and only if $S_\emptyset$ can be obtained from B by successive conformal contractions on rows.*

Proof Let A be the matrix of order $m+1$ obtained from B by inserting a row of all 1's on the bottom. It follows from Theorem 4.1.4 that A is an SNS*-matrix if and only if B is an S-matrix. The matrix $S_\emptyset$ can be obtained from B by successive conformal contractions on rows if and only if $\pm J_1$ can be obtained from A by the corresponding successive conformal contractions. The theorem is now a consequence of Theorem 4.2.4. $\quad\square$

We illustrate Theorem 4.2.5 with two examples. First let

$$B_1 = \begin{bmatrix} 1 & 0 & 0 & -1 & 0 \\ 0 & 1 & -1 & 0 & 0 \\ -1 & 1 & 1 & 0 & 0 \\ 1 & 0 & 1 & 1 & -1 \end{bmatrix}.$$

Then B_1 satisfies the hypotheses of Theorem 4.2.5 and by successive conformal contraction on row 1 we obtain matrices with sign patterns

$$\begin{bmatrix} 1 & -1 & 0 & 0 \\ 1 & 1 & -1 & 0 \\ 0 & 1 & 1 & -1 \end{bmatrix}, \begin{bmatrix} 1 & -1 & 0 \\ 1 & 1 & -1 \end{bmatrix}, \begin{bmatrix} 1 & -1 \end{bmatrix} \text{ and } S_\emptyset.$$

Hence B_1 is an S-matrix. Now let

$$B_2 = \begin{bmatrix} 1 & 0 & 0 & -1 \\ 1 & 0 & -1 & 0 \\ 0 & 1 & 1 & -1 \end{bmatrix}.$$

Conformal contraction on row 1 results in the matrix

$$\begin{bmatrix} 0 & -1 & 1 \\ 1 & 1 & -1 \end{bmatrix},$$

which cannot be further conformally contracted. Hence B_2 is not an S-matrix.

Corollary 4.2.3 and Theorem 4.2.5 imply the validity of the following algorithm for recognizing whether or not a matrix is an S-matrix.

S-matrix-recognition algorithm

Let A be an m by $m + 1$ matrix.

(0) Let $X = A$.
(1) If X is a 0 by 1 matrix, then A is an S-matrix. Otherwise,
(2) if X does not have a balanced row with exactly two nonzero entries, then A is not an S-matrix. Otherwise,
(3) choose a balanced row of X, say row p, with exactly two nonzero entries. Let the nonzero entries in row p be in columns r and s. If the vectors obtained from columns r and s by deleting the entry in row p are not conformal, then A is not an S-matrix. Otherwise,
(4) replace X by the matrix obtained from X by the conformal contraction of columns r and s on row p, and go back to (1).

Theorem 4.1.4 implies the validity of the following algorithm.

SNS^*-matrix-recognition algorithm

Let $B = [b_{ij}]$ be a matrix of order n.

(0) If each row of B contains a 0, then B is not an SNS^*-matrix. Otherwise,
(1) choose a row of B, say row q, each of whose entries is nonzero. Multiply column j of B by sign b_{qj} for each $j = 1, 2, \ldots, n$, and then delete row q to form a matrix A.
(2) Apply the S-matrix recognition algorithm to the matrix A.
(3) If A is an S-matrix, then B is an SNS^*-matrix. Otherwise, B is not an SNS^*-matrix.

Let A be an m by $m + 1$ matrix and suppose that row p of A contains exactly two nonzero entries a_{pr} and a_{ps} where $a_{pr}a_{ps} > 0$. If A is an S^*-matrix, then there exists a strict signing of D such that AD is an S-matrix. Hence, if A is an S^*-matrix, then by Corollary 4.2.3, $a_{ir}a_{is} \leq 0$ for each $i \neq p$, that is, the vectors u and v obtained from columns r and s of A by deleting row p are *anticonformal*. If u and v are anticonformal, then Lemma 4.2.1 implies that the matrix obtained from A by subtracting column r of A from column s and then deleting row p and column r is an S^*-matrix if and only if A is an S^*-matrix. Thus we have the following recognition algorithm for S^*-matrices, which is essentially contained in [5]; see footnote 5.

S^*-matrix-recognition algorithm

Let A be an m by $m + 1$ matrix.

(0) Let $X = A$.
(1) If X is a 0 by 1 matrix, then A is an S^*-matrix. Otherwise,
(2) if X does not have a row with exactly two nonzero entries, then A is not an S^*-matrix. Otherwise,
(3) choose a row of X, say row p, with exactly two nonzero entries. Let the nonzero entries in row p be in columns r and s. If row p is not balanced, then replace column r of X by its negative. If the vectors obtained from columns r and s by deleting the entry in row p are not conformal, then A is not an S^*-matrix. Otherwise,
(4) replace X by the matrix obtained from X by the conformal contraction of columns r and s on row p, and go back to (1).

Because no cancellation occurs in the preceding algorithms, they can be carried out on the sign pattern of A using the qualitative rules $1 + 1 = 1$ and $(-1) + (-1) = -1$ in place of the usual arithmetic rules. Subject to the restrictions on vectors being conformal as specified earlier, the algorithms are basically Gaussian elimination on the sign pattern with complete pivoting as long as the pivot is chosen within a row with two nonzero entries. More precisely, A is an S^*-matrix if and only if A can be reduced in this way to the 0 by 1 matrix $S_\emptyset$. If A is an S^*-matrix, then the sign pattern of the nonzero vectors in the right null spaces of the matrices in $\mathcal{Q}(A)$ can be determined by back substitution.[5]

---

[5]It is tempting to think that the strict signing $D = \operatorname{diag}(d_1, d_2, \ldots, d_{m+1})$ with $d_i = -1$ if and only if column i of A corresponds to a column that is negated in (2) of the S^*-matrix recognition algorithm has the property that AD is an S-matrix. But this need not be true. For example, let

$$A = \begin{bmatrix} -1 & 0 & 0 & -1 \\ 0 & -1 & -1 & 0 \\ -1 & -1 & 1 & 1 \end{bmatrix}.$$

Then successively pivoting on the (1,1)-, (2,2)- and (3,3)-entries of A, we obtain $D = \operatorname{diag}(-1, -1, -1, 1)$. But the strict signings E such that AE is an S-matrix are $E = \pm\operatorname{diag}(1, -1, 1, -1)$. The algorithm as given in [5] contains a forward subalgorithm for determining the sign pattern of the nonzero vectors in the right null spaces of the matrices in $\mathcal{Q}(A)$. Applying this subalgorithm to the matrix A we obtain the sign pattern $(-1, -1, -1, 1)$, and hence the subalgorithm is not correct. The verification of the subalgorithm failed to take into account the fact that both columns of a conformal contraction may have been involved in previous conformal contractions. Since this subalgorithm is incorrect, the algorithm may fail to execute. Referring to the algorithm in [5], one sees there may be no c and d satisfying the specified conditions on the bottom of page 244.

The preceding recognition algorithms for S- and S^*-matrices have time complexity $O(m^2)$. Other polynomial-time recognition algorithms for S- and S^*-matrices have been proposed in [4], [8], and [9]. These algorithms are based on the digraph characterizations of sign-solvable linear systems given in section 3.3.

We now define an operation which is something like an inverse of conformal contraction. For convenience we define it only for matrices each of whose entries is 0, 1, or -1 and thus only for sign patterns of matrices. Let $B = [b_{ij}]$ be an m by n $(0, 1, -1)$-matrix. Let k be an integer with $1 \leq k \leq n$ and let $w = (w_1, w_2, \ldots, w_m)^T$ be the kth column vector of B. Suppose that $u = (u_1, u_2, \ldots, u_m)^T$ and $v = (v_1, v_2, \ldots, v_m)^T$ are $(0, 1, -1)$-vectors each of which is conformal with w. Suppose further that $w_i = 0$ if and only if both $u_i = 0$ and $v_i = 0$. Let A be the $m + 1$ by $n + 1$ $(0, 1, -1)$-matrix obtained from B by replacing column k by v, inserting a new column equal to u, and then inserting a new row with a 1 in the column corresponding to v, a -1 in the column corresponding to u and 0's elsewhere. Then A is obtained from B by the *conformal splitting of column k into the vectors u and v*. Note that B is the sign pattern of the matrix obtained from A by the conformal contraction on the new row. If $k = n$ and the new row and column are inserted on the bottom and right, respectively, then

$$A = \left[\begin{array}{ccc|c|c} & & & & \\ & B[:, \{1, \ldots, n-1\}] & & v & u \\ & & & & \\ \hline 0 & \cdots & 0 & -1 & 1 \end{array} \right].$$

If $w = u = v$, then we say that B is obtained from A by *copying column k*. Lemma 4.2.1 immediately implies the following.

Corollary 4.2.6 *Let B be a square $(0, 1, -1)$-matrix and let A be obtained from B by a conformal splitting of column k into the vectors u and v. Then*

   (i) *A is an SNS-matrix if and only if B is;*
   (ii) *if u and v are nonzero vectors, then A is fully indecomposable if and only if B is.*

The following recursive structure of S-matrices is an immediate consequence of Theorem 4.2.5 and the inverse relationship between conformal contraction and conformal splitting.

Corollary 4.2.7 *The set of all matrices obtained from $S_\emptyset$ by row and column permutations and successive conformal splittings is precisely the set of $(0, 1, -1)$ S-matrices.*

Another recursive structure for S-matrices follows from the discussion at the conclusion of Chapter 3. Let B_i be an m_i by $m_i + 1$ S-matrix with $m_i \geq 0$ ($i = 1, 2, \ldots, k$). For $i \leq j$ let u_{ij} be a 1 by $m_j + 1$ vector such that u_{ii} is nonpositive and u_{ij} is nonnegative for $i \neq j$. Further suppose that each of the matrices

$$\begin{bmatrix} B_i \\ u_{ii} \end{bmatrix} \quad (i = 1, 2, \ldots, k)$$

is fully indecomposable. Let $v = (v_1, \ldots, v_k)^T$ be a column vector such that each entry is nonnegative and $v_i = 1$ whenever $u_{i,i+1} = 0, \ldots, u_{ik} = 0$. Then the matrix

$$A = \left[\begin{array}{ccccccc} B_1 & O & \cdots & O & O & 0 \\ O & B_2 & \cdots & O & O & 0 \\ \vdots & \vdots & \ddots & \vdots & \vdots & \vdots \\ O & O & \cdots & B_{k-1} & O & 0 \\ O & O & \cdots & O & B_k & 0 \\ \hline u_{11} & u_{12} & \cdots & u_{1,k-1} & u_{1,k} & v_1 \\ 0 & u_{22} & \cdots & u_{2,k-1} & u_{2k} & v_2 \\ \vdots & \vdots & \ddots & \vdots & \vdots & \vdots \\ 0 & 0 & \cdots & u_{k-1,k-1} & u_{k-1,k} & v_{k-1} \\ 0 & 0 & \cdots & 0 & u_{k,k} & v_k \end{array} \right]$$

is an S-matrix, and up to row and column permutations and multiplication of rows by -1, every (nonempty) $(0, 1, -1)$ S-matrix can be constructed from S-matrices with fewer rows in this way [6, 7]. Note that the S-matrix H'_{k+1} results in this way by taking $m_1 = \cdots = m_k = 0$ and then permuting rows and columns and multiplying all rows by -1.

4.3 Zero patterns of S-matrices

Let $X = [x_{ij}]$ and $Y = [y_{ij}]$ be two m by n (0,1)-matrices. Then we write $X \leq Y$ provided that $x_{ij} \leq y_{ij}$ for all i and j. The set of all $(0, 1, -1)$-matrices whose zero pattern equals X is denoted by $\mathcal{Z}(X)$. In this section we are concerned with the problem of determining the zero patterns of S^*-matrices. Since a matrix B is an S^*-matrix if and only if there exists a strict signing D such that BD is an S-matrix, it suffices to restrict our attention to S-matrices. We first show that up to multiplication of rows by -1, there is at most one S-matrix with a given zero pattern.

Theorem 4.3.1 *Let A be an m by $m + 1$ $(0, 1)$-matrix and let $\widehat{A}$ and $\widehat{\widehat{A}}$ be S-matrices in $\mathcal{Z}(A)$. Then there exists a unique strict signing D such that $\widehat{A} = D\widehat{\widehat{A}}$.*

Proof We prove the existence of the strict signing D by induction on m. If $m = 1$, then D clearly exists. Assume that $m > 1$. Because of Theorem 4.1.1 we may assume that row 1 of A contains exactly two 1's, which without loss of generality we take in positions $(1,1)$ and $(1,2)$. By Corollary 4.2.3, the conformal contractions $\widehat{B}$ and $\widehat{\widehat{B}}$ of columns 1 and 2 on row 1 of $\widehat{A}$ and $\widehat{\widehat{A}}$, respectively, are S-matrices. By the induction hypothesis there exists a strict signing F such that $\widehat{B} = F\widehat{\widehat{B}}$. Since the vectors obtained from the first two columns of $\widehat{A}$, respectively $\widehat{\widehat{A}}$, by deleting row 1 are conformal, it follows that $\widehat{A}[\overline{\{1\}}, :] = ((I_1 \oplus F)\widehat{\widehat{A}})[\overline{\{1\}}, :]$. Hence $\widehat{A} = D\widehat{\widehat{A}}$ where $D = (\pm I_1) \oplus F$. Since each row of $\widehat{A}$ contains a nonzero entry, D is uniquely determined. $\qquad\square$

Let G be a graph of order n with vertices $1, 2, \ldots, n$ and edges $f_1, f_2, \ldots, f_m$. Then the *edge-vertex incidence matrix of G* is the m by n matrix $B = [b_{ij}]$ where $b_{ij} = 1$ if j is a vertex of f_i and $b_{ij} = 0$ otherwise. The $(0, 1)$-matrix B has exactly two 1's in each row and no two rows are identical. A special case of a well-known theorem of Rado [10] asserts that if $X_1, X_2, \ldots, X_p$ are subsets of the edges of a graph G, then there exist a set of p distinct edges $Y = \{f_{i_1}, f_{i_2}, \ldots, f_{i_p}\}$ of G with $f_{i_k} \in X_k$ $(k = 1, 2, \ldots, p)$ such that Y does not contain a cycle if and only if for each $t = 1, 2, \ldots, p$ and each subset α of $\{1, 2, \ldots, p\}$ with t elements, there are at least $t + 1$ vertices incident with the edges in $\cup_{k \in \alpha} X_k$.

Let A be an m by $m + 1$ $(0, 1)$-matrix. An obvious necessary condition for $\mathcal{Z}(A)$ to contain an S-matrix is that no submatrix of A of order m has an identically zero determinant. Although this condition is not sufficient for $\mathcal{Z}(A)$ to contain an S-matrix, the next theorem implies that it is a necessary and sufficient condition for the existence of a $(0, 1)$-matrix $B \leq A$ such that $\mathcal{Z}(B)$ contains an S-matrix.

Theorem 4.3.2 *Let A be an m by $m + 1$ $(0, 1)$-matrix. Then the following are equivalent:*

    (i) *There exists a $(0, 1)$-matrix $B \leq A$ such that $\mathcal{Z}(B)$ contains an S-matrix.*

   (ii) *No submatrix of A of order m has an identically zero determinant.*

  (iii) *A does not have a p by q zero submatrix for any positive integers p and q with $p + q = m + 1$.*

(iv) *There exists a* $(0, 1)$*-matrix* $C \leq A$ *such that* C *is the edge-vertex incidence matrix of a tree.*

Proof Clearly (i) implies (ii), and it follows from the Frobenius–König theorem that (ii) and (iii) are equivalent. Assume that (iv) holds and let $\widehat{C}$ be a matrix obtained from C by replacing one of the two 1's in each row by a -1. By successive conformal contractions on rows corresponding to pendant edges we obtain the matrix S_0. Hence, by Theorem 4.2.5, $\widehat{C}$ is an S-matrix and hence (i) holds. Now assume that (iii) holds. Then each row of A contains at least two 1's. We define a graph $R(A)$ with vertices $1, 2, \ldots, n$ and an edge $\{j, k\}$ joining distinct vertices j and k if and only if there exists an i such that $a_{ij} = a_{ik} = 1$. The rows of A correspond to certain complete subgraphs of $R(A)$. Let X_i be the set of edges of the complete subgraph corresponding to row i ($i = 1, 2, \ldots, m$). Let α be a subset of $\{1, 2, \ldots, m\}$ of t elements. Let β be the set of vertices incident with the edges in $\cup_{i \in \alpha} X_i$. Then $A[\alpha, \overline{\beta}]$ is a zero matrix. Hence, by (iii), β contains at least $t + 1$ vertices. By Rado's theorem there exist m distinct edges $x_1, x_2, \ldots, x_m$, one from each of the X_i, containing no cycle. Since $R(A)$ has $m + 1$ vertices, these edges are the edges of a spanning tree T of $R(A)$. The edge-vertex incidence matrix C of T satisfies $C \leq A$, and hence (iv) holds. The theorem now follows. ☐

It follows from the equivalence of (i) and (iv) in Theorem 4.3.2 that the $(0,1)$-matrices A such that $\mathcal{Z}(A)$ contains an S-matrix but $\mathcal{Z}(A')$ does not contain an S-matrix for any $(0,1)$-matrix A' with $A' \leq A$ and $A' \neq A$ are precisely the edge-vertex incidence matrices of trees.

The characterization of S-matrices given in Theorem 4.2.5 can be used to obtain a characterization of zero patterns of S-matrices. First we make the following definitions.

Let X and Y be matrices such that Y is obtained from X by the conformal contraction of columns r and s on row p. Then we say that the zero pattern of Y is obtained from the zero pattern of X by a contraction of columns r and s on row p. More precisely, let $A = [a_{ij}]$ be an m by n $(0,1)$-matrix such that row p of A contains exactly two 1's, say $a_{pr} = a_{ps} = 1$ where $r \neq s$. Let u be the m by 1 $(0,1)$ column vector whose ith entry is 1 if and only if $a_{ir} = 1$ or $a_{is} = 1$. Let B be the $m - 1$ by $n - 1$ matrix obtained from A by replacing column s by u and then deleting row p and column r. We say that B is the matrix obtained from A by the *contraction of columns* r *and* s *on row* p. Note that B is the matrix obtained from A by a "symbolic pivot" on a_{pr} and then deletion of row p and column r. If column r of A contains exactly two 1's, say $a_{qr} = 1$ where $q \neq p$, and if $a_{qs} = 0$, then the contraction is called an *elementary contraction*.

The next lemma is a consequence of Lemma 4.2.1 and the equivalence of (ii) and (iii) of Theorem 4.3.2.

Lemma 4.3.3 *Let A be an m by $m + 1$ $(0, 1)$-matrix and let B be a matrix obtained from A by a contraction on a row. Then*

(i) *no submatrix of A of order m has an identically zero determinant if and only if no submatrix of B of order $m - 1$ has an identically zero determinant, and*

(ii) *$\mathcal{Z}(A)$ contains an S-matrix if and only if $\mathcal{Z}(B)$ does.*

Theorem 4.2.5 now implies the following.

Theorem 4.3.4 *The $(0, 1)$-matrices A such that $\mathcal{Z}(A)$ contains an S-matrix are precisely the $(0, 1)$-matrices from which $S_\emptyset$ can be obtained by a sequence of contractions on rows.*

We can also give a forbidden configuration characterization of S-matrices. First we prove the following lemma.

Lemma 4.3.5 *Let C be a k by $k + 1$ $(0, 1)$-matrix such that $J_{2,3}$ can be obtained from C by a sequence of contractions on rows. Then there exists a matrix C' with exactly two 1's in each column and permutation matrices U and V such that $U(C' \oplus I_{k-k'})V \leq C$ and $J_{2,3}$ can be obtained from C' by a sequence of elementary contractions.*

Proof We prove the lemma by induction on k. If $k = 2$, the lemma clearly holds. Now assume that $k > 2$. Using the induction hypothesis, we may assume without loss of generality that row 1 of C has exactly two 1's, and that contracting C on row 1, but not deleting row 1 and column 1, we obtain a matrix of the form

$$\begin{bmatrix} 1 & 0 & \cdots & 0 & 1 & 0 & \cdots & 0 \\ 0 & & & & & & & \\ \vdots & & F & & & O & & \\ 0 & & & & & & & \\ y & & O & & & I_{k-1-k'} & & \end{bmatrix}$$

or of the form

$$
\left[
\begin{array}{c|ccccc|ccc}
1 & 1 & 0 & \cdots & 0 & 0 & \cdots & 0 \\
\hline
x & & F & & & & O & \\
\hline
0 & & & & & & & \\
\vdots & & O & & & & I_{k-1-k'} & \\
0 & & & & & & &
\end{array}
\right]
$$

where $J_{2,3}$ can be obtained from F by a sequence of elementary contractions and F has exactly two 1's in each column. In the first case, there exist permutation matrices U and V such that $U(F \oplus I_{k-k'})V \leq C$. Now consider the second case. The vector x has at most two 1's. By replacing 1's of C in the first column by 0's we may assume that the only row of C that contains 1's in both columns 1 and 2 is row 1. If x has either zero or two 1's, then there exist permutation matrices such that $U(F \oplus I_{k-k'})V \leq C$. If x has only one 1, then the contraction of columns 1 and 2 on row 1 is elementary. Hence the lemma follows by induction. ☐

Theorem 4.3.6 *Let A be an m by $m+1$ $(0, 1)$-matrix such that no submatrix of order m has an identically zero determinant. Then the following are equivalent:*

(i) *$\mathcal{Z}(A)$ does not contain an S-matrix.*

(ii) *There exist a k by $k+1$ $(0, 1)$-matrix C and permutation matrices P and Q such that $P(C \oplus I_{m-k})Q \leq A$ and $J_{2,3}$ can be obtained from C by a sequence of contractions on rows.*

(iii) *There exist a k' by $k'+1$ $(0, 1)$-matrix C' and permutation matrices P' and Q' such that $P'(C' \oplus I_{m-k'})Q' \leq A$ and $J_{2,3}$ can be obtained from C' by a sequence of elementary contractions on rows.*

Proof Clearly (iii) implies (ii) and by Lemma 4.3.5, (ii) implies (iii). Suppose that $J_{2,3}$ can be obtained from C by a sequence of contractions. By Lemma 4.3.3, no submatrix of C of order k has an identically zero determinant and, since $\mathcal{Z}(J_{2,3})$ does not contain an S-matrix, $\mathcal{Z}(C)$ does not contain an S-matrix. Hence each matrix B in $\mathcal{Z}(C \oplus I_{m-k})$ has a submatrix of order m whose standard determinant expansion contains nonzero terms of opposite sign. Thus (ii) implies (i).

Now assume that (i) holds. We prove (ii) holds by induction on the number of 1's contained in A. First suppose that A has a row with exactly two 1's, say row 1. Then by Lemma 4.3.3, the $m-1$ by m matrix obtained from A by contraction on row 1 has no submatrix of order $m - 1$ with an identically zero determinant and is not the zero pattern of an S-matrix, and thus (ii) follows by induction. Now suppose that each row of A contains at least three 1's. We may assume that A is minimal in the sense that each matrix obtained from A by changing a 1 to a 0 either has a submatrix of order m with an identically zero determinant or is not the zero pattern of an S-matrix.

First consider the case that some matrix obtained from A by changing a 1 to a 0 has an identically zero determinant. It follows from Theorem 4.3.2 that we may assume that A has the form

$$
\begin{bmatrix}
A_1 & \begin{matrix} 0 \\ \vdots \\ 0 \end{matrix} & O \\
\hline
u & 1 & 0 \cdots 0 \\
\hline
X & v & A_2
\end{bmatrix}
$$

where A_1 is a p by $p + 1$ matrix. Since each row of A contains at least three 1's, we have $p \geq 1$. Neither A_2 nor a submatrix of A_1 of order p can have an identically zero determinant. Since no row of A_1 can contain exactly two 1's, $\mathcal{Z}(A_1)$ does not contain an S-matrix. Hence (ii) follows by applying the induction hypothesis to A_1. Now consider the case that no matrix obtained from A by changing a 1 to a 0 has an identically zero determinant. Then each column has at least two 1's, and our minimality assumption implies that each submatrix obtained from A by changing a 1 to a 0 is the zero pattern of an S-matrix. It follows from Theorem 4.1.1 that each row of A has exactly three 1's. Since each column has at least two 1's, some column contains exactly two 1's. Without loss of generality we assume that column 1 contains exactly two 1's and that $A[\{1, 2, \}, \{1, 2, 3\}]$ has the form

$$
\begin{bmatrix}
1 & 1 & 1 \\
1 & a & b
\end{bmatrix}.
$$

Let A' be the matrix obtained from A by changing the 1 in position $(1,3)$ to 0. Then $\mathcal{Z}(A')$ contains an S-matrix and by Lemma 4.3.3, the matrix obtained from A' by contraction on row 1 is an S-matrix. Since an S-matrix has a row with two 1's, we conclude that $a = 1$. Similarly, we have $b = 1$. Therefore

$$
A[\{1, 2\}, :] = \begin{bmatrix} J_{2,3} & O \end{bmatrix}.
$$

The matrix $A[\{3, \ldots, m\}, \{4, \ldots, m + 1\}]$ cannot have an identically zero determinant. Hence there exist permutation matrices P and Q such that $P(J_{2,3} \oplus I_{m-2})Q \leq A$, and (ii) holds. □

The forbidden configuration characterization of zero patterns of S-matrices in Theorem 4.3.6 can be formulated in terms of bipartite graphs. The *bipartite graph* of an m by n (0,1)-matrix $A = [a_{ij}]$ is the graph of order $m + n$ with vertex bipartition $\{V, V'\}$ where

$$V = \{1, 2, \ldots, m\} \text{ and } V' = \{1', 2', \ldots, n'\},$$

and where the edges are those pairs $\{i, j'\}$ for which $a_{ij} \neq 0$. Let B be obtained from A by an elementary contraction on a row. Then the bipartite graph of A can be obtained from that of B by subdividing an edge into three edges by the insertion of two new vertices.

Let G be a graph with edges $\alpha_1, \alpha_2, \ldots, \alpha_m$. A *subdivision* of G is a graph H obtained from G by the insertion of $k_i \geq 0$ new vertices on α_i for each $i = 1, 2, \ldots m$. If each of the k_i is even, then H is an *even subdivision* of G. Note that if G is a bipartite graph, then every even subdivision of G is also a bipartite graph. A *matching* of a graph is a set of pairwise vertex disjoint edges. Since the bipartite graph of the matrix $J_{2,3}$ is the complete bipartite graph $K_{2,3}$, a direct translation of the equivalence of (i) and (iii) in Theorem 4.3.6 gives the following.

Corollary 4.3.7 *Let A be an m by $m + 1$ $(0, 1)$-matrix such that no submatrix of A of order m has an identically zero determinant. Then $\mathcal{Z}(A)$ does not contain an S-matrix if and only if the bipartite graph of A contains a spanning subgraph which is the vertex disjoint union of an even subdivision of $K_{2,3}$ and a matching.*

4.4 Maximal S-matrices

In order to more clearly formulate the theorems of this section, we assume without loss of generality that matrices are $(0, 1, -1)$-matrices. We first determine the smallest and largest number of nonzero entries in S-matrices and therefore S^*-matrices. We denote the number of nonzero entries of a matrix A by $\#(A)$. An *oriented edge-vertex incidence matrix* of a graph G is any matrix obtained from an edge-vertex incidence matrix of G by replacing one of the two 1's in each row by -1. Thus each row of an oriented edge-vertex incidence matrix of a graph is balanced.

The inequality (4.3) in the next theorem is contained in [5].

Theorem 4.4.1 *Let A be an m by $m + 1$ $(0, 1, -1)$-matrix. If A is an S-matrix, then*

$$2m \leq \#(A) \leq \frac{m(m+3)}{2}. \tag{4.3}$$

Furthermore,

(i) *if $\#(A) = 2m$, then A is an S-matrix if and only if A is an oriented edge-vertex incidence of a tree;*

(ii) *if $\#(A) = m(m + 3)/2$, then A is an S-matrix if and only if there exist permutation matrices P and Q and a strict signing D such that $DPAQ$ equals H'_{m+1} where H'_{m+1} is the matrix defined in (4.2);*

(iii) *if k is any integer with $2m \leq k \leq m(m + 3)/2$, then there exists an m by $m + 1$ $(0, 1, -1)$ S-matrix B with $\#(B) = k$.*

Proof If A is an oriented edge-vertex incidence matrix of a tree, then $\#(A) = 2m$ and, by successive conformal contractions on a row corresponding to a pendant edge, A can be contracted to the matrix S and hence A is an S-matrix. The matrix H'_{m+1} in (4.2) satisfies $\#(H'_{m+1}) = m(m + 3)/2$ and it has already been noted that H'_{m+1} is an S-matrix.

Now assume that $A = [a_{ij}]$ is an S-matrix. Then each row of A contains at least two nonzero entries and hence $\#(A) \geq 2m$. Suppose that $\#(A) = 2m$. It follows from Lemma 4.3.2 that the zero pattern of A is an edge-vertex incidence matrix of a tree. Since each row of A is balanced, A is an oriented edge-vertex incidence matrix of a tree.

We now use induction to show that $\#(A) \leq m(m + 3)/2$ with equality as specified in the theorem. If $m = 1$, this is clear. Assume that $m > 1$. It follows from Corollary 4.2.7 that up to row and column permutations, A is obtained by a conformal splitting of column 1 of an $m - 1$ by m S-matrix A', where the new column is inserted on the left and the new row is inserted on the top. By induction we have

$$\#(A) \leq \#(A') + m + 1 \leq \frac{(m-1)(m+2)}{2} + m + 1 = \frac{m(m+3)}{2}.$$

Suppose that $\#(A) = m(m + 3)/2$. Then $\#(A') = (m - 1)(m + 2)/2$ and columns 1 and 2 of A contain no zeros. By induction we may assume without loss of generality that either A' is H'_m, or A' is the matrix obtained from H'_m by interchanging its first two columns. In the first case A is H'_{m+1}, and in the second case we multiply row 2 by -1 and interchange columns 2 and 3 of A to obtain H'_{m+1}.

Finally, let k be an integer with $2m \leq k \leq m(m+3)/2$. Let B be a matrix obtained from H'_{m+1} by replacing with 0 any $(m(m+3))/2 - k$ of the -1's below the diagonal running from the $(1, 1)$ position to the (m, m) position. Then $\#(B) = k$, and since B can be contracted to a 1 by 2 S-matrix, B is an S-matrix. □

A $(0, 1, -1)$-matrix A is a *maximal S-matrix* provided A is an S-matrix but no matrix obtained from A by replacing a zero by 1 or -1 is an S-matrix. For example, the matrix

$$\begin{bmatrix} 1 & -1 & 0 & 0 \\ 0 & 0 & 1 & -1 \\ 1 & 1 & -1 & -1 \end{bmatrix}$$

is a maximal S-matrix. It follows from Theorem 4.4.1 that the matrices H'_{m+1} are maximal S-matrices. We use Theorem 4.2.5 to obtain a recursive characterization of maximal S-matrices.

Lemma 4.4.2 *Let* $A = [a_{ij}]$ *be an* m *by* $m + 1$ *S-matrix. Suppose that* $a_{11}a_{12} < 0$ *and the vectors obtained from columns* 1 *and* 2 *of* A *by deleting row* 1 *have the same sign pattern. Then* $a_{13} = \cdots = a_{1,m+1} = 0$.

Proof We prove the lemma by induction on m. If $m = 1$, there is nothing to prove. Assume that $m > 1$. Suppose that there exists a $p \neq 1$ such that row p of A contains exactly two nonzero entries, say $a_{pr}a_{ps} < 0$. The assumptions on A imply that $r, s \geq 3$. By Lemma 4.2.2, the column vectors obtained from columns r and s of A by deleting row p are conformal. Applying the inductive hypothesis to the matrix obtained from A by the conformal contraction of columns r and s on row p, we conclude that the first row of A has exactly two nonzero entries. If each row of A other than row 1 has more than two nonzero entries, then by Theorem 4.1.1 row 1 of A has exactly two nonzero entries. □

Theorem 4.4.3 *Let* A *be an* m *by* $m+1$ $(0, 1, -1)$*-matrix. Then* A *is a maximal S-matrix if and only if* A *can be obtained from* $S_\emptyset$ *by successive copying of columns.*

Proof First assume that A is a maximal S-matrix. It follows from Corollary 4.2.7 and the maximal property of A that A can be obtained from $S_\emptyset$ by successive copying of columns.

Now assume that A can be obtained from $S_\emptyset$ by successive copying of columns. We show by induction on m that A is a maximal S-matrix. If $m = 0$, then clearly A is a maximal S-matrix. Suppose that $m > 0$. Without loss of generality A is obtained from a matrix B by copying column 1 of B where B can be obtained from $S_\emptyset$ by successive copying of columns. By the inductive hypothesis B is a maximal S-matrix. It follows from Lemma 4.2.1 and the maximal property of B that each matrix obtained from A by replacing a zero in rows $2, \ldots, m$ by a nonzero number is not an S-matrix. Lemma 4.4.2 implies that a similar conclusion holds for the zeros in row 1 of A. Hence A is a maximal S-matrix. $\square$

Let $A = [a_{ij}]$ be an m by n $(0, 1, -1)$-matrix. We define the *row multigraph* $RM(A)$ of A as follows. The vertices of $RM(A)$ are $1, 2, \ldots, n$. Let i and j be distinct vertices and let m_{ij} be the number of integers k such that $a_{ki}a_{kj} < 0$. If $m_{ij} = 0$, there is no edge joining i and j. Otherwise, there is an edge of multiplicity m_{ij} joining i and j. Note that corresponding to row i of A there is a *biclique* $F_i(A)$ of $RM(A)$, that is, a complete bipartite graph which is a submultigraph of $RM(A)$.[6]

To each edge of $F_i(A)$ we assign the color i and then speak of the *colored row multigraph* of A.

Theorem 4.4.4 *Let A be an m by $m + 1$ $(0, 1, -1)$-matrix each of whose rows is balanced and nonzero. Then the following are equivalent:*

   (i) *A is an S-matrix.*
  (ii) *Each cycle of the colored multigraph $RM(A)$ contains at least two edges of the same color.*
 (iii) *Any set of m edges of $RM(A)$ of different colors is the set of edges of a spanning tree of $RM(A)$.*

In particular, if A is an S-matrix, then $RM(A)$ is a graph.

Proof First, we assume that (i) holds and prove that (ii) holds. Without loss of generality A is a maximal S-matrix. If $m = 1$, then $RM(A)$ contains exactly one edge. We now assume that $m > 1$ and proceed by induction on m. Theorem 4.4.3 implies that there exists a maximal S-matrix B such that A is obtained from B by copying a column. We assume that A is obtained from B by copying column m of B and inserting the new column on the right and the new row on the bottom. The colored multigraph $RM(A)$ is obtained from the colored multigraph $RM(B)$ by appending the new vertex $m + 1$, an edge of color m joining m and $m + 1$, and an edge of color i joining $m + 1$ and j if and only if there is an edge of color i joining m and j in $RM(A)$. By the inductive

---

[6]This biclique may have no edges and possibly no vertices.

hypothesis and Lemma 4.2.2, each cycle of the colored multigraph $RM(B)$ contains two edges of the same color. It is now easy to verify that (ii) holds.

We now assume that (ii) holds and prove that (i) holds. Let $D = \mathrm{diag}(d_1, d_2, \ldots, d_{m+1})$ be any signing such that each row of AD is balanced. We suppose that $D \neq \pm I_{m+1}$ and obtain a contradiction. If D is not a strict signing, say $D = E \oplus O$ where E is a strict signing of order $k \geq 1$, then since each row of A is balanced, $A(E \oplus (\pm I_{m+1-k}))$ are both strict column signings of A with all rows balanced. It follows that we may assume that D is a strict signing different from $\pm I_{m+1}$. The nonempty sets $X = \{i : d_i = 1\}$ and $Y = \{i : d_i = -1\}$ partition $\{1, 2, \ldots, m+1\}$ into sets of cardinalities p and q, respectively. Since each row of A and AD is balanced and nonzero, for each $i = 1, 2, \ldots, m$ there exists an edge α_i of color i in $RM(A)$ joining either two vertices of X or two vertices of Y. Since $p + q = m + 1$ either at least p of the α_i join two vertices of X or at least q join two vertices of Y. Without loss of generality we assume that at least p of the α_i join two vertices of X. Since X has only p vertices, there is a cycle in $RM(A)$ no two of whose edges have the same color, contradicting our assumption. Hence the only signings D such that each row of AD is balanced are AI_{m+1} and $A(-I_{m+1})$, and we conclude that (i) holds.

Since a set of m edges of $RM(A)$ is the set of edges of a tree if and only if it does not contain the edges of a cycle, (ii) and (iii) are equivalent. $\quad\square$

A *biclique decomposition* of a multigraph G is a partition of the edges of G into bicliques, that is, a coloring of the edges of G such that for each color, the edges of that color are the edges of a biclique of G. The colored multigraph $RM(A)$ of an m by n $(0, 1, -1)$-matrix A is a decomposition of the uncolored multigraph $RM(A)$ into the bicliques $F_1(A), F_2(A), \ldots, F_m(A)$. Conversely, any biclique decomposition of a multigraph of order n into m bicliques corresponds to the colored multigraph $RM(A)$ of some m by n $(0, 1, -1)$-matrix A. It follows from Theorem 4.4.4 that m by $m + 1$ S-matrices are essentially biclique decompositions of graphs of order $m + 1$ into m nonempty bicliques such that any set of m edges of different colors is the set of edges of a tree. The biclique decomposition of the complete graph K_{m+1} of order $m + 1$ in which the edge joining i and j has color $\min\{i, j\}$ is the colored row multigraph of the S-matrix H'_{m+1} defined in (4.2).

Corollary 4.4.5 *Let A be an m by $m + 1$ $(0, 1, -1)$-matrix. Then A is a maximal S-matrix if and only if $RM(A)$ is the complete graph K_{m+1} of order $m + 1$ and any set of m edges of different colors is the set of edges of a spanning tree of $RM(A)$.*

Proof By Theorem 4.4.4, it suffices to show that an S-matrix is a maximal S-matrix if and only if its row multigraph is a complete graph. Let B and C be matrices such that B is obtained from C by copying a column, say column i of C, where the new column is inserted on the right. Then $RM(B)$ is obtained from $RM(C)$ by appending a new vertex v, edges joining v to the vertices joined to i, and an edge joining v to i. The corollary now follows by induction on m using the recursive characterization of maximal S-matrices given in Theorem 4.4.3. $\square$

Any biclique decomposition of the complete graph K_n requires at least $n-1$ bicliques [3]. For each biclique decomposition of K_n into $n-1$ bicliques there exists a spanning tree whose edges have distinct colors [1]. Thus, given any biclique decomposition of K_n into $n-1$ bicliques, there is some choice of $n-1$ edges, one of each color, that gives a spanning tree. By Corollary 4.4.5, the maximal $n-1$ by n S-matrices correspond to the biclique decompositions of K_n which have the special property that every choice of $n-1$ edges, one of each color, gives a spanning tree.

Finally, we determine the minimum number of nonzero entries in maximal S-matrices.

Theorem 4.4.6 *Let $A = [a_{ij}]$ be an m by $m+1$ maximal S-matrix. Then*

$$\#(A) \geq (m+1)\lfloor \log_2(m+1) \rfloor + 2(m+1 - 2^{\lfloor \log_2(m+1) \rfloor}). \qquad (4.4)$$

Equality holds if and only if A can be obtained from $S_\emptyset$ by successively copying a column with the fewest number of nonzero entries. Moreover, if equality holds in (4.4), then each column of A contains either $\lfloor \log_2(m+1) \rfloor$ or $\lfloor \log_2(m+1) \rfloor + 1$ nonzero entries.

Proof We prove the theorem by induction on m. If $m \leq 1$, the theorem holds. Now assume that $m > 1$. A separate induction on m shows that if column j of A has the largest number of nonzero entries, then there exists an i such that $a_{ij} \neq 0$ and row i of A contains exactly two nonzero entries. Without loss of generality we assume that column 1 of A has the largest number of nonzero entries and a_{11} and a_{12} are the only nonzero entries in row 1. The matrix obtained from A by the conformal contraction of columns 1 and 2 on row 1 is the matrix $B = A(\{1\}, \{1\})$. Since A is a maximal S-matrix, B is a maximal S-matrix. First suppose that $m = 2^k$ for some positive integer k. Applying the inductive hypothesis to B, we conclude that either (i) each column of B contains exactly k nonzero entries, or (ii) some column of B contains at least $k+1$ nonzero entries and $\#(B) > mk$. If (i) holds, then (4.4) holds, with equality if and only if as given in the theorem. If (ii) holds, then the strict inequality holds in (4.4).

Now suppose that $2^{k-1} < m < 2^k$ for some positive integer k. By the inductive hypothesis there is a column of B which contains at least ℓ nonzero entries where

$$\ell = \lfloor \log_2 m \rfloor + 1 = \lfloor \log_2(m+1) \rfloor + 1.$$

Thus the first column of A contains at least ℓ nonzero entries. It follows that (4.4) holds with equality if and only if as given in the theorem. $\quad\square$

Suppose that A_r is an m by $m+1$ maximal S-matrix where $m+1 = 2^r$ for some positive integer r. By Theorem 4.4.6, $\#(A_r) \geq r2^r$ with equality only if each column of A_r contains exactly r nonzero entries. Up to row and column permutations the A_r are the matrices defined recursively by

$$A_r = \left[\begin{array}{c|c} A_{r-1} & A_{r-1} \\ \hline I_{2^{r-1}} & -I_{2^{r-1}} \end{array} \right] \quad (r \geq 2) \tag{4.5}$$

where

$$A_1 = \left[\begin{array}{cc} 1 & -1 \end{array} \right].$$

By Theorem 4.4.4, if we choose one edge of each color of the colored multigraph $RM(A_r)$, we obtain a spanning tree of $RM(A_r)$. If we recursively choose one edge of each color in the colored graph $RM(A_r)$, the tree obtained is a spanning tree of the graph of the r-cube.

Bibliography

[1] N. Alon, R.A. Brualdi, and B.L. Shader. Multicolored trees in bipartite decompositions, *J. Combin. Theory Ser. B*, 53:143–8, 1991.

[2] R.A. Brualdi, K.L. Chavey, and B.L. Shader. Rectangular L-matrices, *Linear Alg. Appls.*, 196:37–61, 1994.

[3] R.L. Graham and H.O. Pollak. On the addressing problem for loop switching, *Bell System Tech. J.*, 50:2495–519, 1971.

[4] P. Hansen. Recognizing sign-solvable graphs, *Discrete Appl. Math.*, 6:237–41, 1983.

[5] V. Klee. Recursive structure of S-matrices and an $O(m^2)$ algorithm for recognizing strong sign solvability, *Linear Alg. Appls.*, 96:233–47, 1987.

[6] V. Klee and R. Ladner. Qualitative matrices: Strong sign-solvability and weak satisfiability, in *Computer-assisted Analysis and Model Simplification* (H. Greenberg and J. Maybee, eds.), Academic Press, New York, 293–320, 1981.

[7]   V. Klee, R. Ladner, and R. Manber. Signsolvability revisited, *Linear Alg. Appls.*, 59:131–57, 1984.

[8]   R. Manber. Graph-theoretical approach to qualitative solvability of linear systems, *Linear Alg. Appls.*, 48:457–70, 1982.

[9]   S. Maybee. A method for identifying sign-solvable systems, Master's thesis, Department of Computer Science, University of Colorado, Boulder, 1982.

[10] R. Rado. A theorem on independence relations, *Quart. J. Math. Oxford*, 13:83–9, 1942.

5

Beyond S*-matrices

5.1 Totally L-matrices

Let A be an m by $m + 1$ matrix. By definition A is an S^*-matrix if and only if every submatrix of A of order m is an SNS-matrix. Equivalently, A is an S^*-matrix if and only if every nonzero vector in the right null spaces of the matrices in $\mathcal{Q}(A)$ has only nonzero entries. The first two sections of this chapter are concerned with a generalization of S^*-matrices investigated in [2].

Now let A be an m by n matrix with $m \leq n$. Then A is a *totally L-matrix* provided that each submatrix of A of order m is an SNS-matrix. A matrix of order m is a totally L-matrix if and only if it is an SNS-matrix. An m by $m + 1$ matrix is a totally L-matrix if and only if it is an S^*-matrix. If A is arbitrary and $m < n$, then for each matrix $\widetilde{A}$ in $\mathcal{Q}(A)$ there exists a nonzero vector in the right null space of $\widetilde{A}$ with $m + 1$ or fewer nonzero entries. If $m < n$, then A is a totally L-matrix if and only if each nonzero vector which belongs to the right null space of a matrix in $\mathcal{Q}(A)$ has at least $m + 1$ nonzero entries. It follows from (i) of Theorem 2.1.1 that A is a totally L-matrix if and only if every row signing of A contains at least $n - m + 1$ unsigned columns. The matrices

$$
\begin{bmatrix} 1 & 1 & 1 & 0 \\ 1 & -1 & 0 & 1 \end{bmatrix} \text{ and } \begin{bmatrix} 1 & 1 & 1 & 0 & 0 \\ 1 & -1 & 0 & 1 & 0 \\ 1 & 0 & -1 & 0 & 1 \end{bmatrix} \tag{5.1}
$$

are examples of totally L-matrices. If $m = 1$, then A is a totally L-matrix if and only if each entry of A is nonzero. For $m \geq 2$ we have the following.

Theorem 5.1.1 *Let A be an m by n totally L-matrix with $m \geq 2$. Then $n \leq m + 2$.*

Proof We prove that $n - m \leq 2$ by induction on m. Up to multiplication by -1, there are exactly four nonzero sign patterns of 2 by 1 vectors. Hence a totally L-matrix with two rows has at most four columns. Now assume that $m \geq 3$. First suppose that there exists an integer k with $2 \leq k \leq m - 1$ such that A has a k by $m - k$ zero submatrix. Then there exist permutation matrices P and Q such that

$$PAQ = \left[\begin{array}{cc} A' & O \\ X & Y \end{array} \right]$$

where A' is a k by $k + n - m$ matrix and Y is a matrix of order $m - k$. Since A is a totally L-matrix, A' is a totally L-matrix with at least two rows. Applying the inductive assumption to A' we conclude that $n - m \leq 2$.

Now suppose that A does not contain a k by $m - k$ zero submatrix for any integer k with $2 \leq k \leq m - 1$. Since $m \geq 3$, this implies that A does not contain a column with exactly one nonzero entry. Therefore, by Theorem 4.1.2, every submatrix of A of order m is partly decomposable and contains a 1 by $m - 1$ zero submatrix. Each such 1 by $m - 1$ zero matrix is a submatrix of $n - m + 1$ submatrices of A of order m. Since A is a totally L-matrix, each row of A contains at most one 1 by $m - 1$ zero submatrix. Hence

$$\binom{n}{m} \leq m(n - m + 1),$$

which implies that $n - m \leq 2$. □

It follows from Theorem 5.1.1 that an m by n totally L-matrix with $m \geq 2$ either is an SNS-matrix ($n = m$) or an $S*$-matrix ($n = m + 1$), or has $n = m + 2$ columns. Theorem 4.1.1 implies that an m by $m + 2$ totally L-matrix has a row with at most three, and hence exactly three, nonzero entries. We now give a technique for constructing m by $m + 2$ totally L-matrices.

Let $A = [a_{ij}]$ be an m by $m + 2$ totally L-matrix. Assume that

(i) column t of A has exactly one nonzero entry, say $a_{pt} \neq 0$, and that
(ii) exactly two other entries of row p of A are nonzero, say $a_{pr} \neq 0$ and $a_{ps} \neq 0$.

Let $B = [b_{ij}]$ be the $m + 1$ by $m + 3$ matrix obtained from A by inserting on the right a column of zeros and then inserting on the bottom the row vector all of whose entries are zero except for a 1 in columns r and $m + 3$ and the number $-a_{pr}a_{pt}$ in column t. The matrix B is called the *single-extension* of A on columns t and r. The number $-a_{pr}a_{pt}$ has been chosen so that $B[\{p, m + 1\}, \{r, t\}]$ is an SNS-matrix. Let $C = [c_{ij}]$ be the $m + 2$ by $m + 4$ matrix obtained from B by inserting on the right a column of zeros and then inserting on the bottom the row vector all of whose entries are zero except for a 1 in columns s and $m + 4$ and the number $-a_{ps}a_{pt}$ in column t. The matrix C is the *double-extension* of A on column t. The number $-a_{ps}a_{pt}$ has been chosen so that the matrix $C[\{p, m + 1, m + 2\}, \{r, s, t\}]$ is an SNS-matrix. The second totally L-matrix in (5.1) is the single-extension on columns 1 and 3 of the first totally L-matrix. The double-extension on column 3 of the first matrix in (5.1) is

$$\begin{bmatrix} 1 & 1 & 1 & 0 & 0 & 0 \\ 1 & -1 & 0 & 1 & 0 & 0 \\ 1 & 0 & -1 & 0 & 1 & 0 \\ 0 & 1 & -1 & 0 & 0 & 1 \end{bmatrix}, \tag{5.2}$$

which is easily seen to be a 4 by 6 totally L-matrix. More generally we have the following.

Theorem 5.1.2 *Single-extensions and double-extensions of m by $m + 2$ totally L-matrices are totally L-matrices.*

Proof Let A, B, and C be the matrices described in the paragraph preceding the statement of the theorem. First we prove that B is a totally L-matrix by showing that each row signing of B has at least three unsigned columns. Let $D = D_1 \oplus [d_{m+1}]$ be a signing of order $m + 1$. Suppose that $d_{m+1} = 0$. Then $D_1 A$ is a row signing of A and, since A is a totally L-matrix, $D_1 A$ has at least three unsigned columns. A column of DB corresponding to a unsigned column of $D_1 A$ is also unsigned. Hence DB has at least three unsigned columns. Now suppose that $d_{m+1} \neq 0$. Then column $m + 3$ of DB is unsigned. If $D_1 = O$, then clearly columns r, t, and $m + 3$ are unsigned columns of DB. Suppose $D_1 \neq O$. A column of DB corresponding to a unsigned column of $D_1 A$ other than columns r and t is unsigned. Hence, if $D_1 A$ has two unsigned columns neither of which is column r or column t, then DB has three unsigned columns. Thus we now suppose that at most one column of $D_1 A$ other than columns r and t is unsigned. Then $D_1 A$ has exactly three unsigned columns, and columns r and t of $D_1 A$ are unsigned. Let column q of $D_1 A$ be unsigned where $q \neq r, t$. Since column t has exactly one nonzero entry and $b_{m+1,t} = -a_{pr}a_{pt}$, one of columns r and t of DB is unsigned. Since columns q and $m + 3$ of DB are also unsigned, DB has three unsigned columns. We conclude that B is a totally L-matrix.

We now prove that the double-extension C of A is a totally L-matrix. Let $D = D_1 \oplus [d_{m+1}] \oplus [d_{m+2}]$ be a signing of order $m + 2$. The matrix obtained from C by deleting row $m + 2$ and column $m + 4$ is a single extension of A, as is the matrix obtained from C by deleting row $m + 1$ and column $m + 3$. Hence it follows from the first part of the proof that if either $d_{m+1} = 0$ or $d_{m+2} = 0$, then DC has at least three unsigned columns. Now suppose that both d_{m+1} and d_{m+2} are nonzero. Then columns $m + 3$ and $m + 4$ of DC are unsigned. If $D_1 = O$, then columns r and s of DC are also unsigned and hence DC has at least four unsigned columns. If there is a unsigned column of $D_1 A$ other than columns r, s and t, then it follows that DC has at least three unsigned columns. Now assume that $D_1 \neq O$ and column j of DC is balanced for $j \neq r, s, t$. Since A is a totally L-matrix, columns r, s and t of $D_1 A$ must be unsigned. The only nonzero entry in column t of A is in row p, and hence the pth diagonal entry of D_1 is not zero. Since the row signing of A obtained from $D_1 A$ by multiplying row p by -1 has at least three unsigned columns and since the only nonzero entries in row p of A are in columns r, s, and t, columns r and s of $D_1 A$ each have exactly one nonzero entry. Because the matrix $C[\{p, m + 1, m + 2\}, \{r, s, t\}]$ is an SNS-matrix, at least one of columns r, s, and t of DC is unsigned. Hence DC has at least three unsigned columns, and we conclude that C is a totally L-matrix.   □

Both a single-extension and a double-extension of a totally L-matrix have a column with exactly one nonzero entry where the row containing this nonzero entry has exactly two other nonzero entries, and hence can be further extended. Starting with the 2 by 4 totally L-matrix in (5.1), we can obtain by single-extensions an m by $m + 2$ totally L-matrix for each integer $m \geq 2$.

5.2 Recursive structure of totally L-matrices

In this section we show that up to row and column permutations and multiplication of rows and columns by -1, every m by $m + 2$ $(0, 1, -1)$ totally L-matrix with $m \geq 2$ can be obtained from the 2 by 4 totally L-matrix in (5.1) by a sequence of single- and double-extensions.

Lemma 5.2.1 *Let A be an m by $m + 2$ totally L-matrix with $m \geq 2$. Then at least two columns of A contain exactly one nonzero entry, and every row of A contains exactly three nonzero entries.*

Proof We prove the lemma by induction on m. If $m = 2$, then up to column permutations and multiplication of columns by -1, the sign pattern of A is the 2 by 4 matrix in (5.1). Thus the lemma holds if $m = 2$. Now suppose that $m > 2$.

We first show that at least one column of A contains exactly one nonzero entry. If A has a fully indecomposable submatrix of order m, this follows from Theorem 4.1.2. Now assume that each submatrix of A of order m is partly decomposable. After row and column permutations we may assume that

$$A = \begin{bmatrix} X & O \\ Y & Z \end{bmatrix}$$

where X is a p by $p+2$ matrix and Z is a fully indecomposable matrix of order q. Here p and q are positive integers satisfying $p + q = m$. If $q = 1$, then the last column of A contains exactly one nonzero entry. Assume that $q > 1$. Since A is a totally L-matrix, X is a totally L-matrix and Z is an SNS-matrix. It follows from Theorem 2.2.5 that there exists a strict signing D such that DZ has a column each of whose entries is nonnegative and all other columns of DZ are balanced. Let $v = (v_1, v_2, \ldots, v_{p+2})$ where

$$v_i = \begin{cases} 1 & \text{if column } i \text{ of } DY \text{ is nonzero and has only} \\ & \text{nonnegative entries,} \\ -1 & \text{if column } i \text{ of } DY \text{ is nonzero and has only} \\ & \text{nonpositive entries,} \\ 1 & \text{if column } i \text{ of } DY \text{ is nonzero and balanced,} \\ 0 & \text{if column } i \text{ of } DY \text{ is a zero column.} \end{cases}$$

Consider the $p + 1$ by $p + 3$ matrix

$$M = \begin{bmatrix} X & O \\ v & 1 \end{bmatrix}.$$

We show that M is a totally L-matrix by showing that every row signing of M has at least three unsigned columns. Let $E = \text{diag}(e_1, \ldots, e_p, e_{p+1})$ be a signing of order $p+1$ and consider the signing $U = \text{diag}(e_1, \ldots, e_p) \oplus (e_{p+1}D)$ of order m. Because A is a totally L-matrix, UA has at least three unsigned columns. If $e_{p+1} = 0$, then the number of unsigned columns of EM equals the number of unsigned columns of UA, and hence EM has at least three unsigned columns. Now suppose that $e_{p+1} \neq 0$. Then $e_{p+1}DZ$ has a unique unsigned column and the last column of EM is unsigned. If $1 \leq i \leq p+2$ and column i of UA is unsigned, then the definition of v implies that column i of EM is unsigned. It follows that EM has at least three unsigned columns. Hence M is a totally L-matrix. By the inductive assumption, at least two columns of M contain exactly one nonzero entry. One of these columns, say column j, has a zero in its last position. Since D is a strict signing, the definition of v implies that column j of A contains exactly one nonzero entry.

We may now assume that $q = 1$ and that the only entry of Z is positive. Let $Y = (y_1, y_2, \ldots, y_{m+1})$. We next show that each row of A contains exactly three nonzero entries. By the inductive assumption each row of X contains exactly three nonzero entries. Since A is a totally L-matrix, the matrix

$$\begin{bmatrix} X \\ Y \end{bmatrix}$$

is an S^*-matrix and by Theorem 4.1.1 has a row with exactly two nonzero entries. It follows that Y contains exactly two nonzero entries and every row of A contains exactly three nonzero entries.

Suppose that A has only one column containing exactly one nonzero entry. Then by the inductive assumption we may assume that

$$X = \begin{bmatrix} X_1 & O & O \\ X_2 & a & 0 \\ X_3 & 0 & b \end{bmatrix}$$

where $y_1 = \cdots = y_{m-1} = 0$ and $y_m y_{m-1} \neq 0$. Without loss of generality, each of a, b, y_m, and y_{m+1} is positive. Consider the matrix X' obtained from X by replacing the 0 in position $(m-1, m)$ by -1. Since X' has a row with four nonzero entries, the inductive assumption implies that X' is not a totally L-matrix. Thus there is a signing D such that DX' has at most two unsigned columns. Since DX has at least three unsigned columns, we conclude that DX has exactly three unsigned columns, and the nonzero entries of the last two columns of DX have the same sign. But now either $(D \oplus [1])A$ or $(D \oplus [-1])A$ has only two unsigned columns, contradicting the fact that A is a totally L-matrix. This contradiction shows that A has at least two columns with exactly one nonzero entry, and the theorem follows by induction. ☐

Corollary 5.2.2 *Let A be an m by $m + 2$ $(0, 1, -1)$ totally L-matrix with $m \geq 2$. Let u be any row of A. Then there exists a row v of A such that, up to column permutations and multiplication of columns by -1, the 2 by $m + 2$ submatrix of A determined by u and v equals*

$$\begin{bmatrix} 1 & -1 & 1 & 0 & 0 & \cdots & 0 \\ 1 & 1 & 0 & 1 & 0 & \cdots & 0 \end{bmatrix}.$$

Moreover, each submatrix of A of order 2 either is an SNS-matrix or has an identically zero determinant, and each 3 by 2 submatrix has a zero entry.

Proof The lemma clearly holds when $m = 2$. By Lemma 5.2.1, there are two columns of A which contain exactly one nonzero entry. At least one of these columns has a zero in the row corresponding to u. Deleting such a column and the row containing its nonzero entry, we obtain a totally L-matrix. The first assertion now follows by induction on m.

Consider any submatrix $A[\{i, j\}, \{k, \ell\}]$ of A of order 2 which has no zero entries. Since A is a totally L-matrix, each row signing of A has at least three unsigned columns. Let D be a signing of order m with nonzero entries only in rows i and j. Since each row of A has exactly three nonzero entries, it follows that at least one of columns k and ℓ of DA is unsigned. Hence $A[\{i, j\}, \{k, \ell\}]$ is an SNS-matrix. Thus each submatrix of A of order 2 is an SNS-matrix or has an identically zero determinant. The lemma now follows from the fact that a 3 by 2 matrix with no zero entries has a submatrix of order 2 which is not an SNS-matrix. □

We now obtain a recursive characterization of m by $m + 2$ totally L-matrices.

Theorem 5.2.3 *The set of all matrices which can be obtained from*

$$\begin{bmatrix} 1 & 1 & 1 & 0 \\ 1 & -1 & 0 & 1 \end{bmatrix} \tag{5.3}$$

by a sequence of single- and double-extensions is, up to row and column permutations and multiplication of rows and columns by −1, the set of all m by m + 2 totally L-matrices of 0's, 1's, and −1's with m ≥ 2.

Proof It follows from Theorem 5.1.2 that any matrix obtained as prescribed in the theorem is a totally L-matrix. We prove the converse by induction on m. Let A be an m by $m + 2$ totally L-matrix. If $m = 2$, then (5.3) is a totally L-matrix. If $m = 3$, then using Lemma 5.2.1 and Corollary 5.2.2 we see that, up to row and column permutations and multiplication of rows and columns by −1, A is a single-extension of (5.3). Now assume that $m \geq 4$. By Lemma 5.2.1, A has at least two columns with exactly one nonzero entry. Without loss of generality we assume that A has the form

$$\begin{bmatrix} B & O \\ x & 1 \end{bmatrix}$$

where B is an $m - 1$ by m totally L-matrix. By the inductive assumption, B is either a single- or double-extension of a totally L-matrix C.

First assume that B is a single-extension of C. Without loss of generality we may assume that A has the form

$$\left[\begin{array}{cccccc|c|c}
 & & & & & 0 & 0 & 0 \\
 & & & & & 0 & 0 & 0 \\
 & & & & & \vdots & \vdots & \vdots \\
 & & & & & 0 & 0 & 0 \\
0 & \cdots & 0 & 1 & 1 & 1 & 0 & 0 \\
0 & \cdots & 0 & 0 & 1 & -1 & 1 & 0 \\
x_1 & \cdots & x_{m-3} & x_{m-2} & x_{m-1} & x_m & x_{m+1} & 1
\end{array}\right].$$

If $x_m = 0$ and $x_{m+1} \neq 0$, then using Corollary 5.2.2 we see that A is a single-extension of B on columns $m - 1$ and $m + 1$. If $x_m = 0$ and $x_{m+1} = 0$, then interchanging rows $m - 1$ and m of A we obtain a matrix which is the single-extension of a totally L-matrix. Now suppose that $x_m \neq 0$. By Corollary 5.2.2, either $x_{m-2} \neq 0$ or $x_{m+1} \neq 0$. If $x_{m-2} \neq 0$, A is the double-extension of C on column m. If $x_{m+1} \neq 0$, then A is the single-extension of B on columns m and $m + 1$. Therefore, if B is a single-extension of C, then A is a single- or double-extension of a totally L-matrix.

Now assume that B is a double-extension of a totally L-matrix C. Without loss of generality we assume that A has the form

$$A = \left[\begin{array}{cccccc|cc|c}
 & & & & & 0 & 0 & 0 & 0 \\
 & & & & & \vdots & \vdots & \vdots & \vdots \\
 & & & & & 0 & 0 & 0 & 0 \\
0 & \cdots & 0 & 1 & 1 & 1 & 0 & 0 & 0 \\
0 & \cdots & 0 & 0 & 1 & -1 & 1 & 0 & 0 \\
0 & \cdots & 0 & 1 & 0 & -1 & 0 & 1 & 0 \\
x_1 & \cdots & x_{m-4} & x_{m-3} & x_{m-2} & x_{m-1} & x_m & x_{m+1} & 1
\end{array}\right].$$

If $x_{m-1} = 0$ and both x_m and x_{m+1} equal zero, then we may permute rows $m - 2$, $m - 1$, and m to obtain a matrix which is the double-extension on column $m - 1$ of a totally L-matrix. Otherwise, if $x_{m-1} = 0$, then by Corollary 5.2.2 A is either the single-extension on column m or the single-extension on column $m + 1$. Suppose that $x_{m-1} \neq 0$. By Corollary 5.2.2, either x_m or x_{m+1} is nonzero. In either case, A is a single-extension of a totally L-matrix. $\qquad\square$

Let G be a plane graph with n vertices and m bounded faces. We say that the m by n $(0, 1)$-matrix $B = [b_{ij}]$ is a *face-vertex* incidence matrix of G provided that the sets $V_i = \{j : b_{ij} = 1\}$ $(i = 1, 2, \ldots, m)$ are the vertex sets of the faces of G. A *triangulation of an n-gon* is a plane graph with n vertices, each of whose bounded faces has exactly three vertices and whose unbounded face is a cycle of length n. It is not difficult to show that an m by $m+2$ $(0, 1)$-matrix B with $m \geq 2$ is the zero pattern of a totally L-matrix if and only if B is the face-vertex incidence matrix of a triangulation of an $(m+2)$-gon.

We conclude this section by noting that every m by $m+2$ totally L-matrix A is a submatrix of an SNS*-matrix of order $m+2$. To see this, let B be an $m+1$ by $m+2$ matrix obtained by deleting the last column of a single-extension of A. It follows from Theorem 5.1.2 that B is an S^*-matrix. By Theorem 4.1.4, B and hence A is a submatrix of an SNS*-matrix of order $m+2$.

5.3 Signed compound

Let A be an m by n matrix and let r be a positive integer with $r \leq \min\{m, n\}$. Recall that the rth compound of A is an $\binom{m}{r}$ by $\binom{n}{r}$ matrix whose entries are the numbers

$$a_{\{i_1,i_2,\dots,i_r\},\{j_1,j_2,\dots,j_r\}} = \det A[\{i_1, i_2, \dots, i_r\}, \{j_1, j_2, \dots, j_r\}]$$

where $1 \leq i_1 < i_2 < \cdots < i_r \leq m$ and $1 \leq j_1 < j_2 < \cdots < j_r \leq n$. The matrix A has a *signed rth compound* provided the rth compounds of the matrices in $\mathcal{Q}(A)$ are entirely contained in one qualitative class, that is, every submatrix of A of order r either is an SNS-matrix or has an identically zero determinant. Note that every matrix has a signed first-order compound. Many of the classes of matrices that we have studied can be defined in terms of signed compounds. A square matrix of order n has a signed nth compound if and only if it is an SNS-matrix or has an identically zero determinant. An m by n matrix is a totally L-matrix if and only if it has a signed mth compound with no zero entries. A matrix of order n is an S^2NS-matrix if and only if it has both a signed $(n-1)$th compound and a nonzero signed nth compound.

Consider the matrix

$$F = \begin{bmatrix} 1 & 1 & 0 & \cdots & 0 & 0 \\ 0 & 1 & 1 & \cdots & 0 & 0 \\ 0 & 0 & 1 & \cdots & 0 & 0 \\ \vdots & \vdots & \vdots & \ddots & \vdots & \vdots \\ 0 & 0 & 0 & \cdots & 1 & 1 \\ 1 & 0 & 0 & \cdots & 0 & 1 \end{bmatrix}$$

of order $n \geq 2$ with exactly two 1's in each row and column. Since every submatrix of F of order $r < n$ has at most one nonzero term in its standard determinant expansion, it follows that F has a signed rth compound for each $r < n$. Moreover, F has a signed nth compound if and only if n is odd.

Let A be an SNS-matrix, and let B and C be S^*-matrices. Then the matrix $X = A \oplus B \oplus C^T$ of order, say, n is a square matrix with identically zero determinant. The only submatrices of X of order $n - 1$ which do not have an identically zero determinant are those obtained by deleting a row that intersects C^T and a column that intersects B. Hence X has a signed $(n - 1)$th compound.

Lemma 5.3.1 *Let $A = [a_{ij}]$ be an m by n matrix and let r be an integer with $r \le \min\{m, n\}$. Assume that $a_{mn} \ne 0$ and let*

$$
B = \left[\begin{array}{ccc|c}
 & & & 0 \\
 & A & & \vdots \\
 & & & 0 \\
 & & & a_{mn} \\
\hline
0 & \cdots & 0 \ 1 & -1
\end{array}\right].
$$

(i) *If A has signed $(r - 1)$st and rth compounds, then B has a signed rth compound.*

(ii) *If $m \le n$ and A has a signed mth compound, then B has a signed $(m + 1)$th compound.*

(iii) *If A has signed $(r - 1)$st and rth compounds, then any matrix obtained from A by inserting a new column with exactly one nonzero entry has a signed rth compound.*

Proof First assume that A has signed $(r-1)$th and rth compounds. Consider a submatrix $B[\alpha, \beta]$ of B of order r. If $B[\alpha, \beta]$ does not contain the submatrix of order 2 in the lower right corner of B, then the assumptions imply that $B[\alpha, \beta]$ has a signed determinant. Now assume that $\{m, m+1\} \subseteq \alpha$ and $\{n, n+1\} \subseteq \beta$. If $B[\alpha \setminus \{m + 1\}, \beta \setminus \{n + 1\}]$ has an identically zero determinant, then so does $B[\alpha, \beta]$. Otherwise the assumptions and Theorem 4.1.3 imply that $B[\alpha, \beta]$ is an SNS-matrix and thus has a signed determinant. Hence B has a signed rth compound, and (i) holds.

Now assume that $m \le n$ and A has a signed mth compound. Since A does not have a submatrix of order $m + 1$, it follows as before that B has a signed $(m + 1)$th compound and (ii) holds. Assertion (iii) is easily verified. $\quad\square$

Theorem 5.3.2 *Any m by $m + 2$ matrix obtained from the 2 by 4 matrix given in (5.3) by a sequence of single-extensions has a signed rth compound for each $r = 1, 2, \ldots, m$.*

Proof A single-extension of an m by n matrix A' can be obtained by constructing B, as in Lemma 5.3.1, from an m by $n - 1$ submatrix A of A' and then inserting a new column with exactly one nonzero entry. Thus the theorem follows by induction using Lemma 5.3.1. □

Not every totally L-matrix satisfies the conclusion of Theorem 5.3.2. For example, the totally L-matrix (5.2) obtained from (5.3) by a double-extension does not have a signed 3rd compound. By Corollary 5.2.2, every m by $m + 2$ totally L-matrix has a signed 2nd compound. The zero patterns of m by n matrices ($m \leq n$) which have a signed rth compound for $r = 1, 2, \ldots, m$ are characterized in Corollary 7.2.3.

The *term rank* of an m by n matrix A is the largest order r of a square submatrix of A which does not have an identically zero determinant. A theorem of König [3] asserts that r is the smallest number of rows and columns which cover all the nonzero entries of A. By [3, Theorem 4.2.1] there exist permutation matrices P and Q such that PAQ has the form

$$\begin{bmatrix} X & O \\ Z & Y \end{bmatrix} \tag{5.4}$$

where X is an $m - e$ by f matrix with term rank f, Y is an e by $n - f$ matrix with term rank e, and $r = e + f$. Without loss of generality we assume that A has the form (5.4). The rth compound of A does not depend on the entries of Z, and hence A has a signed rth compound if and only if X has a signed fth compound and Y has a signed eth compound. This reduces the study of signed rth compounds of matrices whose term rank is r to the study of signed pth compounds of p by q matrices whose term rank is p.

Let B be a p by q matrix of term rank p. If the pth compound of B has no zero entries, then B has a signed pth compound if and only if B is a totally L-matrix. Suppose that B has a signed pth compound and that B is not a totally L-matrix. Then some submatrix of B of order p has an identically zero determinant, and hence by the Frobenius–König theorem, B has a $p + 1 - t$ by t zero submatrix for some integer t with $1 \leq t \leq p$. We choose such a zero submatrix with t minimum. After row and column permutations, we may assume that B has the form

$$\begin{bmatrix} E & O \\ F & G \end{bmatrix}$$

where O is a $p + 1 - t$ by t zero matrix. It follows from the minimum property of t that G is a $t - 1$ by t S^*-matrix, possibly a vacuous 0 by 1 matrix. The matrix E is a $p+1-t$ by $q-t$ matrix of term rank $p+1-t$ and E has a signed $(p+1-t)$th compound. Either E is a totally L-matrix or some submatrix of E of order $p+1-t$ has an identically zero determinant. Hence it follows inductively that there exist permutation matrices P and Q such that PBQ has the form

$$\begin{bmatrix} B_1 & O & \cdots & O \\ B_{21} & B_2 & \cdots & O \\ \vdots & \vdots & \ddots & \vdots \\ B_{k1} & B_{k2} & \cdots & B_k \end{bmatrix}$$

where B_1 is a totally L-matrix and $B_2, \ldots, B_k$ are S^*-matrices. Not every matrix satisfying these properties has a signed pth compound. For example,

$$\left[\begin{array}{cc|cc} 1 & 1 & 0 & 0 \\ 1 & 1 & 1 & 1 \end{array} \right]$$

does not have a signed 2nd compound. However, if $q = p + 1$, the converse does hold.

Theorem 5.3.3 *Let B be a p by $p + 1$ matrix of term rank p. Then B has a signed pth compound if and only if there exist permutation matrices P and Q such that PBQ has the form*

$$\begin{bmatrix} B_1 & O \\ B_{21} & B_2 \end{bmatrix} \tag{5.5}$$

where B_1 is an SNS-matrix and B_2 is an S^*-matrix.

Proof First assume that B has a signed pth compound. If $q = p+1$ in the preceding discussion, then either $k = 1$ (and B_1 in (5.5) is vacuous) or $k = 2$ (and B_1 is an SNS-matrix). Now suppose that B has the form (5.5) where B_1 is an SNS-matrix and B_2 is an S^*-matrix. Each matrix obtained from B by deleting a column that intersects B_2 is an SNS-matrix. Each matrix obtained by deleting a column which intersects B_1 has an identically zero determinant. Hence B has a signed pth compound. $\square$

   Taking the p by q matrix B in the preceding discussion to be the matrices X^T and Y in (5.4), we obtain the following.

Theorem 5.3.4 *Let A be a matrix of term rank r such that A has a signed rth compound. Then there exist permutation matrices P and Q such that PAQ has the form*

$$
\begin{bmatrix}
\begin{array}{cccc|cccc}
A_1 & O & \cdots & O & & & & \\
A_{21} & A_2 & \cdots & O & & & & \\
\vdots & \vdots & \ddots & \vdots & & O & & \\
A_{\ell,1} & A_{\ell,2} & \cdots & A_\ell & & & & \\
\hline
 & & & & B_1 & O & \cdots & O \\
 & U & & & B_{21} & B_2 & \cdots & O \\
 & & & & \vdots & \vdots & \ddots & \vdots \\
 & & & & B_{k1} & B_{k2} & \cdots & B_k
\end{array}
\end{bmatrix}
$$

where $A_1^T, \ldots, A_{\ell-1}^T$ are S^*-matrices and A_ℓ^T is a totally L-matrix, B_1 is a totally L-matrix and $B_2, \ldots, B_k$ are S^*-matrices, and U is an e by f matrix with $e + f = r$.

5.4 Sign-central matrices

Let B be an m by n matrix. The matrix B is *central* provided there is a nonnegative vector $x \neq 0$ in its null space. The columns $b^{(1)}, b^{(2)}, \ldots, b^{(n)}$ of B determine a convex polytope

$$
\mathcal{C}(B) = \left\{ c_1 b^{(1)} + c_2 b^{(2)} + \cdots + c_n b^{(n)} : \sum_{i=1}^n c_i = 1, c_i \geq 0 \ (1 \leq i \leq n) \right\},
$$

and the matrix B is central provided that the origin $(0, 0, \ldots, 0)^T$ is contained in $\mathcal{C}(B)$. Necessarily, the columns of a central matrix are linearly dependent. The matrix A is *sign-central* provided that each matrix in $\mathcal{Q}(A)$ is central. A matrix with one row is sign-central if and only if it has at least one zero entry or contains both a positive and a negative entry. Each S-matrix is sign-central, and hence sign-central matrices generalize S-matrices. The idea behind the definition of sign-centrality occurs in [4]. This section is based on the paper [1].

If A is an m by n sign-central matrix, then any m by p matrix which contains A as a submatrix is also a sign-central matrix. A *minimal sign-central matrix* is a sign-central matrix such that each matrix obtained from A by the deletion of a column is not sign-central. Each nonzero row of a minimal sign-central matrix contains both a positive entry and a negative entry and thus has at least two nonzero entries. Every S-matrix is a minimal sign-central matrix. The matrix

$$\begin{bmatrix} 1 & 1 & 1 & 1 & -1 & -1 & -1 & -1 \\ 1 & 1 & -1 & -1 & 1 & 1 & -1 & -1 \\ 1 & -1 & 1 & -1 & 1 & -1 & 1 & -1 \end{bmatrix}$$

is also a minimal sign-central matrix. More generally, for each positive integer m, the m by 2^m matrix E_m such that each m-tuple of 1's and -1's is a column of E_m is a minimal sign-central matrix.

Let A be an m by n matrix, and let P and Q be permutation matrices of orders m and n, respectively. Let D be a strict signing. Then A is a (minimal) sign-central matrix if and only if $PDAQ$ is a (minimal) sign-central matrix. Now assume that A has a zero row, and let A' be a matrix obtained from A by deleting a zero row. Then A is a (minimal) sign-central matrix if and only if A' is a (minimal) sign-central matrix. It follows that zero rows are of no significance in the study of sign-central matrices.

Let $\mathcal{X}$ be a family of nonempty subsets of a set X. Then $\mathcal{X}$ is a *clutter* provided for all α and β in $\mathcal{X}$, $\alpha \subseteq \beta$ implies that $\alpha = \beta$.[1] The *blocker* of $\mathcal{X}$ is the collection $b(\mathcal{X})$ of subsets γ of X such that γ has a nonempty intersection with every set in $\mathcal{X}$ but no proper subset of γ has this property. The blocker $b(\mathcal{X})$ is clearly a clutter. If $\mathcal{X}$ is also a clutter, then it is straightforward to check that $b(b(\mathcal{X})) = \mathcal{X}$.

Let $c = (c_1, c_2, \ldots, c_m)^T$ be a $(0, 1, -1)$ column vector, and let $R(c)$ denote the subset of the set

$$M = \{1, 2, \ldots, m, \overline{1}, \overline{2}, \ldots, \overline{m}\}$$

defined by

$$R(c) = \{i : c_i = 1\} \cup \{\overline{i} : c_i = -1\}.$$

Let $A = [a^{(1)} \ a^{(2)} \ \cdots \ a^{(n)}]$ be an m by n $(0, 1, -1)$-matrix with column vectors $a^{(1)}, a^{(2)}, \ldots, a^{(n)}$. The collection $\mathcal{A}$ of subsets of M consisting of the minimal (with respect to set inclusion) sets among

$$R(a^{(1)}), R(a^{(2)}), \ldots, R(a^{(n)})$$

is called the *clutter of the matrix* A, and $b(\mathcal{A})$ is the *blocker of the matrix* A.[2] Since no entry of A is both 1 and -1, $\{i, \overline{i}\}$ is not a subset of any set in the clutter of A for each $i = 1, 2, \ldots, m$. The blocker of the sign central matrix E_m is $\{\{1, \overline{1}\}, \{2, \overline{2}\}, \ldots, \{m, \overline{m}\}\}$.

Characterizations of sign-central matrices are given in the next theorem.

---

[1] A clutter is an *antichain* in the partially ordered set of all subsets of X.
[2] More properly, $b(\mathcal{A})$ is the blocker of $\mathcal{A}$.

Theorem 5.4.1 *Let A be an m by n* $(0, 1, -1)$-*matrix. Then the following are equivalent:*

(i) *A is a sign-central matrix.*

(ii) *For every strict signing D of order m, the matrix DA has a nonnegative column vector.*

(iii) *For every strict signing D of order m, the matrix DA has a nonpositive column vector.*

(iv) *Each set of the blocker $b(\mathcal{A})$ contains as a subset at least one of the sets $\{1, \overline{1}\}, \{2, \overline{2}\}, \ldots, \{m, \overline{m}\}$.*

(v) *There do not exist permutation matrices P and Q such that*

$$PAQ = \left[\begin{array}{cc} A_1 & X_1 \\ X_2 & A_2 \end{array} \right]$$

where A_1 is a possibly vacuous matrix with at least one 1 in each column and A_2 is a possibly vacuous matrix with at least one -1 in each column.

Proof The equivalence of (ii) and (iii) follows from the fact that D is a strict signing if and only if $-D$ is a strict signing. Assertion (v) is a reformulation of (iv). We next show that (ii) and (iv) are equivalent. Assume that there exists a strict signing D such that each column of DA contains at least one -1. Without loss of generality we assume that the first $k \geq 0$ diagonal entries of D equal 1 and the last $m - k \geq 0$ diagonal entries equal -1. We may also assume that each of the first $p \geq 0$ columns of DA contain at least one -1 in rows $1, \ldots, k$ and each of the last $n - p \geq 0$ columns contain at least one -1 in rows $k + 1, \ldots, m$. Then $\{\overline{1}, \ldots, \overline{k}, k + 1, \ldots, m\}$ has a nonempty intersection with each set $R(a^j)$ and consequently contains a set of $b(A)$ as a subset. Thus (iv) implies (iii). Now assume that (iv) does not hold. Then without loss of generality there exists a set in $b(\mathcal{A})$ of the form $X = \{1, \ldots, k, \overline{k + 1}, \ldots, \overline{l}\}$. Each column of A either contains a 1 in one of rows $1, \ldots, k$ or contains a -1 in one of rows $k + 1, \ldots, l$. Let D be the strict signing such that the first k entries on its main diagonal equal -1 and all other entries on the main diagonal equal 1. Then each column of DA contains a -1 and (ii) does not hold. Thus (ii) and (iv) are equivalent.

We now show that (ii) implies (i) by induction on m. Assume that (ii) and therefore (iii) holds. If the null space of each matrix B in $\mathcal{Q}(A)$ contains a positive vector, then clearly A is a sign-central matrix. Assume that B is a matrix in $\mathcal{Q}(A)$ such that

$$\text{null space}(B) \cap \mathbf{R}_+^n = \emptyset$$

where $\mathbf{R}_+^n$ denotes the open positive orthant in $\mathbf{R}^n$. Applying the separation theorem for convex sets, we conclude that there exists a nonzero, nonnegative vector $v = (v_1, \ldots, v_n)^T$ in $\mathbf{R}^n$ which is orthogonal to the null space of B. Since the orthogonal complement of the null space of B is spanned by the rows of B, it follows that there exists a nonzero vector u in $\mathbf{R}^m$ such that $v = u^T B$. By replacing u by Eu and B by $E^{-1}B$, where E is an invertible diagonal matrix, and then permuting the rows of B and u, we may assume that the first k entries of u are 1 and the last $m - k$ entries are 0 for some $k \geq 2$. Suppose that the first k entries of column j of A are nonpositive. Since the jth entry of v is the sum of the first k entries of column j of B, and since v is nonnegative, it follows that the first k entries of column j of A are zero. Thus we may assume that A has the form

$$\begin{bmatrix} X & O \\ Y & Z \end{bmatrix}$$

where X has k rows and each column of X is nonzero and either balanced or nonnegative. Since (iii) holds, if $k = m$ then A has a column of all 0's, and hence A is a sign-central matrix. Assume that $k < m$. Let D be a strict signing of order $m - k$. By (ii), the matrix $(I_k \oplus D)A$ contains a nonpositive column. The definition of X implies that DZ contains a nonpositive column. Hence, for every strict signing D of order $m - k$, the matrix DZ has a nonpositive column vector. Applying the induction hypothesis, we conclude that Z, and hence A, is a sign-central matrix.

Finally, we prove that (i) implies (ii) by induction on m. If $m = 1$, the assertion is trivially true. We now assume that $m \geq 2$ and that the assertion holds for $m - 1$. If A has a zero row, the conclusion follows easily from the inductive assumption. Assume that each row of A contains a nonzero entry. Without loss of generality we assume that the 1's in the last row of A occur in positions $1, 2, \ldots, k$, the -1's occur in positions $k + 1, \ldots, l$, and the 0's occur in the remaining $n - l$ positions where $0 < k < l \leq n$. Let B be any matrix in $\mathcal{Q}(A)$. Let A' be the $m - 1$ by $n - k$ matrix obtained by deleting the last row of A and columns $1, \ldots, k$, and let A'' be the $m - 1$ by $n - (l - k)$ matrix obtained by deleting the last row of A and columns $k + 1, \ldots, l$. We show that the matrices A' and A'' are sign-central matrices. Let B' be any matrix in $\mathcal{Q}(A')$. Take a matrix B in $\mathcal{Q}(A)$, the corresponding matrix of which in $\mathcal{Q}(A')$ is B'. For each positive integer λ, let $B_\lambda = [b_\lambda^{(1)} \cdots b_\lambda^{(n)}]$ be the matrix in $\mathcal{Q}(A)$ obtained by replacing the positive entries in row m of B with λ. Since A is a sign-central matrix, there exist numbers $c_{j,\lambda} \geq 0$ such that $\sum_{j=1}^n c_{j,\lambda} = 1$ and $\sum_{j=1}^n c_{j,\lambda} b_\lambda^{(j)} = 0$. There exists an infinite sequence of the λ's such that the limits $c_j = \lim_{\lambda \to \infty} c_{j,\lambda}$ exist for each $j = 1, 2, \ldots, n$. We then have

$$\sum_{j=1}^{n} c_j = 1, \; c_j = 0 \; (j = 1, \ldots, k) \text{ and } \sum_{j=k+1}^{n} c_j b^{(j)} = 0.$$

It follows that

$$B'(c_{k+1}, \ldots, c_n)^T = 0.$$

Since this is true for each matrix B' in $\mathcal{Q}(A')$, A' is a sign-central matrix and similarly A'' is a sign-central matrix. By the inductive assumption, for any strict signing D^* of order $m - 1$, some column of $D^* A'$ is nonnegative and some column of $D^* A''$ is nonnegative. Now let D be any strict signing of order m. If the last diagonal entry of D is 1, then one of columns $1, \ldots, k, l + 1, \ldots, n$ is nonnegative. If the last diagonal entry of D is -1, then one of columns $k + 1, \ldots, n$ of DA is nonnegative. Hence (ii) holds and the proof of the theorem is complete. $\qquad\square$

We now consider minimal sign-central matrices. Since a sign-central matrix may have zero rows, a sign-central matrix need not be an L-matrix. However, we have the following theorem.

Theorem 5.4.2 *Let A be an m by n $(0, 1, -1)$ minimal sign-central matrix with no zero rows. Then A is an L-matrix.*

Proof Assume to the contrary that A is not an L-matrix. Then there exists a signing D such that each column of DA is balanced. Without loss of generality assume that $D = D_1 \oplus O$ where D_1 is a strict signing of order k for some integer k with $1 \leq k \leq m$. First suppose that no column of DA is a zero column. Then for each strict signing D' obtained by replacing the 0's, if any, on the main diagonal of D with 1's, $D'A$ does not have a nonnegative column. By Theorem 5.4.1, A is not a sign-central matrix.

Now suppose that some column of DA is a zero column. Without loss of generality we may assume that

$$A = \begin{bmatrix} A_1 & O \\ X & A_2 \end{bmatrix},$$

where A_1 has k rows and where each column of $D_1 A_1$ is balanced and nonzero. Since A is a sign-central matrix, for each strict signing D_2 of order $m - k$, the matrix $(D_1 \oplus D_2)A$ has a nonnegative column. This implies that $D_2 A_2$ has a nonnegative column for all strict signings D_2. Thus, by Theorem 5.4.1, A_2 is a sign-central matrix. The minimality of A now implies that $A = A_2$. Hence DA can have no zero columns and the proof is complete. $\qquad\square$

A sign-central matrix A with no zero rows is not necessarily an L-matrix. For example, the matrix

$$A = \begin{bmatrix} 0 & 1 & -1 \\ 1 & 0 & 0 \\ 1 & 0 & 0 \\ 1 & 0 & 0 \end{bmatrix} \tag{5.6}$$

is a sign-central matrix with no zero rows but is not an L-matrix.

Since a matrix obtained from a (minimal) sign-central matrix by including zero rows is also a (minimal) sign-central matrix, the number of rows of a (minimal) sign-central matrix can be very large in comparison with the number of its columns. In particular, as the matrix (5.6) shows, a sign-central matrix with no zero rows can have more rows than columns. However, we have the following theorem.

Theorem 5.4.3 *Let A be an m by n minimal sign-central matrix with no zero rows. Then $n \geq m + 1$ with equality if and only if A is an S-matrix.*

Proof By Theorem 5.4.2, A is an L-matrix, and hence $n \geq m$. Since an SNS-matrix is not a sign-central matrix, $n \geq m + 1$. Suppose that $n = m + 1$. Now assume that A is a minimal sign-central matrix but some submatrix of A of order m is not an SNS-matrix. Without loss of generality, suppose that the leading submatrix B of A of order m is not an SNS-matrix. Thus there exists a signing D such that each column of DB is balanced. Without loss of generality we assume that $D = I_k \oplus O$ where k is an integer with $1 \leq k \leq m$ and that

$$A = \left[\begin{array}{c|c|c} O & A_1 & u \\ \hline A_2 & A_3 & v \end{array} \right]$$

where A_1 is a k by t matrix, each of whose columns is nonzero and balanced, and A_2 is a $m - k$ by $m - t$ matrix. We have $t \geq 1$ since A has no zero rows and each nonzero row of a minimal sign-central matrix has at least two nonzero entries. We also have $t < m$, for otherwise, either I_m is a strict signing such that $I_m A = A$ has no nonnegative column (if u is either balanced or each nonzero entry of u is negative) or $-I_m$ is a strict signing such that $-I_m A = -A$ has no nonnegative column (if each nonzero entry of u is positive). Finally, we have $k < m$ since $m - t \geq 1$ and since a minimal sign-central with $m + 1 \geq 2$ columns cannot have a zero column. It now follows that A_2 is a nonvacuous matrix. Suppose that u is either balanced or nonpositive. Then since $(I_k \oplus D')A$ has a nonnegative column for all strict signings D' of order $m - k$, we conclude that $D' A_2$ has a nonnegative column for all D', and hence A_2 is a sign-central matrix. If u is nonnegative, we consider the strict signings of the form $-I_k \oplus D'$ and also conclude that A_2 is a sign-central matrix. But this implies that the submatrix of A determined by its first $m - t$ columns is a sign-central matrix, which contradicts the assumption that A is a minimal sign-central matrix. This contradiction completes the proof of the theorem. □

As observed in [1], the problem of deciding if an m by n $(0, 1, -1)$-matrix $A = [a_{ij}]$ is not a sign-central matrix is an NP-complete problem.

Bibliography

[1] T. Ando and R.A. Brualdi. Sign-central matrices, *Linear Alg. Appls.*, 208/209:283–95, 1994.

[2] R.A. Brualdi, K.L. Chavey, and B.L. Shader. Rectangular L-matrices, *Linear Alg. Appls.*, 196:37–61, 1994.

[3] R.A. Brualdi and H.J. Ryser. *Combinatorial Matrix Theory*, Cambridge University Press, New York, 1991.

[4] G.V. Davydov and I.M. Davydov. Solubility of the system $Ax = 0, x \geq 0$ with indefinite coefficients, *Soviet Math. (IZ.VUZ.)*, 34-9:108–12, 1990.

6

SNS-matrices

6.1 Zero patterns of SNS-matrices

Let n be a positive integer. The set $\mathcal{M}_n$ of all $(0,1)$-matrices of order n is a partially ordered set using the entrywise partial order, that is, if $X = [x_{ij}]$ and $Y = [y_{ij}]$ are in $\mathcal{M}_n$, then $X \leq Y$ if and only if $x_{ij} \leq y_{ij}$ for all i and j. In this section we are primarily concerned with identifying the matrices in $\mathcal{M}_n$ which are the zero patterns of SNS-matrices.

Let $A = [a_{ij}]$ be a $(0, 1)$-matrix of order n. The set of all $(0, 1, -1)$-matrices whose zero pattern equals A is denoted by $\mathcal{Z}(A)$. Note that if $\widehat{A}$ is an SNS-matrix in $\mathcal{Z}(A)$, then so is $D_1 \widehat{A} D_2$ for all strict signings D_1 and D_2 of order n. An obvious necessary condition for $\mathcal{Z}(A)$ to contain an SNS-matrix is that A does not have an identically zero determinant.

Assume that A does not have an identically zero determinant. Suppose A is partly decomposable. Then without loss of generality A has the form

$$\begin{bmatrix} B_1 & O \\ B_3 & B_2 \end{bmatrix}$$

where B_1 is a square matrix of order p with $1 \leq p < n$. Let

$$\widehat{A} = \begin{bmatrix} \widehat{B}_1 & O \\ \widehat{B}_3 & \widehat{B}_2 \end{bmatrix}$$

be a matrix in $\mathcal{Z}(A)$ where $\widehat{B}_1$ has order p. Then $\det \widehat{A} = \det \widehat{B}_1 \det \widehat{B}_2$, and it follows that $\widehat{A}$ is an SNS-matrix if and only if both $\widehat{B}_1$ and $\widehat{B}_2$ are SNS-matrices. Hence $\mathcal{Z}(A)$ contains an SNS-matrix if and only if both $\mathcal{Z}(B_1)$ and $\mathcal{Z}(B_2)$ contain an SNS-matrix. More generally, if $A_1, A_2, \ldots, A_k$ are the fully indecomposable components of A, then $\mathcal{Z}(A)$ contains an SNS-matrix if and only if $\mathcal{Z}(A_i)$ contains an SNS-matrix for each $i = 1, 2, \ldots, k$. For this reason, in this section we generally assume that A is a fully indecomposable matrix.

A *signing of a digraph* is a signed digraph obtained by assigning either $+1$ or -1 to each arc of the digraph.

Lemma 6.1.1 *Let A be a $(0, 1)$-matrix of order n with $I_n \leq A$. Then there exists an SNS-matrix in $\mathcal{Z}(A)$ if and only if there is a signing of the digraph of A in which every directed cycle is negative.*

Proof There exists an SNS-matrix in $\mathcal{Z}(A)$ if and only if there exists an SNS-matrix in $\mathcal{Z}(A)$ such that each entry on the main diagonal equals -1. The lemma is thus an immediate consequence of Theorem 3.2.1. □

Let n be a positive integer. We denote by C_n the permutation matrix of order n which has 1's in positions $(1, 2), (1, 3), \ldots, (1, n-1)$, and $(n, 1)$. For instance, we have

$$C_6 = \begin{bmatrix} 0 & 1 & 0 & 0 & 0 & 0 \\ 0 & 0 & 1 & 0 & 0 & 0 \\ 0 & 0 & 0 & 1 & 0 & 0 \\ 0 & 0 & 0 & 0 & 1 & 0 \\ 0 & 0 & 0 & 0 & 0 & 1 \\ 1 & 0 & 0 & 0 & 0 & 0 \end{bmatrix}.$$

In general, the digraph of C_n is the directed cycle

$$1 \rightarrow 2 \rightarrow \cdots \rightarrow n \rightarrow 1$$

of length n.

Corollary 6.1.2 *Let $A = I_n + C_n + C_n^T$ where $n \geq 2$. Then there exists an SNS-matrix in $\mathcal{Z}(A)$ if and only if n is even.*

Proof The digraph $\mathcal{D}(A)$ can be obtained from a cycle of length n by replacing each edge by two oppositely directed arcs. Of the $n+2$ directed cycles of $\mathcal{D}(A)$, two have length n and the other n have length 2. Moreover, each arc is contained in one directed cycle of length n and one of length 2.

First assume that n is even. Let $\widehat{A}$ be the matrix obtained from A by replacing each 1 on and above the main diagonal by -1. Then each directed cycle of the signed digraph $\mathcal{D}(\widehat{A})$ is negative and thus, by Lemma 6.1.1, $\widehat{A}$ is an SNS-matrix in $\mathcal{Z}(A)$.

Now assume that n is odd. Consider any signing of $\mathcal{D}(A)$. Since each arc is contained in exactly two directed cycles, the product of the signs of all the directed cycles is $+1$. Since there is an odd number of directed cycles, at least one directed cycle is positive. Hence, by Lemma 6.1.1, $\mathcal{Z}(A)$ does not contain an SNS-matrix.   □

Let A be a $(0,1)$-matrix of order n where $I_n \leq A$. Suppose that B is a matrix whose determinant is not identically zero such that $\mathcal{Z}(B)$ does not contain an SNS-matrix. If $B \leq A[\alpha]$ for some principal submatrix $A[\alpha]$ of A, then $\mathcal{Z}(A)$ does not contain an SNS-matrix. Hence, if for some principal submatrix $A[\alpha]$ of A of odd order $k > 1$, we have $I_k + C_k + C_k^T \leq A[\alpha]$, then $\mathcal{Z}(A)$ does not contain an SNS-matrix. This implies that an SNS-matrix of order $n \geq 3$ has at least one zero entry.

Any strongly connected digraph can be recursively constructed by choosing a directed cycle and successively inserting a path joining two previously encountered vertices or a directed cycle containing only one previously encountered vertex. More precisely, let $\mathcal{D}$ be a strongly connected digraph of order n. There exists a sequence of strongly connected digraphs

$$\mathcal{D}_1, \mathcal{D}_2, \ldots, \mathcal{D}_k = \mathcal{D} \qquad (6.1)$$

such that $\mathcal{D}_1$ is a digraph consisting of a single vertex and for each $i = 1, \ldots,$ $k - 1$, $\mathcal{D}_{i+1}$ can be obtained from $\mathcal{D}_i$ by inserting a path or directed cycle

$$\alpha_i : x_i \to z_1 \to \cdots \to z_t \to y_i$$

where x_i and y_i are vertices of $\mathcal{D}_i$ and $z_1, \ldots, z_t$ are not. Note that α_i may be an arc joining two vertices of $\mathcal{D}_{i-1}$. We call such a sequence (6.1) of digraphs a *recursive strong construction* of $\mathcal{D}$.

Now let A be a fully indecomposable $(0,1)$-matrix of order n with $I_n \leq A$. Since A is fully indecomposable, the digraph $\mathcal{D}(A)$ is strongly connected. Let (6.1) be a recursive strong construction of $\mathcal{D}(A)$. Without loss of generality we may assume that the vertices of $\mathcal{D}_i$ are $1, 2, \ldots, n_i$ where $n_1 \leq n_2 \leq \cdots \leq n_k = n$ and α_i equals.

$$x_i \to n_i + 1 \to n_i + 2 \to \cdots \to n_{i+1} \to y_i.$$

Let A_i be the $(0,1)$-matrix of order n_i satisfying $I_{n_i} \leq A_i$ and $\mathcal{D}(A_i) = \mathcal{D}_i$. If $n_{i+1} = n_i$, then A_{i+1} can be obtained from A_i by replacing the 0 in position

(x_i, y_i) by a 1. If $n_i < n_{i+1}$, then

$$
A_{i+1} = \left[
\begin{array}{c|cccccc}
A_i & \multicolumn{6}{c}{G_i} \\
\hline
 & 1 & 1 & 0 & \cdots & 0 & 0 \\
 & 0 & 1 & 1 & \cdots & 0 & 0 \\
 & 0 & 0 & 1 & \cdots & 0 & 0 \\
F_i & \vdots & \vdots & \vdots & \ddots & \vdots & \vdots \\
 & 0 & 0 & 0 & \cdots & 1 & 1 \\
 & 0 & 0 & 0 & \cdots & 0 & 1
\end{array}
\right]
\tag{6.2}
$$

where the only 1 in G_i is in the x_i position of its first column and the only 1 in F_i is in the y_i position of its last row. We call the sequence of matrices

$$
I_1 = A_1, A_2, \ldots, A_k = A
\tag{6.3}
$$

a *recursive fully indecomposable construction* of A.

The following theorem [5] (see also [7, 19, 23]) implies that if A is fully indecomposable, then up to multiplication of rows and columns by -1 there is at most one SNS-matrix in $\mathcal{Z}(A)$.

Theorem 6.1.3 *Let A be a fully indecomposable $(0, 1)$-matrix of order n. Let $\widehat{A}$ and $\widehat{\widehat{A}}$ be SNS-matrices in $\mathcal{Z}(A)$. Then there exist strict signings D and E such that $\widehat{A} = D\widehat{\widehat{A}}E$. Moreover, D and E are uniquely determined up to a scalar factor of -1.*

Proof Without loss of generality assume that $I_n \leq A$. Let (6.3) be a recursive fully indecomposable construction of A. If $k = 1$, the theorem clearly holds. We prove the existence of the strict signings D and E by induction on k. Assume that $k > 1$. Let $\widehat{A}_{i+1}$ and $\widehat{\widehat{A}}_{i+1}$ be SNS-matrices in $\mathcal{Z}(A_{i+1})$. Without loss of generality we assume that each entry on the main diagonals of $\widehat{A}_{i+1}$ and $\widehat{\widehat{A}}_{i+1}$ equals -1. Thus each of the nonzero terms in the standard determinant expansions of $\widehat{A}_{i+1}$ and $\widehat{\widehat{A}}_{i+1}$ equals $(-1)^{n_{i+1}}$. First suppose that $n_{i+1} = n_i$. Let $\widehat{A}_i$ and $\widehat{\widehat{A}}_i$ be obtained from $\widehat{A}_{i+1}$ and $\widehat{\widehat{A}}_{i+1}$, respectively, by replacing the (x_i, y_i) entry with 0. Then $\widehat{A}_i$ and $\widehat{\widehat{A}}_i$ are SNS-matrices in $\mathcal{Z}(A_i)$. By the inductive hypothesis there exist strict signings D_i and E_i such that $\widehat{A}_i = D_i\widehat{\widehat{A}}_iE_i$. Corresponding entries of the SNS-matrices $\widehat{A}_{i+1}$ and $D_i\widehat{\widehat{A}}_{i+1}E_i$ are equal, except possibly for the entries in position (x_i, y_i). Since there is a nonzero term of sign $(-1)^{n_{i+1}}$ in the standard determinant expansions of both $\widehat{A}_{i+1}$ and

$D_i \widehat{\widehat{A}}_{i+1} E_i$ containing the entry in position (x_i, y_i), it follows that their entries in position (x_i, y_i) are also equal. Hence $\widehat{A}_{i+1} = D_i \widehat{\widehat{A}}_{i+1} E_i$.

Now suppose that $n_{i+1} > n_i$. It follows from the inductive hypothesis and the structure of A_{i+1} given in (6.2) that there exist strict signings D_{i+1} and E_{i+1} such that the corresponding entries of the SNS-matrices $\widehat{A}_{i+1}$ and $D_{i+1} \widehat{\widehat{A}}_{i+1} E_{i+1}$ are equal except possibly for the entries in position (n_{i+1}, y_i). Since there is a nonzero term of sign $(-1)^{n_{i+1}}$ in the standard determinant expansions of both $\widehat{A}_{i+1}$ and $D_{i+1} \widehat{\widehat{A}}_{i+1} E_{i+1}$ containing the entry in position (n_{i+1}, y_i), their entries in position (n_{i+1}, y_i) are also equal. Hence $\widehat{A}_{i+1} = D_{i+1} \widehat{\widehat{A}}_{i+1} E_{i+1}$. Thus, by induction there exist strict signings D and E such that $\widehat{A} = D \widehat{\widehat{A}} E$.

Suppose that D' and E' are strict signings such that $\widehat{A} = D' \widehat{\widehat{A}} E'$. Then $(DD') \widehat{A} (E'E) = \widehat{A}$. We show that $DD' = I_n$ and $E'E = I_n$, or $DD' = -I_n$ and $E'E = -I_n$. Let β, respectively δ, equal the set of indices i for which the ith diagonal entry of DD', respectively $E'E$, equals 1. Then $\widehat{A}[\beta, \bar{\delta}] = O$ and $\widehat{A}[\bar{\beta}, \delta] = O$. Since $\widehat{A}$ is fully indecomposable, either $\beta = \delta = \emptyset$ or $\beta = \delta = \{1, 2, \ldots, n\}$, and the conclusion follows.   □

The following corollary implies that if A and B are fully indecomposable $(0, 1)$-matrices with $B \leq A$ and $\widehat{B}$ is an SNS-matrix in $\mathcal{Z}(B)$, then $\mathcal{Z}(A)$ contains an SNS-matrix if and only if some matrix obtained from $\widehat{B}$ by replacing some of its zeros with $+1$ or -1 is an SNS-matrix in $\mathcal{Z}(A)$.

Corollary 6.1.4 *Let $A = [a_{ij}]$ be a $(0, 1)$-matrix of order n with $a_{rs} = 1$, and let B be the matrix obtained from A by replacing a_{rs} with zero. Assume that B is fully indecomposable and that $\widehat{B}$ is an SNS-matrix in $\mathcal{Z}(B)$. Then the following are equivalent:*

   (i) *There exists an SNS-matrix in $\mathcal{Z}(A)$.*
   (ii) *Either $\widehat{B} + E_{rs}$ or $\widehat{B} - E_{rs}$ is an SNS-matrix in $\mathcal{Z}(A)$, where E_{rs} denotes the $(0, 1)$-matrix of order n whose only 1 is in position (r, s).*
   (iii) *The matrix $\widehat{B}(\{r\}, \{s\})$ is an SNS-matrix.*

Proof Clearly (ii) implies (i). Moreover, Theorem 1.2.5 and the full indecomposability of B imply that (ii) and (iii) are equivalent. Now assume that (i) holds. Let $\widehat{A}$ be an SNS-matrix in $\mathcal{Z}(A)$. Since B is fully indecomposable, it follows from Theorem 6.1.3 that there exist strict signings D and E such that $\widehat{B}$ and $D\widehat{A}E$ agree except in position (r, s), that is, $D\widehat{A}E = \widehat{A} \pm E_{rs}$. Hence (ii) holds.   □

Corollary 6.1.5 *Let*

$$
A = \begin{bmatrix}
B_1 & U_1 & O & \cdots & O & O \\
O & B_2 & U_2 & \cdots & O & O \\
O & O & B_3 & \cdots & O & O \\
\vdots & \vdots & \vdots & \ddots & \vdots & \vdots \\
O & O & O & \cdots & B_{\ell-1} & U_{\ell-1} \\
U_\ell & O & O & \cdots & O & B_\ell
\end{bmatrix}
$$

where B_i is a fully indecomposable $(0, 1)$-matrix and U_i is a $(0, 1)$-matrix whose only 1 is in position (r_i, s_{i+1}) $(i = 1, 2, \ldots, \ell)$. Let $\widehat{B}_i$ be an SNS-matrix in $\mathcal{Z}(B_i)$ $(i = 1, 2, \ldots, \ell)$. Then the following are equivalent:

(i) *There exists an SNS-matrix in $\mathcal{Z}(A)$.*

(ii) *One of the matrices*

$$
\begin{bmatrix}
\widehat{B}_1 & U_1 & O & \cdots & O & O \\
O & \widehat{B}_2 & U_2 & \cdots & O & O \\
O & O & \widehat{B}_3 & \cdots & O & O \\
\vdots & \vdots & \vdots & \ddots & \vdots & \vdots \\
O & O & O & \cdots & \widehat{B}_{\ell-1} & U_{\ell-1} \\
\pm U_\ell & O & O & \cdots & O & \widehat{B}_\ell
\end{bmatrix}
\tag{6.4}
$$

is an SNS-matrix in $\mathcal{Z}(A)$.

(iii) *Each of the matrices $\widehat{B}_i(\{r_i\}, \{s_i\})$ is an SNS-matrix, with the convention that $s_{\ell+1} = s_1$.*

Proof Clearly (ii) implies (i). Each nonzero term in the standard determinant expansion of $\det A$ which contains the nonzero element of one of $U_1, U_2, \ldots, U_\ell$ contains the nonzero element of each of $U_1, U_2, \ldots, U_\ell$. Since B_i is fully indecomposable, the determinant of $\widehat{B}_i(\{r_i\}, \{s_i\})$ is not identically zero $(i = 1, 2, \ldots, \ell)$. Hence (ii) implies (iii). Moreover, if (iii) holds, then we can choose U_ℓ or $-U_\ell$ so that all nonzero terms in the standard determinant expansion of the matrix (6.4) have the same sign and thus (ii) holds. Therefore (ii) and (iii) are equivalent.

Now assume that (i) holds. Let $\widehat{A}$ be an SNS-matrix in $\mathcal{Z}(A)$. By Theorem 6.1.3, there exist strict signings $D_1, D_2, \ldots, D_\ell$ and $E_1, E_2, \ldots, E_\ell$ such that with $D = D_1 \oplus D_2 \oplus \cdots \oplus D_\ell$ and $E = E_1 \oplus E_2 \oplus \cdots \oplus E_\ell$, corresponding entries of $D\widehat{A}E$ and $\widehat{B}_1 \oplus \widehat{B}_2 \oplus \cdots \oplus \widehat{B}_\ell$ are equal except possibly for the entries in positions (r_i, s_{i+1}) $(i = 1, 2, \ldots, \ell)$. Replacing some of the D_i by $-D_i$ and the corresponding E_i by $-E_i$, we obtain strict signings D and E such that $D\widehat{A}E$ has the form (6.4), and hence (ii) holds. $\qquad\square$

Corollaries 6.1.4 and 6.1.5 and Theorem 3.2.2 imply that the following algorithm [5] either constructs an SNS-matrix with a given zero pattern or correctly concludes that there is no such SNS-matrix.

Algorithm for the construction of an SNS-matrix with a prescribed zero pattern

Let A be a fully indecomposable $(0,1)$-matrix with $I_n \leq A$. Let $\mathcal{D}_1, \mathcal{D}_2, \ldots, \mathcal{D}_k = \mathcal{D}(A)$ be a recursive strong construction of the digraph of A.

(0) Let $i = 1$ and let $\mathcal{D}'_1$ be $\mathcal{D}_1$. (Since $\mathcal{D}_1$ has exactly one vertex and no arcs, we may view $\mathcal{D}'_1$ as a signed digraph.)

(1) If $i = k$, then the $(0, 1, -1)$-matrix $\widehat{A}$ each of whose main diagonal elements equals -1 and whose signed digraph equals $\mathcal{D}'_k$ is an SNS-matrix in $\mathcal{Z}(A)$. Otherwise,

(2) let $\mathcal{D}_{i+1}$ be obtained from $\mathcal{D}_i$ by inserting a path or directed cycle

$$\alpha_i : x_i \to z_1 \to \cdots \to z_t \to y_i$$

where x_i and y_i, but not $z_1, \ldots, z_t$, are vertices of $\mathcal{D}_i$. If there exists both a positive path and a negative path in $\mathcal{D}'_i$ from y_i to x_i, then there does not exist an SNS-matrix in $\mathcal{Z}(A)$. Otherwise,

(3) let $\mathcal{D}'_{i+1}$ be the signed digraph obtained from $\mathcal{D}'_i$ by signing all but the last arc of α_i with $+1$ and signing the last arc with $-a$ where a is the common sign of all the paths in $\mathcal{D}'_i$ from y_i to x_i. (Note that if $x_i = y_i$, then the only path from y_i to x_i is the path of length 0 and hence $a = 1$.) Increment i by 1 and go back to (1).

This algorithm is a nonbacktracking algorithm which attempts to construct a signing of the strongly connected digraph $\mathcal{D}(A)$ with the property that all directed cycles are negative by sequentially constructing signings $\mathcal{D}'_i$ of the $\mathcal{D}_i$ with this property.[1] If it is successful, then by Theorem 3.2.1, the matrix $\widehat{A}$ is an SNS-matrix. Otherwise, it follows from Corollaries 6.1.4 and 6.1.5 that no signing of $\mathcal{D}(A)$ has only negative directed cycles and hence that $\mathcal{Z}(A)$ does not contain an SNS-matrix. If in step (2) of the algorithm $x_i \neq y_i$, then it is required to decide whether or not all paths from y_i to x_i in $\mathcal{D}'_i$ have the same sign. This decision problem is polynomial-time equivalent to the even cycle recognition problem for digraphs and hence its complexity is unknown.

Suppose that in applying the algorithm one knows in advance that $\mathcal{Z}(A)$ contains an SNS-matrix. Then in step (2) all paths from y_i to x_i in $\mathcal{D}'_i$ must have the same sign and hence we need only compute the sign a of one path from

---

[1]That such an algorithm exists is implicit in [13] and [20].

y_i to x_i. Hence there is a polynomial-time algorithm for finding an SNS-matrix in $\mathcal{Z}(A)$, provided it is known in advance that $\mathcal{Z}(A)$ contains an SNS-matrix.[2]

Similarly, if one knows that $\mathcal{Z}(A)$ contains an SNS-matrix and $\widehat{A}$ is a matrix in $\mathcal{Z}(A)$, then one can determine in polynomial time whether or not $\widehat{A}$ is an SNS-matrix. The reason is as follows. Suppose in a recursive strong construction of $\mathcal{D}(A)$ we sign each arc with the sign of the corresponding entry of $\widehat{A}$. Then by Corollaries 6.1.4 and 6.1.5, $\widehat{A}$ is an SNS-matrix if and only if the sign of some path (and hence all paths) from y_i to x_i in $\mathcal{D}_i$ is the negative of the sign of α_i for each $i = 1, 2, \ldots, k - 1$. It follows that the problem of determining whether $\mathcal{Z}(A)$ contains an SNS-matrix is polynomial-time equivalent to the problem of whether a matrix B in $\mathcal{Z}(A)$ is an SNS-matrix [27].

We conclude this section with two examples illustrating the SNS-matrix construction algorithm.

Let

$$A = \begin{bmatrix} 1 & 1 & 0 & 0 \\ 0 & 1 & 1 & 1 \\ 1 & 1 & 1 & 1 \\ 1 & 0 & 0 & 1 \end{bmatrix}.$$

Then one way to carry out the algorithm is illustrated as follows:

$$\begin{bmatrix} -1 & 0 & 0 & 0 \\ 0 & -1 & 0 & 0 \\ 0 & 0 & -1 & 0 \\ 0 & 0 & 0 & -1 \end{bmatrix}, \quad \begin{bmatrix} -1 & 1 & 0 & 0 \\ 0 & -1 & 1 & 0 \\ -1 & 0 & -1 & 0 \\ 0 & 0 & 0 & -1 \end{bmatrix},$$

$$\begin{bmatrix} -1 & 1 & 0 & 0 \\ 0 & -1 & 1 & 0 \\ -1 & -1 & -1 & 0 \\ 0 & 0 & 0 & -1 \end{bmatrix}, \quad \begin{bmatrix} -1 & 1 & 0 & 0 \\ 0 & -1 & 1 & 1 \\ -1 & -1 & -1 & 0 \\ -1 & 0 & 0 & -1 \end{bmatrix},$$

$$\widehat{A} = \begin{bmatrix} -1 & 1 & 0 & 0 \\ 0 & -1 & 1 & 1 \\ -1 & -1 & -1 & 1 \\ -1 & 0 & 0 & -1 \end{bmatrix}.$$

The matrix $\widehat{A}$ is an SNS-matrix in $\mathcal{Z}(A)$.

---

[2]For instance, suppose $\widehat{A}$ is a fully indecomposable $(0, 1, -1)$ SNS-matrix of order n. Let A be the $(0, 1)$-matrix obtained from $\widehat{A}$ by deleting all the minus signs. Then in polynomial time we can reconstruct $\widehat{A}$, up to left and right multiplication by strict signings.

Now let

$$A = \begin{bmatrix} 1 & 1 & 1 \\ 1 & 1 & 1 \\ 1 & 1 & 1 \end{bmatrix}.$$

Then one way to carry out the algorithm is illustrated as follows:

$$\begin{bmatrix} -1 & 0 & 0 \\ 0 & -1 & 0 \\ 0 & 0 & -1 \end{bmatrix}, \quad \begin{bmatrix} -1 & 1 & 0 \\ 0 & -1 & 1 \\ -1 & 0 & -1 \end{bmatrix},$$

$$\begin{bmatrix} -1 & 1 & 0 \\ -1 & -1 & 1 \\ -1 & 0 & -1 \end{bmatrix}, \quad \begin{bmatrix} -1 & 1 & 0 \\ -1 & -1 & 1 \\ -1 & -1 & -1 \end{bmatrix}.$$

The signed digraph $\mathcal{D}'_4$ of the last matrix contains both a positive and negative path from vertex 3 to vertex 1. Hence there is no SNS-matrix in $\mathcal{Z}(A)$.

6.2 Maximal SNS-matrices

Let $B = [b_{ij}]$ be a $(0, 1, -1)$ SNS-matrix of order n. Then B is a *maximal SNS-matrix* [15] provided that no matrix obtained from B by replacing a 0 by a 1 or a -1 is an SNS-matrix. By Theorem 1.2.5, B is a maximal SNS-matrix if and only if for each $b_{ij} = 0$ there are nonzero terms of opposite sign in the standard determinant expansion of the matrix $B(\{i\}, \{j\})$. If B has a negative main diagonal, then by Corollary 3.2.2, B is a maximal SNS-matrix if and only if for each ordered pair of distinct vertices i and j, either there is an arc from i to j or there exist both a positive path and a negative path from j to i in the signed digraph $\mathcal{D}(B)$. In contrast to maximal SNS-matrices, S^2NS-matrices have the property that any single 0 can be replaced by either a 1 or a -1 to obtain an SNS-matrix. Since an SNS-matrix of order $n \geq 3$ has at least one zero, an S^2NS-matrix of order $n \geq 3$ is not a maximal SNS-matrix.

It can be verified that the following matrices are maximal SNS-matrices:

$$\begin{bmatrix} -1 & 1 & 0 \\ -1 & -1 & 1 \\ -1 & -1 & -1 \end{bmatrix}, \quad \begin{bmatrix} -1 & 1 & 0 & 1 \\ -1 & -1 & 1 & 0 \\ 0 & -1 & -1 & 1 \\ -1 & 0 & -1 & -1 \end{bmatrix}, \qquad (6.5)$$

$$I_7 + C_7 + C_7^3 = \begin{bmatrix} 1 & 1 & 0 & 1 & 0 & 0 & 0 \\ 0 & 1 & 1 & 0 & 1 & 0 & 0 \\ 0 & 0 & 1 & 1 & 0 & 1 & 0 \\ 0 & 0 & 0 & 1 & 1 & 0 & 1 \\ 1 & 0 & 0 & 0 & 1 & 1 & 0 \\ 0 & 1 & 0 & 0 & 0 & 1 & 1 \\ 1 & 0 & 1 & 0 & 0 & 0 & 1 \end{bmatrix}. \tag{6.5}$$

Each of 24 nonzero terms in the standard determinant expansion of the matrix $I_7 + C_7 + C_7^3$ is positive, and hence the matrix is an SNS-matrix. Because $I_7 + C_7 + C_7^3$ is a circulant, in verifying that it is a maximal SNS-matrix it suffices to show that there are nonzero terms of opposite sign in the standard determinant expansion of each of the four matrices of order 6 obtained by deleting row 1 and one of the columns 3, 5, 6, and 7. It is interesting to note that $I_7 + C_7 + C_7^3$ is an incidence matrix of the projective plane of order 2.

The following lemma [5] (see also [12, 16]) shows how to construct a maximal SNS-matrix of order $n + 1$ from a maximal SNS-matrix of order n. The construction used is that of column copying defined in section 4.2.

Lemma 6.2.1 *Let B be a $(0, 1, -1)$-matrix of order n, and let C be obtained from B by copying column k. Then B is a fully indecomposable, maximal SNS-matrix if and only if C is.*

Proof By Corollary 4.2.6, B is a fully indecomposable SNS-matrix if and only if C is, and if C is a maximal SNS-matrix, then B is a maximal SNS-matrix. Now assume that B is a fully indecomposable, maximal SNS-matrix. Without loss of generality we assume that

$$C = \left[\begin{array}{ccc|c} & B & & w \\ \hline 0 & \cdots & 0 \ \ 1 & -1 \end{array} \right]$$

where w is equal to the last column vector of B. It follows from Corollary 4.2.6 that no matrix obtained from C by replacing a 0 in its first n rows with ± 1 is an SNS-matrix. Since C is fully indecomposable, none of the matrices $C(\{n+1\}, \{j\})$ has an identically zero determinant. For $j = 1, 2, \ldots, n-1$, the matrix $C(\{n+1\}, \{j\})$ has two equal columns and hence is not an SNS-matrix. We conclude that C is a maximal SNS-matrix. $\qquad \square$

Copying a row of a matrix is defined in an analogous way to copying a column. It follows from Lemma 6.2.1 that a matrix obtained from a fully indecomposable, maximal SNS-matrix by successively copying rows and columns is a fully indecomposable, maximal SNS-matrix. Starting with the matrix

$$B = \begin{bmatrix} -1 & 1 \\ -1 & -1 \end{bmatrix}$$

and alternately copying the last column and last row, we obtain fully indecomposable maximal SNS-matrices of the form

$$\begin{bmatrix} B & E & O & \cdots & O & O \\ F & B & E & \cdots & O & O \\ O & F & B & \cdots & O & O \\ \vdots & \vdots & \vdots & \ddots & \vdots & \vdots \\ O & O & O & \cdots & B & E \\ O & O & O & \cdots & F & B \end{bmatrix} \tag{6.6}$$

where

$$E = \begin{bmatrix} 1 & 0 \\ -1 & 0 \end{bmatrix} \text{ and } F = \begin{bmatrix} 0 & 1 \\ 0 & 1 \end{bmatrix}.$$

We next describe methods to combine two SNS-matrices to obtain an SNS-matrix of larger order. The first method [16] constructs a fully indecomposable SNS-matrix of order $m + n - 1$ from a fully indecomposable SNS-matrix of order m and a fully indecomposable SNS-matrix of order n.

Theorem 6.2.2 *Let*

$$A = \begin{bmatrix} A' \\ \hline u^T \end{bmatrix}$$

be a fully indecomposable SNS-matrix of order $m \geq 2$ where A' is an $m - 1$ by m matrix. Let

$$B = \begin{bmatrix} v & | & B' \end{bmatrix}$$

be a fully indecomposable SNS-matrix of order $n \geq 2$ where B' is an n by $n - 1$ matrix. Suppose that $W = [w_{ij}]$ is an n by m matrix obtained from vu^T by replacing certain nonzero entries by zeros, and that a row (respectively, a column) of W is nonzero if and only if the corresponding row (respectively, column) of vu^T is nonzero. Then the matrix

$$M = \begin{bmatrix} A' & O \\ W & B' \end{bmatrix}$$

is a fully indecomposable SNS-matrix of order $m + n - 1$.

Proof It is easy to verify that the full indecomposability of A and B and the assumptions on W imply that M does not contain a p by q zero submatrix for any positive integers p and q with $p + q = m + n - 1$. Thus M is a fully indecomposable matrix.

Let

$$u^T = \begin{bmatrix} u_1 & u_2 & \cdots & u_m \end{bmatrix} \text{ and } v = \begin{bmatrix} v_1 \\ v_2 \\ \vdots \\ v_n \end{bmatrix}.$$

We show that M is an SNS-matrix, by showing that the sign of each nonzero term in its standard determinant expansion equals the sign of det A det B. Since the $m - 1$ by $n - 1$ submatrix in the upper right corner of M is a zero submatrix, each nonzero term in the determinant expansion contains exactly one nonzero element of W. Suppose that $w_{ij} \neq 0$. Then $u_j \neq 0$ and $v_i \neq 0$. In addition, $M(\{m - 1 + i\}, \{j\})$ is partly decomposable, and det $M(\{m - 1 + i\}, \{j\}) = $ det $A(\{m\}, \{j\})$ det $B(\{i\}, \{1\})$. Thus the sign of each nonzero term containing w_{ij} in the standard determinant expansion of M equals the sign of

$$(-1)^{m-1+i+j} w_{ij} \det A(\{m\}, \{j\}) \det B(\{i\}, \{1\}). \tag{6.7}$$

Since A is a fully indecomposable SNS-matrix and $u_j \neq 0$, $A(\{m\}, \{j\})$ is an SNS-matrix with

$$\text{sign det } A = \text{sign}((-1)^{m+j} u_j \det A(\{m\}, \{j\})). \tag{6.8}$$

Similarly, $B(\{i\}, \{1\})$ is an SNS-matrix with

$$\text{sign det } B = \text{sign}((-1)^{1+i} v_i \det B(\{i\}, \{1\})). \tag{6.9}$$

Since $w_{ij} = v_i u_j$, it follows from (6.8) and (6.9) that (6.7) equals

$$\text{sign(det } A \text{ det } B).$$

Hence M is an SNS-matrix. □

The matrix obtained from M in Theorem 6.2.2 by replacing W by $-W$ is also an SNS-matrix. This follows by applying the theorem to the matrix A and the matrix obtained from B by negating its first column.

Corollary 6.2.3 *Let A, B, W, and M be matrices satisfying the hypotheses of Theorem 6.2.2. Then M is a maximal SNS-matrix if and only if both A and B are maximal SNS-matrices and $W = vu^T$.*

Proof If M is a maximal SNS-matrix, then it follows from Theorem 6.2.2 that $W = vu^T$ and that both A and B are maximal SNS-matrices. Now assume that A and B are maximal SNS-matrices and that $W = vu^T$. Then a matrix obtained from M by changing any zero belonging to one of the submatrices A', B', and W to a nonzero results in a matrix which is not an SNS-matrix. Let k and ℓ be integers with $1 \le k \le m - 1$ and $m + 1 \le \ell \le m + n - 1$. Let M' be a matrix obtained from M by replacing the (k, ℓ) entry by a nonzero. Then each nonzero term t_σ in the standard determinant expansion of M' which contains the (k, ℓ) entry contains exactly two nonzero entries, say $w_{i_1 j_1}$ and $w_{i_2 j_2}$, of W. Since $M = vu^T$, by replacing $w_{i_1 j_1}$ by $w_{i_1 j_2}$, and $w_{i_2 j_2}$ by $w_{i_2 j_1}$, we obtain a nonzero term in the standard determinant expansion of M' whose sign is opposite that of t_σ. Hence M' is not an SNS-matrix, and we conclude that M is a maximal SNS-matrix.   □

If in Theorem 6.2.2 we take

$$A = \begin{bmatrix} -1 & 1 \\ 1 & 1 \end{bmatrix}$$

and $W = vu^T$, then M is the matrix obtained from B by copying its first column. Similarly, if we take

$$B = \begin{bmatrix} 1 & 1 \\ 1 & -1 \end{bmatrix}$$

and $W = vu^T$, then M is the matrix obtained from A by copying its last row. Thus Lemma 6.2.1 is also a consequence of Corollary 6.2.3. The following provides a converse to Theorem 6.2.2.

Theorem 6.2.4 *Let*

$$M = \begin{bmatrix} A' & O \\ W & B' \end{bmatrix}$$

be a fully indecomposable SNS-matrix of order $m + n - 1$ where A' is an $m - 1$ by m matrix, B' is an n by $n - 1$ matrix, and $W = [w_{ij}]$ is an n by m matrix. If *i and j are integers such that $w_{ij} \neq 0$, then the matrix A'_j obtained from A' by deleting its column j and the matrix B'_i obtained from B' by deleting its row i are SNS-matrices. Let $u^T = (u_1, u_2, \ldots, u_m)$ be defined by*

$$u_j = \begin{cases} sign((-1)^{m+j} \det A'_j \det M) & \text{if column } j \text{ of } W \text{ is nonzero} \\ 0 & \text{if column } j \text{ of } W \text{ is a zero column,} \end{cases}$$

and let $v = (v_1, v_2, \ldots, v_n)^T$ be defined by

$$v_i = \begin{cases} sign((-1)^{i+1} \det B'_i) & \text{if row } i \text{ of } W \text{ is nonzero} \\ 0 & \text{if row } i \text{ of } W \text{ is a zero row.} \end{cases}$$

Then the matrices

$$A = \left[\begin{array}{c} A' \\ \hline u^T \end{array} \right] \quad and \quad B = \left[\begin{array}{c|c} v & B' \end{array} \right]$$

are fully indecomposable SNS-matrices and W can be obtained from vu^T by replacing certain nonzero entries by zeros.

Proof First suppose that $w_{ij} \neq 0$. Since M is fully indecomposable, $M(\{m - 1 + i\}, \{j\})$ is an SNS-matrix. The form of M implies that

$$\det M(\{m - 1 + i\}, \{j\}) = \det A'_j \det B'_i. \tag{6.10}$$

Since M is an SNS-matrix,

$$\text{sign } w_{ij} = \text{sign}((-1)^{m-1+i+j} \det M(\{m - 1 + 1\}, \{j\}) \det M). \tag{6.11}$$

It follows from (6.10) and (6.11) that W can be obtained from vu^T by replacing certain zeros by nonzeros.

Equation (6.10) also implies that A'_j and B'_i are both SNS-matrices. By the Laplace expansion of the determinant of A along the last row we see that the sign of each nonzero term in the standard determinant expansion of A equals $sign(\det M)$. Hence A is an SNS-matrix. A similar argument shows that B is an SNS-matrix. Suppose that A contains a p by q zero submatrix for some positive integers p and q with $p + q = m + 1$. If the zero submatrix intersects the last row of A, then M contains a $p + n - 1$ by q zero submatrix, which contradicts the full indecomposability of M. Otherwise M contains a p by $q + n - 1$ submatrix, again contradicting the full indecomposability of M. We conclude that A is fully indecomposable and in a similar way we conclude that B is also fully indecomposable. $\square$

An immediate and useful consequence of Theorem 6.2.4 is the following.

Corollary 6.2.5 *Let A be a fully indecomposable SNS-matrix of order n. If $A[\alpha, \beta]$ is a p by q zero submatrix of A with $p + q = n - 1$ for which the complementary submatrix $A(\alpha, \beta)$ contains a submatrix X of order 2 each of whose entries is nonzero, then X is not an SNS-matrix.*

For zero patterns we have the following characterization.

Corollary 6.2.6 *Let*

$$M = \begin{bmatrix} A' & O \\ W & B' \end{bmatrix}$$

be a fully indecomposable (0, 1)-matrix of order $m + n - 1$ where A' is an $m - 1$ by m matrix and B' is an n by $n - 1$ matrix. Let A be the matrix of order m obtained from A' by appending on the bottom the (0, 1)-vector u^T, and let B be the matrix obtained from B' by appending on the left the (0, 1)-vector v, where u^T has a 1 in column j if and only if column j of W is nonzero, and v has a 1 in row i if and only if row i of W is nonzero. Then

   (i) *$Z(M)$ contains an SNS-matrix if and only if both $Z(A)$ and $Z(B)$ contain SNS-matrices, and*

  (ii) *$Z(M)$ contains a maximal SNS-matrix if and only if both $Z(A)$ and $Z(B)$ contain maximal SNS-matrices and $W = vu^T$.*

We now give a method for combining two SNS-matrices for which there exists a maximal SNS-matrix which is a submatrix of both. Let

$$B = \begin{bmatrix} B_1 & B_2 \\ B_3 & M \end{bmatrix} \quad \text{and} \quad C = \begin{bmatrix} M & C_1 \\ C_2 & C_3 \end{bmatrix} \tag{6.12}$$

be partitioned matrices each of which has a submatrix equal to M. The matrix

$$B \star C = \begin{bmatrix} B_1 & B_2 & O \\ B_3 & M & C_1 \\ O & C_2 & C_3 \end{bmatrix}$$

is called the *join of the matrices B and C at M*. If M has no zero entries, then $B \star C$ is a 1-*join* if M has order 1 and is a 2-*join* if M has order 2.

Suppose that B and C are fully indecomposable SNS-matrices and each has a negative main diagonal. It follows from Theorem 6.2.2 that the 1-join $B \star C$ of B and C at $M = [-1]$ is an SNS-matrix. This can also be argued directly as follows. The signed digraph of $B \star C$ is obtained by identifying a vertex of the signed digraph of B with a vertex of the signed digraph of C. Thus every directed cycle of $\mathcal{D}(B \star C)$ can be viewed as a directed cycle of $\mathcal{D}(B)$ or of $\mathcal{D}(C)$. It follows that every directed cycle of $\mathcal{D}(B \star C)$ is negative and hence by Theorem 3.2.1, $B \star C$ is an SNS-matrix. The next result is a generalization of a theorem in [5].

Theorem 6.2.7 *Let B and C be $(0, 1, -1)$ SNS-matrices of the form (6.12) each having a negative main diagonal. If M is a maximal SNS-matrix, then $B \star C$ is an SNS-matrix.*

Proof Assume that M is a maximal SNS-matrix. By Theorem 3.2.1, each directed cycle of the signed digraphs $\mathcal{D}(B)$, $\mathcal{D}(C)$, and $\mathcal{D}(M)$ is negative. In order to prove the theorem we show that each directed cycle of the signed digraph $\mathcal{D}(B \star C)$ is negative. Each of the signed digraphs $\mathcal{D}(B)$, $\mathcal{D}(C)$, and $\mathcal{D}(M)$ can be viewed as signed subdigraphs of the signed digraph $\mathcal{D}(B \star C)$. Let the set of vertices of $\mathcal{D}(B \star C)$ be $X \cup Z \cup Y$ where the set of vertices of $\mathcal{D}(M)$ is Z, the set of vertices of $\mathcal{D}(B)$ is $X \cup Z$, and the set of vertices of $\mathcal{D}(C)$ is $Z \cup Y$.

Let

$$\gamma = v_1 \rightarrow v_2 \rightarrow \cdots \rightarrow v_k \rightarrow v_1$$

be a directed cycle of $\mathcal{D}(B \star C)$. We show that sign $\gamma = -1$ by induction on the length of γ. If γ is a directed cycle of $\mathcal{D}(B)$ or of $\mathcal{D}(C)$, then sign $\gamma = -1$. Each directed cycle of $\mathcal{D}(B \star C)$ of length 2 is a directed cycle either of $\mathcal{D}(B)$ or of $\mathcal{D}(C)$ and hence is negative. Assume that γ is not in $\mathcal{D}(B)$ or in $\mathcal{D}(C)$. There exists a subpath α of γ of length at least 2 whose internal vertices all lie in X or all lie in Y and whose initial vertex u and terminal vertex v lie in Z. Each path δ from v to u in $\mathcal{D}(M)$ can be combined with α to obtain a directed cycle either in $\mathcal{D}(B)$ or in $\mathcal{D}(C)$. Since all directed cycles in $\mathcal{D}(B)$ and $\mathcal{D}(C)$ are negative, all such paths δ have the same sign. Corollary 3.2.2 and the maximality of M imply that the arc (u, v) belongs to $\mathcal{D}(M)$. Since this is true for each such subpath α, it follows that if $v_{i_1}, v_{i_2}, \ldots, v_{i_\ell}$ are the vertices of Z that lie on γ where $i_1 < i_2 < \cdots < i_\ell$, then

$$v_{i_1} \rightarrow v_{i_2} \rightarrow \cdots \rightarrow v_{i_\ell} \rightarrow v_{i_1}$$

is a directed cycle of $\mathcal{D}(M)$. Hence the vertices of γ that belong to Z lie in the same strong component of $\mathcal{D}(M)$. In particular, u and v belong to the same strong component of $\mathcal{D}(M)$.

Since u and v belong to the same strong component of $\mathcal{D}(M)$, there exists a path δ from v to u in $\mathcal{D}(M)$. We may combine δ with the arc (u, v) to obtain a directed cycle $\delta \rightarrow v$ in $\mathcal{D}(M)$. We may also combine the subpath, β, of γ from v to u with the arc (u, v) to obtain the directed cycle $\beta \rightarrow v$ in $\mathcal{D}(B \star C)$. The directed cycle $\beta \rightarrow v$ has length strictly less than that of γ and hence, by the inductive hypothesis, is negative. We have

$$\text{sign } \gamma = \frac{(\text{sign } \delta \text{ sign } \alpha)\text{sign } (\beta \to v)}{\text{sign } (\delta \to v)} = \frac{(-1)(-1)}{-1} = -1.$$

Hence the theorem follows by induction.   □

The following example shows that Theorem 6.2.7 is not true without the maximality assumption on M. Let

$$B = \left[\begin{array}{c|ccc} -1 & 0 & 1 & 0 \\ \hline 1 & -1 & 0 & -1 \\ 0 & 0 & -1 & -1 \\ 0 & 1 & 1 & -1 \end{array}\right], \quad C = \left[\begin{array}{ccc|c} -1 & 0 & -1 & 0 \\ 0 & -1 & -1 & 1 \\ 1 & 1 & -1 & 0 \\ \hline 1 & 0 & 0 & -1 \end{array}\right]$$

and

$$M = \left[\begin{array}{ccc} -1 & 0 & -1 \\ 0 & -1 & -1 \\ 1 & 1 & -1 \end{array}\right].$$

Then $B, C,$ and M are SNS-matrices, but the matrix

$$B \star C = \left[\begin{array}{ccccc} -1 & 0 & 1 & 0 & 0 \\ 1 & -1 & 0 & -1 & 0 \\ 0 & 0 & -1 & -1 & 1 \\ 0 & 1 & 1 & -1 & 0 \\ 0 & 1 & 0 & 0 & -1 \end{array}\right]$$

is not an SNS-matrix.

Corollary 6.2.8 *Let B and C be $(0, 1)$-matrices of the form (6.12) with a positive main diagonal. Assume that M is fully indecomposable and that $\mathcal{Z}(M)$ contains a maximal SNS-matrix. Then $\mathcal{Z}(B \star C)$ contains an SNS-matrix if and only if both $\mathcal{Z}(B)$ and $\mathcal{Z}(C)$ contain SNS-matrices.*

Proof Assume that both $\mathcal{Z}(B)$ and $\mathcal{Z}(C)$ contain SNS-matrices. Since M is fully indecomposable, it follows from Theorem 6.1.3 that there exist an SNS-matrix $\widehat{B}$ in $\mathcal{Z}(B)$ and an SNS-matrix $\widehat{C}$ in $\mathcal{Z}(C)$ such that their submatrices corresponding to M are equal. By Theorem 6.2.7, $\widehat{B} \star \widehat{C}$ is an SNS-matrix in $\mathcal{Z}(B \star C)$. The converse is immediate.   □

If, in Theorem 6.2.7, B and C are maximal SNS-matrices, then it need not be the case that $B \star C$ is a maximal SNS-matrix. For example, let

$$B = C = \left[\begin{array}{cc} -1 & 1 \\ -1 & -1 \end{array}\right] \text{ and } M = [\ -1\ ].$$

Then B and C are maximal SNS-matrices, but

$$B \star C = \begin{bmatrix} -1 & 1 & 0 \\ -1 & -1 & 1 \\ 0 & -1 & -1 \end{bmatrix}$$

is not a maximal SNS-matrix. In fact, it follows from Corollary 6.2.3 that the matrix obtained from $B \star C$ by changing the zero in its lower left corner to a -1 is an SNS-matrix.

The following theorem gives another way to construct maximal SNS-matrices.

Theorem 6.2.9 *Let M be a maximal SNS-matrix and let $B = \begin{bmatrix} M & E \\ F & M \end{bmatrix}$ be a fully indecomposable maximal SNS-matrix with a negative main diagonal. If $B \star B$ is a maximal SNS-matrix, then the matrix*

$$B^{\star k} = \overbrace{B \star \cdots \star B}^{k \, factors} = \begin{bmatrix} M & E & O & \cdots & O & O \\ F & M & E & \cdots & O & O \\ O & F & M & \cdots & O & O \\ \vdots & \vdots & \vdots & \ddots & \vdots & \vdots \\ O & O & O & \cdots & M & E \\ O & O & O & \cdots & F & M \end{bmatrix}$$

is a fully indecomposable maximal SNS-matrix for each $k \geq 2$.

Proof The full indecomposability of B implies the full indecomposability of $B^{\star k}$ for each k. Assume that $B \star B$ is a maximal SNS-matrix (thus B is also a maximal SNS-matrix). We prove the corollary by induction on k. Let k be an integer with $k > 2$ and assume that $B^{\star(k-1)}$ is a maximal SNS-matrix. By Theorem 6.2.7, $B^{\star k}$ is an SNS-matrix. Partition the vertices of $\mathcal{D}(B)$ into $Z_1, Z_2, \ldots, Z_{k+1}$ where Z_i is the set of vertices corresponding to the ith diagonal block of $B^{\star k}$. Suppose that the entry in the (u, v) position of $B^{\star k}$ equals 0. If u is in Z_i and v is in Z_i, then since M is a maximal SNS-matrix, neither of the matrices obtained from $B^{\star k}$ by replacing the (u, v) entry with 1 and with -1 is an SNS-matrix. If u is not in Z_{k+1} and v is not in Z_{k+1}, or if u is not in Z_1 and v is not in Z_1, then it follows from the inductive hypothesis that neither of the matrices obtained from $B^{\star k}$ by replacing the (u, v) entry with 1 and with -1 is an SNS-matrix. Now suppose that u and v are vertices in Z_1 and Z_{k+1}, respectively. Since B is fully indecomposable, there exists a path from v to a vertex w in Z_k whose internal vertices lie in Z_{k+1}. Since $k > 2$, the (u, w) entry of $B^{\star k}$ equals 0. Hence, by the inductive hypothesis there exist both a positive path and a negative path from w to u in $\mathcal{D}(B^{\star(k-1)})$ and thus both a positive path and a negative path from v to u in $\mathcal{D}(B^{*k})$. By Corollary 3.2.3, neither of the matrices obtained from B^{*k} by replacing the (u, v) entry with 1 and with -1 is an SNS-matrix. The same conclusion holds if u is in Z_{k+1} and v is in Z_1. Therefore B^{*k} is a maximal SNS-matrix, completing the induction.   □

The matrices B^{*k} where

$$
B = \left[\begin{array}{cc|cc}
-1 & 1 & 1 & 0 \\
-1 & -1 & -1 & 0 \\ \hline
0 & 1 & -1 & 1 \\
0 & 1 & -1 & -1
\end{array} \right]
$$

are the maximal SNS-matrices defined in (6.6).

Let A be a matrix of order $n \geq 2$. Then A is *doubly indecomposable* provided A has no p by q zero submatrix for any positive integers p and q with $p + q = n - 1$. Equivalently, A is a doubly indecomposable matrix if and only if each of its submatrices of order $n - 1$ is fully indecomposable. By the Frobenius–König theorem, A is a doubly indecomposable matrix if and only if no submatrix of order $n - 2$ has an identically zero determinant. A doubly indecomposable matrix of order $n \geq 3$ has at least 3 nonzero entries in each row and column. A digraph $\mathcal{D}$ with at least three vertices is *strongly 2-connected* provided each digraph obtained from $\mathcal{D}$ by removing a vertex is strongly connected. Thus a strongly 2-connected digraph has the property that given any three distinct vertices u, v, and w, there exists a path from u to v not containing w. It is a simple observation that a $(0, 1)$-matrix A with $I_n \leq A$ is a doubly indecomposable matrix if and only if $\mathcal{D}(A)$ is a strongly 2-connected digraph.

Clearly, any matrix which is the 1-join of two matrices is not doubly indecomposable. Theorems 6.2.2 and 6.2.4 characterize the SNS-matrices which are not doubly indecomposable. Corollary 6.2.6 characterizes the zero patterns of SNS-matrices which are not doubly indecomposable. In the remainder of this section we consider doubly indecomposable SNS-matrices. The second and third matrices in (6.5) are doubly indecomposable, SNS-matrices.

Lemma 6.2.10 *Let A be a doubly indecomposable $(0, 1)$-matrix of order n such that $J_{2,3}$ or $J_{3,2}$ is a submatrix of A. Then $\mathcal{Z}(A)$ does not contain an SNS-matrix.*

Proof Let B be a doubly indecomposable SNS-matrix. Consider a submatrix $B[\alpha, \beta]$ of B of order 2. Since B is doubly indecomposable, $B(\alpha, \beta)$ does not have an identically zero determinant. Thus, since B is an SNS-matrix, $B[\alpha, \beta]$

either is an SNS-matrix or has an identically zero determinant. We conclude that each submatrix of order 2 of a doubly indecomposable SNS-matrix either is an SNS-matrix or has an identically zero determinant. The result now follows from the fact that every 2 by 3 $(1, -1)$-matrix has a submatrix of order 2 which is not an SNS-matrix. □

The next theorem shows how to use the join to construct doubly indecomposable SNS-matrices.

Theorem 6.2.11 *Let B and C be doubly indecomposable SNS-matrices of the form* (6.12) *where M is a maximal SNS-matrix of order* 2. *Then the following hold:*

(i) *$B \star C$ is a doubly indecomposable SNS-matrix.*

(ii) *Let X be a maximal SNS-matrix obtained from $B \star C$ by replacing certain zero entries with nonzeros. Then X is the join at M of two maximal SNS-matrices B and C obtained from B and C, respectively, by replacing certain zero entries with nonzeros.*

(iii) *$B \star C$ is a maximal SNS-matrix if and only if B and C are maximal SNS-matrices.*

Proof Since B and C are doubly indecomposable, we may assume that B and C each have a negative main diagonal. By Theorem 6.2.7, $B \star C$ is an SNS-matrix. To show that $B \star C$ is doubly indecomposable, we consider the digraph $\mathcal{D}(B \star C)$. Let v be a vertex of $\mathcal{D}(B \star C)$ and let $\mathcal{D}_v$ be the digraph obtained from $\mathcal{D}(B \star C)$ by removing vertex v. If v is also a vertex of $\mathcal{D}(B)$, then since B is doubly indecomposable the digraph obtained from $\mathcal{D}(B)$ by removing v is strongly connected. Similarly, if v is a vertex of $\mathcal{D}(C)$, then the digraph obtained from $\mathcal{D}(C)$ by removing v is strongly connected. It follows that $\mathcal{D}_v$ is obtained by identifying either one or two vertices of two strongly connected digraphs. Hence $\mathcal{D}_v$ is strongly connected, and we conclude that $B \star C$ is doubly indecomposable.

Let m and n be, respectively, the order of B and of C. Let i and j be integers with $m < i \leq m+n-2$ and $1 \leq j \leq m-2$. Then $(B \star C)(\{i\}, \{j\})$ is a matrix of order $m+n-3$ which contains an $m-2$ by $n-2$ zero submatrix in its upper right corner. The submatrix of $(B \star C)(\{i\}, \{j\})$ which is complementary to this zero matrix contains a maximal SNS-matrix of order 2. Corollary 6.2.5 implies that $B \star C(\{i\}, \{j\})$ is not an SNS-matrix. A similar argument shows that $B \star C(\{i\}, \{j\})$ is not an SNS-matrix if $1 \leq i \leq m-2$ and $m < j \leq m+n-2$. Assertions (ii) and (iii) now follow. □

Fig. 6.1.

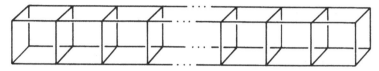

Fig. 6.2.

Let

$$X = \begin{bmatrix} 1 & 1 & 1 & 0 \\ 1 & 1 & 0 & 1 \\ 1 & 0 & 1 & 1 \\ 0 & 1 & 1 & 1 \end{bmatrix}.$$

Then X is a doubly indecomposable matrix and $\mathcal{Z}(X)$ contains a maximal SNS-matrix. Let Y be a doubly indecomposable $(0,1)$-matrix of order n such that $\mathcal{Z}(Y)$ contains a maximal SNS-matrix and J_2 is a submatrix of Y. Then we can form a 2-join $Y \star X$ of Y and X. By Theorem 6.2.11, $Y \star X$ is the zero pattern of a doubly indecomposable, maximal SNS-matrix. The bipartite graph of X is the graph of a 3-dimensional cube, as in Figure 6.1. Thus the bipartite graph of $Y \ast X$ is obtained from that of Y by identifying a cycle of length four in Y with a cycle of length four in X. For example, the bipartite graph can be obtained from the cube by adjoining cubes, as in Figure 6.2. This figure is the bipartite graph of the matrix

$$\begin{bmatrix} J_2 & I_2 & O & \cdots & O \\ I_2 & J_2 & O & \cdots & O \\ \vdots & \vdots & \ddots & \vdots & \vdots \\ O & O & \cdots & J_2 & I_2 \\ O & O & \cdots & I_2 & J_2 \end{bmatrix}.$$

The next four results give sufficient conditions for a doubly indecomposable SNS-matrix to be a 2-join of SNS-matrices.

Lemma 6.2.12 *Let $A = [a_{ij}]$ be a doubly indecomposable $(0, 1, -1)$-matrix* *such that the first row and column of A each contain exactly three 1's, $a_{11} = a_{12} = a_{13} = 1$, $a_{11} = a_{21} = a_{31} = 1$, $a_{22} = 1$, $a_{33} = 0$. Let B be the matrix obtained from A by replacing a_{33} by a 1. Then $\mathcal{Z}(A)$ contains an SNS-matrix if and only if $\mathcal{Z}(B)$ does.*

Proof Clearly, if $\mathcal{Z}(B)$ contains an SNS-matrix, then so does $\mathcal{Z}(A)$. Conversely, assume that $\widehat{A} = [\hat{a}_{ij}]$ is an SNS-matrix in $\mathcal{Z}(A)$. Since $\widehat{A}$ is a doubly indecomposable SNS-matrix, $\widehat{A}[\{1, 2\}]$ is an SNS-matrix, and $\hat{a}_{2,3} = \hat{a}_{32} = 0$. Thus, without loss of generality we may assume that

$$\widehat{A}[\{1, 2, 3\}] = \begin{bmatrix} 1 & -1 & 1 \\ 1 & 1 & 0 \\ 1 & 0 & 0 \end{bmatrix}.$$

Let $\widehat{B} = [\hat{b}_{ij}]$ be the matrix obtained from $\widehat{A}$ by replacing the $(3, 3)$-entry by -1. Each nonzero term t_σ in the standard determinant expansion of $\widehat{B}$ that contains $\hat{b}_{33}$ either contains $\hat{b}_{11}$, or contains both $\hat{b}_{12}$ and $\hat{b}_{31}$. In the first case, if in t_σ we replace $\hat{b}_{33}$ and $\hat{b}_{11}$ by $\hat{b}_{13}$ and $\hat{b}_{31}$, then we obtain a nonzero term in the standard determinant expansion of A with the same sign as t_σ. In the second case, we obtain the same conclusion by considering the entries $\hat{b}_{13}, \hat{b}_{22}$, and $\hat{b}_{31}$. Hence $\widehat{B}$ is an SNS-matrix. □

Theorem 6.2.13 *Let A be a doubly indecomposable, maximal SNS-matrix whose first and second rows and columns each contain exactly three nonzero entries such that $A[\{1, 2\}]$ has no zero entries. Then up to row and column permutations and multiplication by strict signings, A is a 2-join of a maximal SNS-matrix and the maximal SNS-matrix (6.5) of order 4.*

Proof Since A is doubly indecomposable, we may assume without loss of generality that

$$A[\{1, 2\}] = \begin{bmatrix} 1 & -1 \\ 1 & 1 \end{bmatrix},$$

and by Lemma 6.2.10 that the other nonzero entries in the first two rows and columns of A are a_{13}, a_{24}, a_{31}, and a_{42}. When we apply Lemma 6.2.12 to each of the four entries of $A[\{1, 2\}]$, the maximality of A implies that

$$A[\{3, 4\}] = \begin{bmatrix} -1 & -1 \\ 1 & -1 \end{bmatrix}.$$

The corollary now follows from Theorem 6.2.11. □

Theorem 6.2.14 *Let $A = [a_{ij}]$ be a doubly indecomposable SNS-matrix of order $n \geq 2$ such that $A[\{1, 2\}]$ has no zero entries and $A(\{1, 2\})$ is not fully indecomposable. Then A is the 2-join of two doubly indecomposable SNS-matrices at $A[\{1, 2\}]$.*

Proof Since A is a doubly indecomposable SNS-matrix, $A[\{1, 2\}]$ is an SNS-matrix of order 2. Since $A(\{1, 2\})$ is not fully indecomposable, there exist integers p and q with $p + q = n - 2$ and a zero submatrix $A[\alpha, \beta]$ of $A(\{1, 2\})$. Let i and j be integers with $i \notin \alpha \cup \{1, 2\}$ and $j \notin \beta \cup \{1, 2\}$. Since $A[\alpha, \beta]$ is a zero submatrix of $A(\{i\}, \{j\})$, and since $A[\{1, 2\}]$ is a submatrix of $A(\alpha \cup \{i\}, \beta \cup \{j\})$, it follows from Corollary 6.2.5 that the fully indecomposable matrix $A(\{i\}, \{j\})$ is not an SNS-matrix. Thus $A(\alpha \cup \{1, 2\}, \beta \cup \{1, 2\}) = O$. This implies that A is the join of $A[\alpha \cup \{1, 2\}, \overline{\beta}]$ and $A[\overline{\alpha}, \beta \cup \{1, 2\}]$ at $A[\{1, 2\}]$. That these two matrices are doubly indecomposable follows from the double indecomposability of A. □

Theorem 6.2.15 *Let A be a doubly indecomposable SNS-matrix of order n for which there exists $\alpha \subseteq \{1, 2, \ldots, n\}$ of cardinality $m \geq 3$ such that $M = A[\alpha]$ is a maximal SNS-matrix and $A(\alpha)$ does not have an identically zero determinant. Then M is doubly indecomposable and A is the 2-join of two doubly indecomposable SNS-matrices at a submatrix of M.*

Proof Since A is doubly indecomposable and $m \geq 3$, we have $n \geq 4$. Since $A(\alpha)$ does not have an identically zero determinant, we may assume that A has a negative main diagonal. Suppose that u and v are vertices in α and that γ is a path from u to v whose interior vertices are not in α. Then every path from v to u in $\mathcal{D}(M)$ can be combined with γ to obtain a directed cycle in $\mathcal{D}(A)$. Hence the sign of every path in $\mathcal{D}(M)$ from v to u has the same sign. Since M is a maximal SNS-matrix, (u, v) is an arc of $\mathcal{D}(M)$. This implies that $\mathcal{D}(M)$ is strongly connected, and hence M is fully indecomposable. Since A is doubly indecomposable, every submatrix of A of order 2 is an SNS-matrix or has an identically zero determinant. By Theorem 6.2.2, every maximal SNS-matrix which is not doubly indecomposable and is of order at least 3 contains a matrix of order 2 which is not an SNS-matrix and which does not have an identically zero determinant. Hence M is doubly indecomposable.

Let w be a vertex of $\mathcal{D}(A)$ which is not in α. Since $\mathcal{D}(A)$ is strongly 2-connected, there exist vertices u_1 and u_2 and paths in $\mathcal{D}(A)$ from u_1 to w and from u_2 to w none of whose interior vertices is in α. Similarly, there exist vertices v_1 and v_2 and paths in $\mathcal{D}(A)$ from w to v_1 and w to v_2 none of whose interior vertices is in α. The maximality of M implies that $A[\{u_1, u_2\}, \{v_1, v_2\}]$ is a matrix of order 2 each of whose entries is nonzero. Since A is doubly indecomposable, $A[\{u_1, u_2\}, \{v_1, v_2\}]$ is an SNS-matrix. By permuting rows and columns of A which intersect M, we may assume without loss of generality that $u_1 = v_1$ and $u_2 = v_2$.

Let y be a vertex in α different from u_1 and u_2. If there is a path from y to w whose interior vertices are outside α, then $A[\{u_1, u_2, u_3\}, \{v_1, v_2\}]$ is a 3 by 2 submatrix of A each of whose entries is nonzero. By Lemma 6.2.10, every 3 by 2 submatrix of a doubly indecomposable SNS-matrix contains at least one 0. We conclude that the digraph obtained from $\mathcal{D}(A)$ by removing the vertices u_1 and u_2 is not strongly connected. It follows that $A(\{u_1, u_2\}, \{v_1, v_2\})$ is not fully indecomposable. Therefore, by Theorem 6.2.14, A is the 2-join of doubly indecomposable SNS-matrices at $A[\{u_1, u_2\}, \{v_1, v_2\}]$. $\square$

Theorem 6.2.16 *Let B and C be SNS-matrices with a negative main diagonal of the form* (6.12) *where M is a maximal SNS-matrix. Assume that $B \star C$ is doubly indecomposable. Then $B \star C$ is a 2-join of two doubly indecomposable SNS-matrices.*

Proof Since a 2-join is doubly indecomposable if and only if each of the matrices being joined is doubly indecomposable, we need only show that $B \star C$ is a 2-join of two SNS-matrices. The proof is by induction on the order k of M. Since $B \star C$ is doubly indecomposable, $k \neq 1$. If k equals 2, then each entry of M is nonzero and the result holds. Assume that M has order at least 3. By Theorem 6.2.15, $B \star C$ is the 2-join $B_1 \star B_2$ at a maximal SNS-matrix N of order 2 which is a submatrix of M. Let M_1 and M_2 be the submatrices of M contained in B_1 and B_2, respectively. Then M is the 2-join of M_1 and M_2 at N. By Theorem 6.2.11, both M_1 and M_2 are doubly indecomposable, maximal SNS-matrices. Since $k \geq 3$, either M_1 or M_2 has order less than k. Without loss of generality assume that the order of M_1 is less than k. The matrix $B \star C$ can be formed by joining B_2 and C at M_2 to get $B_2 \star C$, and then joining B_1 and $B_2 \star C$ at M_1. Thus $B \star C$ is the join of two matrices at M_1. Since M_1 has order less than k, it follows by induction that $B \star C$ is a 2-join of two SNS-matrices. By Theorem 6.2.11, the matrices are doubly indecomposable. $\square$

We conclude this section with a construction for doubly indecomposable SNS-matrices which are not the join of two SNS-matrices at a maximal SNS-matrix of order 2.

Lemma 6.2.17 *Let $B = [b_{ij}]$ be a doubly indecomposable $(0, 1)$-matrix of order n which has exactly three nonzero entries in column $n - 1$ such that*

$$B[\{n - 2, n - 1, n\}] = \begin{bmatrix} 1 & 1 & 0 \\ 1 & 1 & 1 \\ 0 & 1 & 1 \end{bmatrix}.$$

Let B' be the matrix obtained from B by replacing $b_{n-1,n-1}$ by 0, and let

$$A = [a_{ij}] = \begin{bmatrix} & & & 0 \\ & B' & & \vdots \\ & & & 1 \\ & & & 1 \\ & & & 0 \\ \hline 0 & \cdots & 0 & 1 & 1 & 1 \end{bmatrix}.$$

Then A is a doubly indecomposable matrix of order $n + 1$, and $\mathcal{Z}(A)$ contains an SNS-matrix if and only if $\mathcal{Z}(B)$ does. Moreover, if $\mathcal{Z}(B)$ contains a maximal SNS-matrix, then so does $\mathcal{Z}(A)$, and if B is not a 2-join, then neither is A.

Proof Let $\mathcal{O}$ be a p by q zero submatrix of A. Since $a_{n+1,n+1} = 1$, $\mathcal{O}$ does not intersect both row $n + 1$ and column $n + 1$. If $\mathcal{O}$ does not intersect row $n + 1$ or column $n + 1$, then B contains a p by $q - 1$ zero submatrix. If $\mathcal{O}$ intersects row $n + 1$, then either $p = 1$ and so $p + q \leq n - 1$, or B contains a $p - 1$ by q zero submatrix. Similarly, if $\mathcal{O}$ intersects column $n + 1$, then either $q = 1$ and $p + q \leq n - 1$, or B contains a p by $q - 1$ zero submatrix. Since B is doubly indecomposable, it follows that $p + q \leq n - 1$ and hence that A is doubly indecomposable.

Since B can be obtained from A by replacing $a_{n+1,n}$ by 0 and then contracting, Theorem 4.2.6 implies that if $\mathcal{Z}(A)$ contains an SNS-matrix then so does $\mathcal{Z}(B)$.

Now assume that $\widetilde{B} = [\tilde{b}_{ij}]$ is an SNS-matrix in $\mathcal{Z}(B)$. Since B is doubly indecomposable, we may assume without loss of generality that

$$\widetilde{B}[\{n - 2, n - 1, n\}] = \begin{bmatrix} -1 & -1 & 0 \\ 1 & -1 & -1 \\ 0 & -1 & 1 \end{bmatrix}$$

and that det $\widetilde{B} > 0$. Let $\widetilde{B}'$ be the SNS-matrix in $\mathcal{Z}(B')$ obtained from $\widetilde{B}$ by replacing $\tilde{b}_{n-1,n-1}$ by 0, and let

$$
\widetilde{A} = [\tilde{a}_{ij}] = \left[\begin{array}{c|c} & \begin{matrix} 0 \\ \vdots \\ 1 \\ 1 \\ 0 \end{matrix} \\ B' & \\ \hline 0 \ \cdots \ 0 \ \ 1 \ \ 1 & 1 \end{array} \right].
$$

Then $\widetilde{A}$ is in $\mathcal{Z}(A)$, and we now show that $\widetilde{A}$ is an SNS-matrix.

Since $\widetilde{B}$ is a fully indecomposable SNS-matrix with positive determinant, $\widetilde{B}(\{n-1\},\{n\})$ and $\widetilde{B}(\{n-2\},\{n-1\})$ are SNS-matrices with positive determinants, and $\widetilde{B}(\{n-1\},\{n-1\})$ is an SNS-matrix with negative determinant. Since $\widetilde{B}$ is a doubly indecomposable SNS-matrix, $\widetilde{B}(\{n-2,n\},\{n-1,n\})$ is an SNS-matrix with positive determinant. Let t_σ be a nonzero term in the standard determinant expansion of $\widetilde{A}$. If t_σ contains $\tilde{a}_{n+1,n+1}$, then

$$
\operatorname{sign} t_\sigma = \operatorname{sign} \det \widetilde{B}' = 1.
$$

If t_σ contains both $\tilde{a}_{n+1,n}$ and $\tilde{a}_{n-1,n+1}$, then

$$
\operatorname{sign} t_\sigma = \operatorname{sign} \det \widetilde{B}'(\{n-1\},\{n\}) = \operatorname{sign} \det \widetilde{B}(\{n-1\},\{n\}) = 1.
$$

Similarly if t_σ contains both $\tilde{a}_{n+1,n-1}$ and $\tilde{a}_{n-1,n+1}$, or both $\tilde{a}_{n+1,n-1}$ and $\tilde{a}_{n-2,n+1}$, then $\operatorname{sign} t_\sigma = 1$. If t_σ contains both $\tilde{a}_{n+1,n}$ and $\tilde{a}_{n-2,n+1}$, then t_σ also contains $\tilde{a}_{n,n-1}$ and hence

$$
\begin{aligned}
\operatorname{sign} t_\sigma &= \operatorname{sign} \det \widetilde{B}'(\{n-2,n\},\{n-1,n\}) \\
&= \operatorname{sign} \det \widetilde{B}(\{n-2,n\},\{n-1,n\}) = 1.
\end{aligned}
$$

Thus each nonzero term in the standard determinant expansion of $\widetilde{A}$ is positive and $\widetilde{A}$ is an SNS-matrix.

Next assume that $\widetilde{B}$ is a maximal SNS-matrix. Let i and j be integers such that $\tilde{a}_{ij} = 0$. First suppose that $i \le n$, $j \le n$ and $(i, j) \ne (n-1, n-1)$. The matrix $\widetilde{B}(\{i\},\{j\})$ can be obtained from $\widetilde{A}(\{i\},\{j\})$ by possibly replacing $\tilde{a}_{n+1,n-1}$ or $\tilde{a}_{n-2,n+1}$ by zero and conformally contracting. Since $\widetilde{B}$ is a maximal SNS-matrix and $\tilde{b}_{ij} = 0$, neither $\widetilde{B}(\{i\},\{j\})$ nor $\widetilde{A}(\{i\},\{j\})$ is an SNS-matrix: A similar conclusion holds if $i = n+1$ and $\tilde{b}_{nj} = 0$, or if $j = n+1$ and $\tilde{b}_{in} = 0$. Now suppose that $i = j = n-1$, $i = n+1$, and $\tilde{b}_{nj} \ne 0$, or $j = n+1$

and $\tilde{b}_{in} \neq 0$. Then $\tilde{a}_{ij}$ is the only zero entry in either a 2 by 3 or a 3 by 2 submatrix of $\tilde{A}$. By Lemma 6.2.10, each 2 by 3 and each 3 by 2 submatrix of a doubly indecomposable SNS-matrix contains a zero entry, and hence a matrix obtained from $\tilde{A}$ by replacing $\tilde{a}_{ij}$ by a nonzero is not an SNS-matrix. Therefore we conclude that $\tilde{A}$ is a maximal SNS-matrix.

Finally, assume that A is a 2-join of two matrices U and V at a submatrix $A[\alpha, \beta] = J_2$. Since A is doubly indecomposable, U and V are doubly indecomposable. Since the last row and the last column of A contain exactly three nonzero entries, and since U and V are doubly indecomposable, $n + 1 \notin \alpha$ and $n + 1 \notin \beta$. Without loss of generality we may assume that $a_{n+1,n+1}$ is an entry of U. It now follows that B is the 2-join of U' and V where U' is the matrix obtained from U by replacing the entry corresponding to $a_{n+1,n}$ by a 0 and then contracting. Hence, if B is not a 2-join, then neither is A.   □

The preceding construction can be used to construct doubly indecomposable, maximal SNS-matrices of each order $n \geq 4$ which are not 2-joins. The matrices constructed have the property that they each contain an $n - 2$ by n totally L-matrix which can be obtained from the 2 by 4 totally L-matrix in (5.1) by a sequence of single extensions.

Theorem 6.2.18 *Let X be the zero pattern of an $n - 2$ by n totally L-matrix which can be obtained from a 2 by 4 totally L-matrix by a sequence of single extensions. Assume that columns $n - 1$ and n of X have exactly two nonzero entries. Then there exist $(0, 1)$-vectors $u^T = (u_1, u_2, \ldots, u_{n-2})$ and $v^T = (v_1, v_2, \ldots, v_{n-2})$ such that $u_i + v_i = 1$ for $i = 1, \ldots, n - 2$ and the matrix*

$$\left[\begin{array}{c|cc} u^T & 1 & 1 \\ v^T & 1 & 1 \\ \hline & X & \end{array} \right] \tag{6.13}$$

is the zero pattern of a doubly indecomposable, maximal SNS-matrix which is not a 2-join.

Proof The proof is by induction on n. If $n = 4$, then

$$\begin{bmatrix} 1 & 0 & 1 & 1 \\ 0 & 1 & 1 & 1 \\ 1 & 1 & 1 & 0 \\ 1 & 1 & 0 & 1 \end{bmatrix}$$

is a matrix satisfying the conclusions of the theorem. Suppose that $n > 4$. We may assume that

$$X = \left[\begin{array}{ccccccc|c} & & & & & & & 0 \\ & & & Y & & & & \vdots \\ & & & & & & & 0 \\ \hline 0 & \cdots & 0 & 1 & 1 & 0 & & 1 \end{array}\right]$$

where Y is an $n - 3$ by $n - 1$ totally L-matrix with a one in position $(n - 3, n - 2)$ and whose only nonzero entries in its last two columns are 1's in entries $(n - 4, n - 1)$ and $(n - 3, n - 2)$. By the induction hypothesis there exist $(0, 1)$-vectors $u^T = (u_1, \ldots, u_{n-3})$ and $v^T = (v_1, \ldots, v_{n-3})$ such that $u_i + v_i = 1$ for $i = 1, 2, \ldots, n - 3$ and

$$B = \left[\begin{array}{c|cc} u^T & 1 & 1 \\ v^T & 1 & 1 \\ \hline & Y & \end{array}\right]$$

is the zero pattern of a doubly indecomposable, maximal SNS-matrix which is not a 2-join. It follows from Lemma 6.2.10 that $B[\{1, 2, n - 3\}, \{n - 3, n - 2, n - 1\}]$ is either

$$\left[\begin{array}{ccc} 1 & 1 & 1 \\ 0 & 1 & 1 \\ 1 & 1 & 0 \end{array}\right] \quad \text{or} \quad \left[\begin{array}{ccc} 0 & 1 & 1 \\ 1 & 1 & 1 \\ 1 & 1 & 0 \end{array}\right].$$

Without loss of generality we may assume that

$$B[\{1, 2, n - 3\}\{n - 3, n - 2, n - 1\}] = \left[\begin{array}{ccc} 1 & 1 & 1 \\ 0 & 1 & 1 \\ 1 & 1 & 0 \end{array}\right].$$

Since column $n - 2$ of B contains exactly three 1's, Lemma 6.2.17 can be applied using the submatrix $B[\{1, 2, n - 3\}, \{n - 3, n - 2, n - 1\}]$. Let B' be the matrix obtained from B by replacing the 1 in position $(1, n - 2)$ by 0, and let A be the matrix obtained from B by inserting a new column on the right with 1's rows 1 and 2 and 0's elsewhere and then inserting a row on the bottom with 1's in columns $n - 3$, $n - 2$, and n and 0's elsewhere. It follows from Lemma 6.2.17 that A is the zero pattern of a doubly indecomposable, maximal SNS-matrix which is not a 2-join of SNS-matrices and that A has the form (6.13).   □

6.3 Zero patterns of partly decomposable maximal SNS-matrices

Consider the subposet of $\mathcal{M}_n$ consisting of the zero patterns of SNS-matrices of order n. The minimal elements are the permutation matrices of order n. The maximal elements are the $(0,1)$-matrices A of order n such that $\mathcal{Z}(A)$ contains an SNS-matrix and every SNS-matrix in $\mathcal{Z}(A)$ is a maximal SNS-matrix. Let A be a $(0, 1)$-matrix of order n, and let $\widehat{A}$ and $\widehat{\widehat{A}}$ be SNS-matrices in $\mathcal{Z}(A)$. If A is fully indecomposable, then it follows from Theorem 6.1.3 that $\widehat{A}$ is a maximal SNS-matrix if and only if $\widehat{\widehat{A}}$ is a maximal SNS-matrix. Thus the zero pattern of a fully indecomposable, maximal SNS-matrix is a maximal element of this subposet; however, the zero pattern of a partly decomposable, maximal SNS-matrix need not be a maximal element. This is because there exist partly decomposable SNS-matrices with the same zero pattern exactly one of which is a maximal SNS-matrix. For example, each of the matrices

$$
\left[\begin{array}{ccc|ccc}
-1 & 1 & 0 & 0 & 0 & 0 \\
-1 & -1 & 1 & 0 & 0 & 0 \\
-1 & -1 & -1 & 0 & 0 & 0 \\
\hline
1 & -1 & -1 & -1 & 1 & 0 \\
-1 & -1 & -1 & -1 & -1 & 1 \\
-1 & -1 & -1 & -1 & -1 & -1
\end{array}\right]
\text{ and }
\left[\begin{array}{ccc|ccc}
-1 & 1 & 0 & 0 & 0 & 0 \\
-1 & -1 & 1 & 0 & 0 & 0 \\
-1 & -1 & -1 & 0 & 0 & 0 \\
\hline
-1 & -1 & -1 & -1 & 1 & 0 \\
-1 & -1 & -1 & -1 & -1 & 1 \\
-1 & -1 & -1 & -1 & -1 & -1
\end{array}\right]
$$

$$(6.14)$$

is an SNS-matrix. It is easy to verify that the matrix on the left is a maximal SNS-matrix. The matrix on the right is not a maximal SNS-matrix, since replacing the $(3,4)$ entry by 1 gives the SNS-matrix H_6 defined in (1.12).

In this section we characterize maximal partly decomposable SNS-matrices and the maximal zero patterns of partly decomposable SNS-matrices.

Lemma 6.3.1 *Let*

$$
C = [c_{ij}] = \left[\begin{array}{ccc}
C_1 & O & O \\
C_{21} & C_2 & O \\
C_{31} & C_{32} & C_3
\end{array}\right]
$$

where $C_1, C_2,$ and C_3 are square matrices of orders $n_1, n_2,$ and n_3, respectively. Assume that C_1 is an SNS-matrix, each entry of C_{21} is nonzero, and

$$
\left[\begin{array}{cc}
C_2 & O \\
C_{32} & C_3
\end{array}\right]
\tag{6.15}
$$

is a maximal SNS-matrix. Let r and s be integers with $1 \leq r \leq n_1$ and $n_1 + n_2 + 1 \leq s \leq n_1 + n_2 + n_3$. Then the matrices obtained from C by replacing c_{rs} by ± 1 are not SNS-matrices.

Proof Without loss of generality we assume that C has a negative main diagonal. Since the matrix in (6.15) is maximal, C_{32} has no zero entries. Let t be any integer with $n_1 + 1 \leq t \leq n_1 + n_2$. It follows from Corollary 3.2.3 and the maximality of (6.15) that there exist both a positive path and a negative path in $\mathcal{D}(C)$ from s to t which does not contain r. Combining these paths with the arc from t to r, we obtain both a positive path and a negative path in $\mathcal{D}(C)$ from s to r, and the lemma follows. $\qquad\square$

Corollary 6.3.2 *Let*

$$C = \begin{bmatrix} C_1 & O & O \\ C_{21} & C_2 & O \\ C_{31} & C_{32} & C_3 \end{bmatrix}$$

where C_1, C_2, and C_3 are square matrices. Assume that

$$\begin{bmatrix} C_1 & O \\ C_{21} & C_2 \end{bmatrix} and \begin{bmatrix} C_2 & O \\ C_{32} & C_3 \end{bmatrix} \qquad (6.16)$$

are maximal SNS-matrices. Then C is a maximal SNS-matrix if and only if each entry of C_{31} is nonzero.

Proof If C is a maximal SNS-matrix, then clearly each entry of C_{31} is nonzero. Conversely, the maximality of the matrices in (6.16) implies that each entry of C_{21} and C_{32} is nonzero, and none of their 0's can be replaced by ± 1 to obtain an SNS-matrix. The corollary now follows from Lemma 6.3.1. $\qquad\square$

Let B be a partly decomposable $(0, 1, -1)$-matrix of order n whose determinant is not identically zero. Without loss of generality we assume that B has the form

$$\begin{bmatrix} B_1 & O & \cdots & O \\ B_{21} & B_2 & \cdots & O \\ \vdots & \vdots & \ddots & \vdots \\ B_{k1} & B_{k2} & \cdots & B_k \end{bmatrix} \quad (B_1, B_2, \ldots, B_k \text{ fully indecomposable})$$

$$(6.17)$$

where $k \geq 2$. If B is a maximal SNS-matrix, then each of the fully indecomposable components B_i of B is a maximal SNS-matrix and each entry of the matrices B_{ij} is nonzero. We now identify the sign patterns of partly decomposable, maximal SNS-matrices [21].

Theorem 6.3.3 *Let $B = [b_{ij}]$ be a partly decomposable $(0, 1, -1)$-matrix of* *the form* (6.17). *Then B is a maximal SNS-matrix if and only if each of the following conditions is satisfied:*

   (i) *Each B_i is a maximal SNS-matrix.*
   (ii) *Each entry of the matrices B_{ij} is nonzero.*
   (iii) *There does not exist an i with $1 \le i \le k - 1$ such that $B_{i+1,i} = \pm uv^T$ for some row v^T of B_i and some column u of B_{i+1}.*

Proof First assume that B is a maximal SNS-matrix. Then (i) and (ii) hold. Suppose that there exists an integer i such that $B_{i+1,i} = \epsilon uv^T$ where $\epsilon = \pm 1$ and v^T is a row of B_i and u is a column of B_{i+1}. By (ii) each entry of v^T and u is nonzero. Let C be the matrix obtained from B by replacing the zero in the row corresponding to v^T and column corresponding to u by $-\epsilon$. A nonzero term t_σ in the standard determinant expansion of C which contains the new entry $-\epsilon$ contains exactly one entry $u_p v_q$ from $B_{i+1,i}$, and all other entries lie in the diagonal blocks. By replacing $-\epsilon$ and $u_p v_q$ by u_p and v_q we obtain a nonzero term $t_{\sigma'}$ in the standard determinant expansion of B which has the same sign. Hence all terms in the standard determinant expansion of C have the same sign and thus C is an SNS-matrix. This contradicts the maximality of B, and we conclude that (iii) holds.

Conversely, assume that (i), (ii), and (iii) hold. Clearly, B is an SNS-matrix. First suppose that $k = 2$, and let m be the order of B_1. By (i) no zero entry of B_1 or B_2 can be replaced by ± 1 to obtain an SNS-matrix. Now consider a zero entry $b_{r,m+s}$ of B. If row r of B_1 contains a zero, then by the maximality of B_1, the matrices obtained from B by replacing $b_{r,m+s}$ by ± 1 are not SNS-matrices. Similarly, if the column s of B_2 contains a zero, then the matrices obtained from B by replacing $b_{r,m+s}$ by ± 1 are not SNS-matrices. Thus we may assume that row r of B_1 and column s of B_2 do not contain any zero entries. Since (iii) holds, there exist integers p and q with $p > m$ and $q \le m$ such that (a) $b_{pq} = b_{rq}b_{p,s+m}$, and there exist integers p' and q' with $p' > m$ and $q' \le m$ such that (b) $b_{p'q'} = -b_{rq'}b_{p',s+m}$. Property (a) implies that the matrix obtained from B by replacing $b_{r,m+s}$ by $+1$ is not an SNS-matrix, and property (b) implies that the matrix obtained from B by replacing $b_{r,m+s}$ by -1 is not an SNS-matrix. Therefore, if $k = 2$, B is a maximal SNS-matrix. If $k > 2$, then the maximality of B follows by induction using Corollary 6.3.2. $\qquad\square$

The SNS-matrix on the right in (6.14) has $k = 2$ fully indecomposable components and satisfies conditions (i) and (ii) of Theorem 6.3.3 but does not satisfy (iii), since $B_{21} = -uv^T$ where u is the first column of B_2 and v^T is the last row of B_1. Thus this matrix is not a maximal SNS-matrix. In the next corollary we identify the zero patterns of partly decomposable, maximal SNS-matrices.

Corollary 6.3.4 *Let A be a partly decomposable $(0, 1)$-matrix of the form*

$$
\begin{bmatrix}
A_1 & O & \cdots & O \\
A_{21} & A_2 & \cdots & O \\
\vdots & \vdots & \ddots & \vdots \\
A_{k1} & A_{k2} & \cdots & A_k
\end{bmatrix}
\tag{6.18}
$$

where $A_1, A_2, \ldots, A_k$ are fully indecomposable matrices of order $n_1, n_2, \ldots, n_k$, respectively. Then there exists a maximal SNS-matrix in $\mathcal{Z}(A)$ if and only if

(i) $A_{ij} = J \ (1 \le j < i \le k)$,

(ii) $\mathcal{Z}(A_i)$ *contains a maximal SNS-matrix* $(i = 1, 2, \ldots, k)$, *and*

(iii) $n_i + n_{i+1} \ge 4 \ (i = 1, 2, \ldots, k - 1)$.

Proof First assume that (i), (ii), and (iii) hold. Let B_i be a maximal SNS-matrix in $\mathcal{Z}(A_i)$ $(i = 1, 2, \ldots, k)$. Let p be an integer with $1 \le p \le k - 1$. If $n_p \ge 2$ and $n_{p+1} \ge 2$, then there exists an n_{p+1} by n_p $(1, -1)$-matrix $B_{p+1,p}$ whose rank is at least 2. If $n_p = 1$, then since an SNS-matrix has at most two columns without zero entries, there exists an n_{p+1} by 1 $(1, -1)$ column vector $B_{p+1,p}$ such that neither it nor its negative is a column of B_{p+1}. Similarly, if $n_{p+1} = 1$, then there exists a 1 by n_p $(1, -1)$ row vector $B_{p+1,p}$ such that neither it nor its negative is a row of B_p. Let $B_{pj} = A_{pj}$ for $j < p - 1$. Then by Theorem 6.3.3, the matrix (6.17) is a maximal SNS-matrix in $\mathcal{Z}(A)$.

Conversely, let B be a maximal SNS-matrix in $\mathcal{Z}(A)$. Then clearly (i) and (ii) hold. A maximal SNS-matrix X of order 2 has no zero entries and each 2 by 1 $(1, -1)$ column vector is a column of either X or $-X$. Thus if there exists an i with $n_i = 1$ and $n_{i+1} = 2$, then we contradict (iii) of Theorem 6.3.3. A similar contradiction results if $n_i = 2$ and $n_i = 1$, or $n_i = n_{i+1} = 1$. Hence (iii) also holds. $\qquad\square$

We now identify the maximal elements of the partially ordered set of zero patterns of SNS-matrices of order n.

Corollary 6.3.5 *Let A be a partly decomposable $(0, 1)$-matrix of the form (6.18) where $A_1, A_2, \ldots, A_k$ are fully indecomposable matrices. Assume that there is an SNS-matrix in $\mathcal{Z}(A)$. Then every SNS-matrix in $\mathcal{Z}(A)$ is a maximal SNS-matrix if and only if*

(i) $A_{ij} = J \ (1 \le j < i \le k)$,

(ii) $\mathcal{Z}(A_i)$ *contains a maximal SNS-matrix* $(i = 1, 2, \ldots, k)$, *and*

(iii) *for each* $i = 1, 2, \ldots, k - 1$, *either each row of* A_i *contains a zero or each column of* A_{i+1} *contains a zero.*

Proof First assume that (i), (ii), and (iii) hold. Let B be an SNS-matrix in $\mathcal{Z}(A)$ of the form (6.17) where B_i is in $\mathcal{Z}(A_i)$. By (ii) and the full indecomposability of the A_i, the B_i are maximal SNS-matrices. Since each entry of $B_{i+1,i}$ is nonzero, $B_{i+1,i} \neq uv^T$ for any row vector v^T of B_i and column vector u of B_{i+1}. Hence, by Theorem 6.3.3, B is a maximal SNS-matrix.

Conversely, assume that every SNS-matrix in $\mathcal{Z}(A)$ is a maximal SNS-matrix. Then clearly (i) and (ii) hold. Let B_i be a maximal SNS-matrix in $\mathcal{Z}(A_i)$. Suppose to the contrary that there exists an integer ℓ such that B_ℓ has a row v^T with no zero entries and $B_{\ell+1}$ has a column u with no zero entries. By Theorem 6.3.3 the SNS-matrix in $\mathcal{Z}(A)$ of the form (6.17) where $B_{ij} = uv^T$ if $i = \ell$ and $j = \ell + 1$, and $B_{ij} = A_{ij}$ otherwise, is not a maximal SNS-matrix. Hence (iii) also holds.   □

For example, let

$$A = \left[\begin{array}{c|cccc} 1 & 0 & 0 & 0 & 0 \\ \hline 1 & 1 & 1 & 0 & 0 \\ 1 & 1 & 1 & 1 & 1 \\ 1 & 0 & 0 & 1 & 1 \\ 1 & 1 & 1 & 1 & 1 \end{array} \right].$$

Then by Corollary 6.3.5 every SNS-matrix in $\mathcal{Z}(A)$ is a maximal SNS-matrix. The SNS-matrices in $\mathcal{Z}(A)$ with -1's on the main diagonal are the sixteen matrices

$$\left[\begin{array}{c|cccc} -1 & 0 & 0 & 0 & 0 \\ \hline \pm 1 & -1 & 1 & 0 & 0 \\ \pm 1 & -1 & -1 & 1 & -1 \\ \pm 1 & 0 & 0 & -1 & -1 \\ \pm 1 & 1 & 1 & 1 & -1 \end{array} \right]$$

and any matrix obtained from one of these 16 matrices by pre- and post-multiplying by a strict signing.

6.4 SNS-matrices and the permanent

Let $X = [x_{ij}]$ be a matrix of order n. The *permanent* of X is defined by

$$\operatorname{per} X = \sum_\sigma x_{1i_1} x_{2i_2} \cdots x_{ni_n}$$

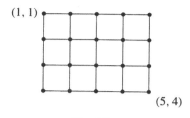

(1, 1)

(5, 4)

Fig. 6.3.

where the summation extends over all permutations $\sigma = (i_1, i_2, \ldots, i_n)$ of $\{1, 2, \ldots, n\}$. Note that

$$x_{1i_1} x_{2i_2} \cdots x_{ni_n} \longleftrightarrow \operatorname{sgn}(\sigma) x_{1i_1} x_{2i_2} \cdots x_{ni_n}$$

is a one-to-one correspondence between the terms in the permanent of X and the terms in the standard determinant expansion of X. The permanent is a multilinear function of the rows and of the columns of a matrix and is invariant under transposition and arbitrary row and column permutations. In contrast to the determinant, the computation of the permanent is apparently an intractable problem. In fact, Valiant [26] has shown that if there is a polynomial-time algorithm to compute the permanent of a $(0,1)$-matrix, then there are polynomial-time algorithms to compute the number of solutions to any NP-complete problem.

Recall that a matching of a graph is a set of pairwise vertex disjoint edges. A *perfect matching* is a matching such that each vertex is incident to exactly one edge of M. Let $\mathcal{G}$ be the bipartite graph of X with bipartition $\{V, V'\}$ where

$$V = \{1, 2, \ldots, n\} \text{ and } V' = \{1', 2', \ldots, n'\}.$$

If X is a $(0,1)$-matrix, then there is a one-to-one correspondence between the perfect matchings of $\mathcal{G}$ and the permutation matrices P of order n such that $P \le X$, and hence the permanent of X equals the number of perfect matchings of $\mathcal{G}$.

The essential idea of an SNS-matrix arose independently in the work of Kasteleyn [9, 10, 11] on the dimer problem of statistical mechanics. Consider a p by q rectangular grid $R_{p,q}$ of points (representing atoms of a molecule) in which consecutive points in a row or a column are joined by an edge (representing an interaction between the corresponding atoms). Thus, for example, if $p = 4$ and $q = 5$, we have the grid shown in Figure 6.3. We may regard $R_{p,q}$ as a graph of order pq. The graph $R_{p,q}$ is a bipartite graph where

$$V = \{(i, j) : 1 \le i \le p, \ 1 \le j \le q, \ i + j \text{ is even}\}$$

and

$$V' = \{(i, j) : 1 \le i \le p, \ 1 \le j \le q, \ i + j \text{ is odd}\}.$$

We choose our notation so that the vertex $(1, 1)$ is the point of the grid in the northwest corner and then order the vertices of V and then those of V' in the order in which they are encountered by traversing the grid, starting with the vertex $(1, 1)$ and going down the first column, up the second column, down the third column, and so forth. The *dimer problem* asks for the number of perfect matchings of the graph $R_{p,q}$ (the number of ways that the atoms can pair themselves up through their interactive forces). If p and q are both odd, there are no perfect matchings in $R_{p,q}$. We henceforth assume that p is even. Let $A_{p,q}$ be the (0,1)-matrix whose associated bipartite graph is $R_{p,q}$. Then $A_{p,q}$ is the matrix of order $pq/2$ defined by

$$
A_{p,q} = \begin{bmatrix}
N_p & U_p & O & \cdots & O & O \\
U_p & N_p & U_p & \cdots & O & O \\
O & U_p & N_p & \cdots & O & O \\
\vdots & \vdots & \vdots & \ddots & \vdots & \vdots \\
O & O & O & \cdots & N_p & U_p \\
O & O & O & \cdots & U_p & N_p
\end{bmatrix},
$$

where N_p and U_p are the matrices of order $p/2$ given by

$$
N_p = \begin{bmatrix}
1 & 0 & 0 & \cdots & 0 & 0 \\
1 & 1 & 0 & \cdots & 0 & 0 \\
0 & 1 & 1 & \cdots & 0 & 0 \\
\vdots & \vdots & \vdots & \ddots & \vdots & \vdots \\
0 & 0 & 0 & \cdots & 1 & 0 \\
0 & 0 & 0 & \cdots & 1 & 1
\end{bmatrix}
\quad \text{and} \quad
U_p = \begin{bmatrix}
0 & 0 & \cdots & 0 & 1 \\
0 & 0 & \cdots & 1 & 0 \\
\vdots & \vdots & \ddots & \vdots & \vdots \\
0 & 1 & \cdots & 0 & 0 \\
1 & 0 & \cdots & 0 & 0
\end{bmatrix}.
$$

The dimer problem is equivalent to determining the permanent of $A_{p,q}$. Kasteleyn solved the dimer problem by first showing how to replace some of the 1's in $A_{p,q}$ with -1's in order that the determinant of the resulting matrix $B_{p,q}$ equals the permanent of $A_{p,q}$ (thus he found an SNS-matrix in $\mathcal{Z}(A_{p,q})$).[3] He then calculated the eigenvalues of $B_{p,q}$ and obtained the following explicit formula for the determinant of $B_{p,q}$:

$$
\det B_{p,q} = 2^{pq/2} \prod_{k=1}^{p} \prod_{\ell=1}^{q} \left(\cos^2\left(\frac{\pi k}{p+1}\right) + \cos^2\left(\frac{\pi \ell}{q+1}\right) \right)^{1/4}
$$
$$
\approx e^{.29pq}.
$$

We now describe the relationship between SNS-matrices and the permanent

[3] More precisely, Kasteleyn used the adjacency matrix of order $p + q$ of the graph $R_{p,q}$ and worked with the pfaffian (see section 11.4).

in more detail [2]. Let $A = [a_{ij}]$ be a $(0,1)$-matrix of order n and let $B = [b_{ij}]$ be a matrix in $\mathcal{Z}(A)$. Since the number of nonzero terms in the standard determinant expansion of B equals per A, we have $|\det B| \leq$ per A with equality if and only if B is an SNS-matrix. Assume that B is an SNS-matrix. Then

$$\text{per } A = \pm \det B$$

implies that if we view the entries of a matrix X of order n as indeterminates,

$$\text{per } A * X = \pm \det B * X \qquad (6.19)$$

is a polynomial identity.[4] Here $*$ denotes the *Hadamard* (i.e., entrywise) product. Therefore an SNS-matrix B in $\mathcal{Z}(A)$ is a prescription for converting the permanent function, on the set of matrices which have zeros wherever A does, into the determinant function by affixing minus signs to certain of the entries. It is for this reason that a $(0,1)$-matrix A for which there exists an SNS-matrix in $\mathcal{Z}(A)$ is sometimes called a *convertible matrix*, and an SNS-matrix in $\mathcal{Z}(A)$ is called a *conversion* of A.

For example, suppose that $n = 4$, and let

$$A = \begin{bmatrix} 1 & 1 & 0 & 0 \\ 1 & 1 & 1 & 1 \\ 0 & 0 & 1 & 1 \\ 1 & 1 & 1 & 1 \end{bmatrix}.$$

The matrix

$$B = \begin{bmatrix} -1 & 1 & 0 & 0 \\ -1 & -1 & 1 & -1 \\ 0 & 0 & -1 & -1 \\ 1 & 1 & 1 & -1 \end{bmatrix}$$

is an SNS-matrix in $\mathcal{Z}(A)$. Thus we have the polynomial identity

$$\text{per} \begin{bmatrix} x_{11} & x_{12} & 0 & 0 \\ x_{21} & x_{22} & x_{23} & x_{24} \\ 0 & 0 & x_{33} & x_{34} \\ x_{41} & x_{42} & x_{43} & x_{44} \end{bmatrix} = \det \begin{bmatrix} -x_{11} & x_{12} & 0 & 0 \\ -x_{21} & -x_{22} & x_{23} & -x_{24} \\ 0 & 0 & -x_{33} & -x_{34} \\ x_{41} & x_{42} & x_{43} & -x_{44} \end{bmatrix} =$$

$$x_{11}x_{22}x_{33}x_{44} + x_{11}x_{22}x_{34}x_{43} + x_{11}x_{23}x_{34}x_{42} + x_{11}x_{24}x_{33}x_{42} +$$

$$x_{12}x_{21}x_{33}x_{44} + x_{12}x_{21}x_{34}x_{43} + x_{12}x_{23}x_{34}x_{41} + x_{12}x_{24}x_{33}x_{41}.$$

---

[4]Thus, if there exists an SNS-matrix in $\mathcal{Z}(A)$, then the permanent of any matrix whose zero pattern equals A can be computed in polynomial time.

It follows from Lemma 6.1.2 that, as polynomials,

$$\det \begin{bmatrix} \pm x_{11} & \pm x_{12} & \pm x_{13} \\ \pm x_{21} & \pm x_{22} & \pm x_{23} \\ \pm x_{31} & \pm x_{32} & \pm x_{33} \end{bmatrix} \neq x_{11}x_{22}x_{33} + x_{12}x_{23}x_{31} +$$

$$x_{13}x_{21}x_{32} + x_{13}x_{22}x_{31} + x_{12}x_{21}x_{31} + x_{11}x_{23}x_{32}$$

for any of the 2^9 choices of signs. The question of whether or not there was a choice of signs such that a polynomial identity is obtained was posed by Polyá [18, 24]. Thus the origins of SNS-matrices can be traced back to an innocent exercise from 1913.

The bipartite graph $R_{p,q}$ is clearly a planar graph. We now show that if A is any $(0,1)$-matrix whose determinant is not identically zero such that the bipartite graph associated with A is planar, then $\mathcal{Z}(A)$ contains an SNS-matrix. The proof we give follows the proof of Kasteleyn's theorem given in [14].

Let $\mathcal{G}$ be a connected plane graph with v vertices, e edges, and f faces (including the unbounded face). By Euler's formula

$$v - e + f = 2.$$

In order that the boundary of each face is a cycle, we assume that each edge of $\mathcal{G}$ is contained in at least one cycle.[5] Let γ be a cycle of $\mathcal{G}$ enclosing the bounded region F_γ. The *clockwise orientation* of γ is the orientation of γ in which the region F_γ is on the right. By assigning a direction to each edge of $\mathcal{G}$ we obtain an *orientation* of $\mathcal{G}$. Let $\overrightarrow{\mathcal{G}}$ be an orientation of $\mathcal{G}$. A *forward edge* of γ is an edge of γ whose direction in $\overrightarrow{\mathcal{G}}$ agrees with the clockwise orientation of γ.

Lemma 6.4.1 *Let $\mathcal{G}$ be a connected plane graph such that each edge is contained in a cycle, and let $\overrightarrow{\mathcal{G}}$ be an orientation of $\mathcal{G}$. Assume that every cycle of $\mathcal{G}$ that encloses a bounded face has an odd number of forward edges. Let γ be any cycle of $\mathcal{G}$. Then the number r of vertices of $\mathcal{G}$ in the interior of F_γ and the number s of forward edges of γ have opposite parity.*

Proof Suppose that F_γ contains k faces of $\mathcal{G}$, and let a_i be the number of forward edges of the cycle γ_i bounding the ith of these faces. Let p be the length of γ and let q be the number of edges in the interior of F_γ. By the hypothesis, each a_i is odd and hence

$$k \equiv \sum_{i=1}^{k} a_i \pmod 2.$$

[5]That is, the edge-connectivity of $\mathcal{G}$ is at least two.

Each edge in the interior of F_γ is a forward edge of exactly one of the γ_i. Hence $s + q = \sum_{i=1}^{k} a_i \equiv k \pmod{2}$. Applying Euler's formula, we obtain

$$(r + p) - (q + p) + (k + 1) = 2,$$

and hence $r + s \equiv 1 \pmod{2}$. ☐

Lemma 6.4.2 *Let $\mathcal{G}$ be a connected plane graph. Then there exists an orientation of $\mathcal{G}$ such that every cycle of $\mathcal{G}$ that encloses a bounded face has an odd number of forward edges.*

Proof We prove the lemma by induction. If $\mathcal{G}$ is a tree, then any orientation works. Assume $\mathcal{G}$ is not a tree. Let $\mathcal{G}'$ be the connected plane graph obtained from $\mathcal{G}$ by removing an edge α which lies on the boundary of the unbounded face and a bounded face F_γ. The lemma follows by applying the inductive hypothesis to $\mathcal{G}'$ and choosing the direction of α so that γ contains an odd number of forward arcs. ☐

The following theorem is a reformulation in terms of SNS-matrices of a special case of a theorem of Kasteleyn [10] (see also section 11.4).

Theorem 6.4.3 *Let $A = [a_{ij}]$ be a fully indecomposable $(0, 1)$-matrix of order $n \geq 2$. Assume that the bipartite graph $\mathcal{G}$ of A is planar. Then there exists an SNS-matrix in $\mathcal{Z}(A)$.*

Proof We assume that $\mathcal{G}$ is embedded in the plane as a plane graph. The full indecomposability of A implies that $\mathcal{G}$ is a connected graph in which each edge belongs to a cycle. Without loss of generality we also assume that $I_n \leq A$, and thus that

$$\{\{1, 1'\}, \{2, 2'\}, \ldots, \{n, n'\}\} \tag{6.20}$$

is a perfect matching in $\mathcal{G}$. By Lemmas 6.4.1 and 6.4.2 there exists an orientation $\overrightarrow{\mathcal{G}}$ such that for each cycle γ of $\mathcal{G}$, the number r of vertices in the interior of the bounded region F_γ and the number s of forward edges of γ have opposite parity. Let $B = [b_{ij}]$ be the matrix in $\mathcal{Z}(A)$ where

$$b_{ij} = \begin{cases} 0 & \text{if } a_{ij} = 0, \\ 1 & \text{if } a_{ij} = 1 \text{ and the edge } \{i, j'\} \text{ is directed from } i \text{ to } j', \\ -1 & \text{if } a_{ij} = 1 \text{ and the edge } \{i, j'\} \text{ is directed from } j' \text{ to } i. \end{cases}$$

Consider a nonzero term $t_\sigma = \text{sgn}(\sigma) b_{1k_1} b_{2k_2} \cdots b_{nk_n}$ in the standard determinant expansion of B. We prove that B is an SNS-matrix by showing that $t_\sigma = b_{11} b_{22} \cdots b_{nn}$. Consider the spanning subgraph $\mathcal{G}_\sigma$ of $\mathcal{G}$ with edges

$$\{\{1, 1'\}, \{2, 2'\}, \ldots, \{n, n'\}\} \cup \{\{1, k_1'\}, \{2, k_2'\}, \ldots, \{n, k_n'\}\}.$$

The connected components of $\mathcal{G}_\sigma$ are cycles γ_j of length $2e_j$ ($j = 1, 2, \ldots, \ell$) and edges. It follows that

$$\text{sgn}(\sigma) = \prod_{j=1}^{\ell} (-1)^{e_j - 1}.$$

Let γ be one of the cycles γ_j of length $2e = 2e_j$ and let the vertices of γ in clockwise order be

$$p_1, p_1', p_2, p_2', \ldots, p_e, p_e'.$$

Since $\mathcal{G}_\sigma$ has a perfect matching of $\mathcal{G}$, the number of vertices r in the interior of F_γ is even. By Lemma 6.4.1, r and s have opposite parity and hence γ has an odd number of forward edges. If $\{p_i, p_i'\}$ is a forward edge of γ, then $b_{p_i p_i} = +1$; otherwise, $b_{p_i p_i} = -1$. If $\{p_i', p_{i+1}\}$ is a forward edge of γ, then $b_{p_i p_{i+1}} = -1$; otherwise, $b_{p_i p_{i+1}} = +1$. Since γ has an odd number of forward edges, it follows that

$$b_{p_1 p_1} b_{p_2 p_1} b_{p_2 p_2} b_{p_3 p_2} \cdots b_{p_e p_e} b_{p_1 p_e} = (-1)^{e-1}.$$

Since this is true for each γ_j, we have

$$\prod_{i=1}^{n} b_{ik_i} \prod_{i=1}^{n} b_{ii} = \prod_{j=1}^{\ell} (-1)^{e_j - 1} = \text{sgn}(\sigma),$$

and hence

$$t_\sigma = b_{11} b_{22} \cdots b_{nn}.$$

Therefore all the nonzero terms in the standard determinant expansion of B have the same sign, and B is an SNS-matrix. $\qquad\square$

An orientation $\overrightarrow{R_{p,q}}$ of $R_{p,q}$ that satisfies the hypothesis of Lemma 6.4.1 is illustrated for $p = 4$ and $q = 5$ in Figure 6.4.

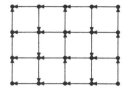

Fig. 6.4.

The SNS-matrix $B_{4,5}$ constructed from the orientation $\overrightarrow{R_{4,5}}$ in the manner described in the proof of Theorem 6.4.3 is

$$
\left[
\begin{array}{cc|cc|cc|cc|cc}
-1 & 0 & 0 & -1 & 0 & 0 & 0 & 0 & 0 & 0 \\
1 & -1 & -1 & 0 & 0 & 0 & 0 & 0 & 0 & 0 \\
\hline
0 & 1 & -1 & 0 & 0 & -1 & 0 & 0 & 0 & 0 \\
1 & 0 & 1 & -1 & -1 & 0 & 0 & 0 & 0 & 0 \\
\hline
0 & 0 & 0 & 1 & -1 & 0 & 0 & -1 & 0 & 0 \\
0 & 0 & 1 & 0 & 1 & -1 & -1 & 0 & 0 & 0 \\
\hline
0 & 0 & 0 & 0 & 0 & 1 & -1 & 0 & 0 & -1 \\
0 & 0 & 0 & 0 & 1 & 0 & 1 & -1 & -1 & 0 \\
\hline
0 & 0 & 0 & 0 & 0 & 0 & 0 & 1 & -1 & 0 \\
0 & 0 & 0 & 0 & 0 & 0 & 1 & 0 & 1 & -1
\end{array}
\right] .
$$

Since $\det B_{4,5} = 95$, $\operatorname{per}(A_{4,5}) = 95$, and the bipartite graph $R_{4,5}$ has exactly 95 perfect matchings. More generally, the matrix obtained from $A_{p,q}$ by replacing the 1's on and above the main diagonal by -1's is the SNS-matrix constructed from the orientation $\overrightarrow{R_{p,q}}$.

We conclude this section with a brief remark concerning a permanent analogue of SNS-matrices. Let A be a fully indecomposable matrix of order n. Then A is an SNS-matrix if and only if the nonzero terms in its standard determinant expansion have the same sign. It is proved in [1] and [6] that the nonzero terms in the permanent of A have the same sign if and only if there are permutation matrices P and Q such that

$$
PAQ = \left[
\begin{array}{cc}
A_{11} & A_{12} \\
A_{21} & A_{22}
\end{array}
\right]
$$

where $A_{11} \geq O$, $A_{22} \geq O$, $A_{12} \leq O$, and $A_{21} \leq O$. In particular, matrices for which the nonzero terms in the permanent have the same sign have a very simple structure.

6.5 Characterizations of zero patterns of SNS-matrices

In this section we obtain two characterizations of (0, 1)-matrices A for which $\mathcal{Z}(A)$ contains an SNS-matrix. One of these was obtained by Little [13] in the context of converting the permanent of a matrix into the determinant. The other was obtained by Seymour and Thomassen [20] in the context of characterizing those digraphs for which there exists a signing such that every directed cycle is negative. The proof that we give of these characterizations is based on those given in [22]. The first characterization is in terms of forbidden configurations in the bipartite graph of A; the second is in terms of forbidden configurations in the digraph of A, assuming the rows of A have been permuted to bring 1's to the main diagonal of A. The bipartite graph characterization identifies the (0,1)-matrices A such that $\mathcal{Z}(A)$ does not contain an SNS-matrix but $\mathcal{Z}(A')$ contains an SNS-matrix for every matrix A' obtained from A by changing a 1 to a 0. The digraph characterization identifies the (0, 1)-matrices A with 1's on the main diagonal such that $\mathcal{Z}(A)$ does not contain an SNS-matrix but $\mathcal{Z}(A')$ contains an SNS-matrix for every matrix A' obtained from A by changing a 1 *off the main diagonal* to a 0. If P is a permutation matrix, then the bipartite graphs of A and PA are isomorphic, but the digraphs of A and PA need not be isomorphic, even if both A and PA have only 1's on their main diagonals. Thus one would expect that there are fewer minimal forbidden configurations in the bipartite characterization than in the digraph characterization. For example, let $A = I_5 + C_5 + C_5^T$. By Corollary 6.1.2, $\mathcal{Z}(A)$ does not contain an SNS-matrix. Changing any 1 off the main diagonal to a 0 we obtain a matrix A' for which $\mathcal{Z}(A')$ contains an SNS-matrix. For instance, if we change the 1 in position (5,4) to a 0, then

$$\begin{bmatrix} -1 & 1 & 0 & 0 & -1 \\ -1 & -1 & 1 & 0 & 0 \\ 0 & -1 & -1 & 1 & 0 \\ 0 & 0 & -1 & -1 & -1 \\ 1 & 0 & 0 & 0 & -1 \end{bmatrix}$$

is an SNS-matrix in $\mathcal{Z}(A')$. Thus the digraph of A is a minimal forbidden digraph configuration for SNS-matrices. However, the matrix

$$B = \begin{bmatrix} 0 & -1 & 0 & 0 & 1 \\ -1 & -1 & -1 & 0 & 0 \\ 0 & -1 & -1 & 1 & 0 \\ 0 & 0 & 1 & 0 & -1 \\ 1 & 0 & 0 & -1 & 0 \end{bmatrix}$$

is a fully indecomposable SNS-matrix, such that the submatrix $B(\{4\})$ is not an SNS-matrix. It follows from (iii) of Corollary 6.1.4 that the matrix obtained from A by changing the 1's in positions $(1,1)$ and $(5,5)$ to 0's is not the zero pattern of an SNS-matrix. Hence, although the bipartite graph of A is a forbidden bipartite configuration for SNS-matrices, it is not a minimal forbidden bipartite configuration.

Let $A = [a_{ij}]$ be an m by n $(0,1)$-matrix. In Chapter 4 we defined the contraction of A on a row with exactly two 1's. The contraction of A on a column with exactly two 1's is defined in a similar way. Suppose that $a_{pr} = 1$ and that either row p or column r of A contains exactly two 1's. If row p contains exactly two 1's, then let B be the the $m - 1$ by $n - 1$ matrix obtained from A by a contraction on row p. If column r contains exactly two 1's, then let B be the matrix obtained from A by a contraction on column r. Note that if both row p and column r each contain exactly two 1's, then there is no ambiguity in the definition of B. We say that B is the matrix obtained from A by the *contraction on a_{pr}*. If row p and column r of A each contain exactly two 1's, say $a_{ps} = a_{qr} = 1$ where $s \neq r$ and $q \neq p$, and if $a_{qs} = 0$, then the contraction on a_{pr} is called an *elementary contraction*. If A is a square matrix and $p = r$, then we say that B is obtained from A by a *principal contraction* on a_{pp}.

The matrix A can be *contracted* to a matrix C provided that $A = C$ or C can be obtained from A by a sequence of contractions. Note that if A can be contracted to C, then C can be obtained from A by a sequence of symbolic pivots and row and column deletions using, as pivots, entries whose row or column contains exactly two 1's. If C can be obtained from A by a sequence of principal contractions, then the pivots used are all on the main diagonal of A.

The following lemma is a consequence of Lemmas 4.2.1 and 4.2.2 and Corollary 4.2.6.

Lemma 6.5.1 *Let A be a fully indecomposable $(0, 1)$-matrix and assume that A can be contracted to B. Then B is a fully indecomposable matrix, and $\mathcal{Z}(A)$ contains an SNS-matrix if and only if $\mathcal{Z}(B)$ contains an SNS-matrix.*

The main results of this section can now be stated as follows. Up to contraction, J_3 (i.e., $I_3 + C_3 + C_3^T$) is the only fully indecomposable $(0,1)$-matrix A such that $\mathcal{Z}(A)$ does not contain an SNS-matrix but $\mathcal{Z}(A')$ contains an SNS-matrix for each matrix A' obtained from A by changing a 1 to a 0. Up to principal contraction and simultaneous row and column permutations, the matrices $I_n + C_n + C_n^T$ ($n = 3, 5, 7, \dots$) are the only fully indecomposable $(0,1)$-matrices A with $I_n \leq A$ such that $\mathcal{Z}(A)$ does not contain an SNS-matrix but $\mathcal{Z}(A')$ contains an SNS-matrix for each matrix A' obtained from A by changing a 1 off the main diagonal to a 0.

A fully indecomposable $(0,1)$-matrix A of order $n \geq 2$ has at least two 1's in each row. If some row contains exactly two 1's, then A can be contracted. Otherwise we have the following lemma, which is a generalization of a result in [3].

Lemma 6.5.2 *Let* $A = [a_{ij}]$ *be a fully indecomposable* $(0, 1)$-*matrix of order* n *with at least three 1's in each row. Then there exist integers* k *and* ℓ *such that* $a_{k\ell} = 1$ *and* $A(\{k\}, \{\ell\})$ *is fully indecomposable. Indeed, if* $P = [p_{ij}]$ *is a permutation matrix with* $P \leq A$, *then* k *and* ℓ *can be chosen so that* $p_{k\ell} = 1$.

Proof Let P be a permutation matrix with $P \leq A$ and without loss of generality assume that $P = I_n$. We consider the strongly connected digraph $\mathcal{D}(A)$ and a maximal proper subset α of $\{1, 2, \ldots, n\}$ such that $\mathcal{D}(A[\alpha])$ is a strongly connected digraph. Since $\mathcal{D}(A)$ is strongly connected, there exist vertices u and v in α and a path or a directed cycle

$$u \to w_1 \to \cdots \to w_t \to v \quad (t \geq 1)$$

from u to v none of whose interior vertices belong to α. It follows from the maximality of α that $\bar{\alpha} = \{w_1, \ldots, w_t\}$ and that there is no arc from w_1 to any of the vertices $w_3, \ldots, w_t$. Since A has at least three 1's in each row, there exists an arc from w_1 to some vertex in α. Hence $\mathcal{D}(A[\alpha \cup \{w_1\}])$ is strongly connected, and by the maximality of α we have $\alpha \cup \{w_1\} = \{1, 2, \ldots, n\}$. Thus $A(\{w_1\}, \{w_1\})$ is fully indecomposable. $\square$

Lemma 6.5.3 *Let*

$$
A = \left[
\begin{array}{ccc|c}
 & & & u_1 \\
 & B & & \vdots \\
 & & & u_{n-1} \\
\hline
v_1 & \cdots & v_{n-1} & 1
\end{array}
\right]
$$

be a fully indecomposable $(0, 1)$-*matrix of order* n *where* B *is also fully indecomposable. Let*

$$\alpha = \{i : u_i = 1\} \text{ and } \beta = \{v_j : v_j = 1\}.$$

Assume that there exists an SNS-matrix $\widehat{B}$ *in* $\mathcal{Z}(B)$. *Then there exists an SNS-matrix in* $\mathcal{Z}(A)$ *if and only if*

(i) *the matrices* $\widehat{B}(\{i\}, \{j\})$, $(i \in \alpha, j \in \beta)$ *are SNS-matrices, and*

(ii) *there exists* $\hat{u}_i = \pm 1$ $(i \in \alpha)$ *and* $\hat{v}_j = \pm 1$ $(j \in \beta)$ *such that*

$$\hat{u}_i \hat{v}_j = (-1)^{i+j} \text{sign}(\det \widehat{B}) \, \text{sign}(\det \widehat{B}(\{i\}, \{j\})) \quad (i \in \alpha, j \in \beta).$$

Moreover, if (i) *holds and* B' *is a fully indecomposable* $(0, 1)$-*matrix with* $B' \leq B$, *then* $\mathcal{Z}(A)$ *contains an SNS-matrix if and only if* $\mathcal{Z}(A')$ *contains an SNS-matrix where*

$$
A' = \left[
\begin{array}{ccc|c}
 & & & u_1 \\
 & B' & & \vdots \\
 & & & u_{n-1} \\
\hline
v_1 & \cdots & v_{n-1} & 1
\end{array}
\right].
$$

Proof It follows from Theorem 6.1.3, applied to the fully indecomposable matrix B, that there exists an SNS-matrix in $\mathcal{Z}(A)$ if and only if there exists an SNS-matrix in $\mathcal{Z}(A)$ of the form

$$
\left[
\begin{array}{ccc|c}
 & & & \hat{u}_1 \\
 & \widehat{B} & & \vdots \\
 & & & \hat{u}_{n-1} \\
\hline
\hat{v}_1 & \cdots & \hat{v}_{n-1} & -1
\end{array}
\right].
\tag{6.21}
$$

Let $\widehat{A}$ be a matrix of the form (6.21). Then

$$
\det(\widehat{A}) = -\det \widehat{B} + \sum_{j \in \beta} \sum_{i \in \alpha} (-1)^{i+j-1} \hat{u}_i \hat{v}_j \det \widehat{B}(\{i\}, \{j\}).
$$

Since B is fully indecomposable, none of the matrices $\widehat{B}(\{i\}, \{j\})$ has an identically zero determinant. That there exists an SNS-matrix in $\mathcal{Z}(A)$ if and only if (i) and (ii) hold now follows from the fact that $\widehat{A}$ is an SNS-matrix if and only if all of the nonzero terms in its standard determinant expansion have the same sign.

Let B' be a fully indecomposable $(0,1)$-matrix with $B' \leq B$. Clearly, if $\mathcal{Z}(A)$ contains an SNS-matrix, then so does $\mathcal{Z}(A')$. Conversely, suppose that $\mathcal{Z}(A')$ contains an SNS-matrix and that $\widehat{B}$ satisfies (i). The matrix $\widehat{B'} = B' * \widehat{B}$ is an SNS-matrix in $\mathcal{Z}(B')$. By first part of the theorem, $\widehat{B'}$ satisfies (i) and (ii). Since $\widehat{B}$ satisfies (i) and B' is fully indecomposable, we have

$$
\text{sign}(\det \widehat{B}(\{i\}, \{j\})) = \text{sign}(\det \widehat{B'}(\{i\}, \{j\})) \quad (i \in \alpha, j \in \beta),
$$

and hence $\widehat{B}$ satisfies (ii). By the first part of the theorem, $\mathcal{Z}(A)$ contains an SNS-matrix. $\qquad\square$

A fully indecomposable matrix X is *nearly decomposable* provided that each matrix obtained from X by replacing a nonzero entry by zero is partly decomposable. Clearly, a nearly decomposable $(0,1)$-matrix of order $n \geq 2$ has at least

$2n$ ones. If $n \geq 2$ the matrix $I_n + C_n$ is a nearly decomposable $(0,1)$-matrix with exactly $2n$ ones. The matrix

$$
\begin{bmatrix}
0 & 1 & 1 & \cdots & 1 \\
1 & & & & \\
1 & & & & \\
\vdots & & & I_{n-1} & \\
1 & & & &
\end{bmatrix}
$$

is a nearly decomposable $(0,1)$-matrix with $3(n-1)$ ones for $n \geq 3$. It is a standard result in combinatorial matrix theory [4] that a nearly decomposable $(0,1)$-matrix of order $n \geq 3$ has at most $3(n-1)$ ones.

Theorem 6.5.4 *The fully indecomposable $(0,1)$-matrices A with the property that $\mathcal{Z}(A)$ does not contain an SNS-matrix but $\mathcal{Z}(A')$ contains an SNS-matrix for each matrix A' obtained from A by changing a 1 to a 0 are the matrices that can be contracted to J_3 using only elementary contractions.*

Proof Let A be a $(0,1)$-matrix of order n such that A can be contracted to J_3 using only elementary contractions. By (ii) of Lemma 4.2.1, A is fully indecomposable. By Corollary 6.1.2 and Lemma 6.5.1, $\mathcal{Z}(A)$ does not contain an SNS-matrix. It follows by a straightforward induction using Lemma 6.5.1 and the fact that a $(0,1)$-matrix of order 3 with exactly one zero is the zero pattern of an SNS-matrix that $\mathcal{Z}(A')$ contains an SNS-matrix for each matrix A' obtained from A by changing a 1 to a 0.

We prove the converse by induction on n. Let A be a fully indecomposable $(0,1)$-matrix of order n such that $\mathcal{Z}(A)$ does not contain an SNS-matrix but $\mathcal{Z}(A')$ contains an SNS-matrix for each matrix A' obtained from A by changing a 1 to a 0. The matrices J_1 and J_2 are zero patterns of SNS-matrices and hence $n \geq 3$. First suppose that A has a row or column with exactly two 1's. Without loss of generality we assume that the last row of A has exactly two 1's and that $a_{n,n-1} = a_{nn} = 1$. Let $B = [b_{ij}]$ be the matrix obtained from A by the contraction on a_{nn}. By Lemma 6.5.1, B is fully indecomposable and $\mathcal{Z}(B)$ does not contain an SNS-matrix. If $p \neq n$ is an integer such that $a_{p,n-1} = a_{pn} = 1$, then the matrix A' obtained from A by changing a_{pn} to 0 can be contracted to B, and hence by Lemma 6.5.1, $\mathcal{Z}(A')$ does not contain an SNS-matrix contrary to our assumptions. Hence $a_{p,n-1}a_{pn} = 0$ for all $p = 1, 2, \ldots, n-1$. Now let p and q be integers such that $b_{pq} = 1$ and let B' be the matrix obtained from B by changing b_{pq} to 0. If $q \neq n-1$, then $a_{pq} = 1$ and the matrix A' obtained from A by changing a_{pq} to 0 can be contracted to B'. If

$q = n - 1$, $a_{p,n-1} = 1$, and $a_{pn} = 0$, then the matrix A' obtained from A by changing $a_{p,n-1}$ to 0 can be contracted to B'. If $q = n - 1$, $a_{p,n-1} = 0$, and $a_{pn} = 1$, then the matrix A' obtained from A by changing a_{pn} to 0 can be contracted to B'. Since $\mathcal{Z}(A')$ contains an SNS-matrix, it follows from Lemma 6.5.1 that in each case $\mathcal{Z}(B')$ contains an SNS-matrix. Therefore B satisfies the inductive hypothesis. Hence B can be contracted to J_3 using only elementary contractions and, in particular, each column of B contains either two or three 1's. Since $a_{p,n-1}a_{pn} = 0$ for all $p \leq n - 1$, either column $n - 1$ or n of A contains exactly two 1's. Hence the contraction of A on either $a_{n,n-1}$ or a_{nn} is an elementary contraction which gives B, and thus A can be contracted to J_3 using only elementary contractions.

Now suppose that each row and column of A has at least three 1's. We apply Lemma 6.5.2 and assume without loss of generality that $a_{nn} = 1$ and that $A(\{n\})$ is a fully indecomposable matrix. We first show that row n of A contains exactly three 1's. Assume to the contrary that $a_{nr} = a_{ns} = a_{nt} = 1$ where $1 < r < s < t < n$. Let C be the matrix obtained from A by changing both a_{nr} and a_{ns} to 0. Then $\mathcal{Z}(C)$ contains an SNS-matrix $\widehat{C}$. Using the fact that $a_{nt} = 1$ and the full indecomposability of $A(\{n\})$, we conclude that C is fully indecomposable. It follows from our assumptions and Corollary 6.1.4 that for some choice of $\epsilon = \pm 1$, the matrix obtained from $\widehat{C}$ by replacing the zero in position (n, r) by ϵ is an SNS-matrix. Similarly, for some choice of $\delta = \pm 1$, the matrix obtained from $\widehat{C}$ by replacing the zero in position (n, s) with δ is an SNS-matrix. Let $\widehat{A}$ be the matrix in $\mathcal{Z}(A)$ obtained from $\widehat{C}$ by replacing the zeros in positions (n, r) and (n, s) by ϵ and δ, respectively. Then all nonzero terms in the standard determinant expansion of $\widehat{A}$ have the same sign and hence $\widehat{A}$ is an SNS-matrix, contrary to assumption. We conclude that row n of A, and similarly column n of A, contains exactly three 1's. Without loss of generality we assume that $a_{n,n-2} = a_{n,n-1} = 1$. Let u and v be integers with $1 \leq u < v < n$ such that $a_{u,n} = a_{v,n} = 1$.

Let $M = A(\{n\})$. It follows from our assumptions that there is an SNS-matrix $\widehat{M}$ in $\mathcal{Z}(M)$. Let A' be the matrix obtained from A by replacing $a_{u,n}$ by 0. By assumption there is an SNS-matrix $\widehat{A'}$ in $\mathcal{Z}(A')$, and by the full indecomposability of M and Theorem 6.1.3 we may assume that $\widehat{A'}(\{n\}) = \widehat{M}$. The full indecomposability of M now implies that $M(\{u\}, \{n - 1\})$ and $M(\{v\}, \{n-1\})$ are SNS-matrices. Similarly, $M(\{u\}, \{n-2\})$ and $M(\{v\}, \{n-2\})$ are SNS-matrices. It now follows from Lemma 6.5.3 and the choice of A that M is a nearly decomposable matrix. Since A has at least three 1's in each row, M contains at least $3n - 5$ ones. Since a nearly decomposable matrix of order $n - 1 \geq 3$ contains at most $3(n - 2)$ ones, we conclude that $n = 3$ and hence $A = J_3$. Thus the theorem follows by induction. $\square$

Theorem 6.5.5 *Let A be a $(0, 1)$-matrix of order n such that A does not have an identically zero determinant. Then the following are equivalent:*

   (i) *$\mathcal{Z}(A)$ does not contain an SNS-matrix.*

  (ii) *There exists a $(0, 1)$-matrix C of order k and permutation matrices P and Q such that $P(C \oplus I_{n-k})Q \le A$ and C can be contracted to J_3.*

 (iii) *There exists a $(0, 1)$-matrix C' of order k' and permutation matrices P' and Q' such that $P'(C' \oplus I_{n-k'})Q' \le A$ and C' can be contracted to J_3 using only elementary contractions.*

Proof Clearly, (iii) implies (ii). Now assume that (ii) holds. Then by Corollary 6.1.2 and Lemma 6.5.1, $\mathcal{Z}(C \oplus I_{n-k})$ does not contain an SNS-matrix, and hence (i) holds. Finally, assume that (i) holds.

There exists a $(0,1)$-matrix B with $B \le A$ such that B does not have an identically zero determinant, $\mathcal{Z}(B)$ does not contain an SNS-matrix, and each matrix B' obtained from B by changing a 1 to a 0 either has an identically zero determinant or $\mathcal{Z}(B')$ contains an SNS-matrix. The choice of B implies that there exist permutation matrices U and V such that $UBV = C \oplus I_{n-k}$ where C is a fully indecomposable matrix of order k which is not the zero pattern of an SNS-matrix but each matrix obtained from C by changing a 1 to a 0 is. Hence, by Theorem 6.5.4, (iii) holds. $\square$

Elementary contraction clearly preserves the permanent of a $(0,1)$-matrix. Hence it follows from Theorem 6.5.5 that if A is a $(0,1)$-matrix with per $A > 0$, then $\mathcal{Z}(A)$ contains an SNS-matrix if and only if $\mathcal{Z}(A')$ contains an SNS-matrix for all $A' \le A$ with per $A' = 6$.

We now give an example to illustrate Theorem 6.5.5. Let

$$A = \begin{bmatrix} 1 & 1 & 0 & 0 & 1 & 1 \\ 0 & 1 & 1 & 1 & 0 & 1 \\ 1 & 0 & 1 & 1 & 0 & 0 \\ 0 & 1 & 1 & 1 & 0 & 0 \\ 1 & 0 & 0 & 0 & 1 & 0 \\ 0 & 0 & 0 & 0 & \mathbf{1} & 1 \end{bmatrix}.$$

By successively contracting on the boldface 1's, we obtain

$$\begin{bmatrix} 1 & 1 & 0 & 0 & 1 \\ 0 & 1 & 1 & 1 & 1 \\ 1 & 0 & 1 & 1 & 0 \\ 0 & 1 & 1 & 1 & 0 \\ 1 & 0 & 0 & 0 & \mathbf{1} \end{bmatrix}, \begin{bmatrix} \mathbf{1} & 1 & 0 & 0 \\ 1 & 1 & 1 & 1 \\ 1 & 0 & 1 & 1 \\ 0 & 1 & 1 & 1 \end{bmatrix}, \begin{bmatrix} 1 & 1 & 1 \\ 1 & 1 & 1 \\ 1 & 1 & 1 \end{bmatrix}.$$

Hence (ii) of Theorem 6.5.5 holds, and $\mathcal{Z}(A)$ does not contain an SNS-matrix. Let

$$
B = \begin{bmatrix}
1 & 1 & 0 & 0 & 0 & 0 \\
0 & 1 & 1 & 1 & 0 & 0 \\
1 & 0 & 1 & 1 & 0 & 0 \\
0 & 1 & 1 & 1 & 0 & 0 \\
\hline
0 & 0 & 0 & 0 & 1 & 0 \\
0 & 0 & 0 & 0 & 0 & 1
\end{bmatrix}.
$$

Then $B \leq A$ and the matrix J_3 can be obtained from $B[\{1, 2, 3, 4\}]$ by the elementary contraction on the 1 in position (1,1), and hence (iii) of Theorem 6.5.5 holds.

Since the bipartite graph of J_3 is the complete bipartite graph $K_{3,3}$, a direct translation of the equivalence of (i) and (iii) in Theorem 6.5.5 gives the following characterization of zero patterns of SNS-matrices in terms of bipartite graphs [13].

Corollary 6.5.6 *Let A be a $(0, 1)$-matrix of order n such that A does not have an identically zero determinant. Then $\mathcal{Z}(A)$ does not contain an SNS-matrix if and only if there is a spanning subgraph of the bipartite graph of A which is the vertex disjoint union of an even subdivision of $K_{3,3}$ and a matching.*

Since a subdivision of $K_{3,3}$ is not a planar graph, Corollary 6.5.6 implies Theorem 6.4.3.

An irreducible matrix X is *nearly reducible* provided that each matrix obtained from X by replacing a nonzero entry by zero is reducible. Each diagonal entry of a nearly reducible matrix is zero. A nearly reducible $(0,1)$-matrix of order $n \geq 2$ has at least n ones. The matrix C_n is a nearly reducible matrix with exactly n ones. The adjacency matrix of a tree of order n is a nearly reducible $(0,1)$-matrix of order n with $2(n-1)$ ones. It is a standard result in combinatorial matrix theory [4] that a nearly reducible $(0,1)$-matrix of order n has at most $2(n-1)$ ones with equality if and only if it is the adjacency matrix of a tree.

By Theorem 6.5.4, up to contraction, indeed up to elementary contraction, there is only one fully indecomposable $(0,1)$-matrix which is minimal with respect to the property of not being the zero pattern of an SNS-matrix. However, there exist fully indecomposable matrices A of order n which are minimal with respect to the properties that A is not the zero pattern of an SNS-matrix and $I_n \leq A$, such that A can be contracted by a principal contraction but not by an elementary, principal contraction. For example, the only principal contractions of the matrix

$$\begin{bmatrix} 1 & 1 & 0 & 0 \\ 0 & 1 & 1 & 1 \\ 1 & 0 & 1 & 1 \\ 1 & 0 & 1 & 1 \end{bmatrix}$$

are the nonelementary contractions on the 1 in position $(1, 1)$ and the 1 in position $(2, 2)$. We now define a type of contraction, called pseudo-elementary, which by the next theorem has the property that matrices of the preceding type which can be contracted by a principal contraction can also be contracted by a pseudo-elementary, principal contraction.

Let $A = [a_{ij}]$ be an m by n $(0,1)$-matrix such that $a_{pr} = 1$ and either row p or column r, say row p, of A contains exactly two 1's. Let s be the integer such that $s \neq r$ and $a_{ps} = 1$. The contraction on a_{pr} is *pseudo-elementary* provided that either it is elementary or the contraction on a_{ps} is elementary. For example, the contraction on the 1 in the $(1,1)$ position of the preceding matrix of order 4 is pseudo-elementary but not elementary.

Theorem 6.5.7 *The fully indecomposable $(0, 1)$-matrices A with $I_n \leq A$ which have the property that $\mathcal{Z}(A)$ does not contain an SNS-matrix but $\mathcal{Z}(A')$ contains an SNS-matrix for each matrix A' obtained from A by changing a 1 off the main diagonal to a 0 are, up to simultaneous row and column permutations, the matrices that can be contracted to $I_k + C_k + C_k^T$ for some odd integer $k \geq 3$ using only pseudo-elementary, principal contractions.*

Proof Assume that A can be contracted to $I_k + C_k + C_k^T$ for some integer $k \geq 3$ using only pseudo-elementary, principal contractions. By Corollary 6.1.2, the matrix $I_k + C_k + C_k^T$ for odd $k \geq 3$ is not the zero pattern of an SNS-matrix. Let X be a matrix obtained from $I_k + C_k + C_k^T$ by changing a 1 off the main diagonal to a 0. Without loss of generality, assume that the 1 in position $(1, k)$ is changed to 0. Then $X * H_k$ is an SNS-matrix in $\mathcal{Z}(X)$ where H_k is the SNS-matrix in (1.12). Thus changing any 1 off the main diagonal of $I_k + C_k + C_k^T$ to a 0 results in the zero pattern of an SNS-matrix. It now follows by a straightforward induction using Lemma 6.5.1 that $\mathcal{Z}(A)$ does not contain an SNS-matrix, but $\mathcal{Z}(A')$ contains an SNS-matrix for each matrix A' obtained from A by changing a 1 off the main diagonal to a 0.

We prove the converse by induction on n. Since J_1 and J_2 are zero patterns of SNS-matrices, we have $n \geq 3$. First suppose that A has a row or column with exactly two 1's. Without loss of generality we assume that the last row of A has exactly two 1's and that $a_{n,n-1} = 1$. Let $B = [b_{ij}]$ be the matrix obtained from A by the principal contraction on a_{nn}. By Lemma 6.5.1, B is fully indecomposable, $I_{n-1} \le B$, and $\mathcal{Z}(B)$ does not contain an SNS-matrix. If $p \ne n$ is an integer such that $a_{p,n-1} = a_{pn} = 1$, then the result of the principal contraction on a_{nn} of the matrix A' obtained from A by changing a_{pn} to 0 is B and hence, by Lemma 6.5.1, $\mathcal{Z}(A')$ does not contain an SNS-matrix, contrary to our assumptions. Hence $a_{p,n-1}a_{pn} = 0$ for all $p = 1, 2, \ldots, n-1$. Now let p and q be distinct integers such that $b_{pq} = 1$ and let B' be the matrix obtained from B by changing b_{pq} to 0. Then as in the proof of Theorem 6.5.4, there exists a matrix A' obtained from A by replacing a 1 off the main diagonal with a 0 such that B' is the result of the principal contraction on a_{nn}. Since $\mathcal{Z}(A')$ contains an SNS-matrix, it follows from Lemma 6.5.1 that $\mathcal{Z}(B')$ contains an SNS-matrix. Therefore B satisfies the inductive hypothesis. Hence there exist an odd integer $k \ge 3$ and a permutation matrix P such that B can be contracted $P(I_k + C_k + C_k^T)P^T$ using only pseudo-elementary, principal contractions. This implies that each column of B contains either two or three 1's. Since $a_{p,n-1}a_{pn} = 0$ for all $p \le n-1$, either column $n-1$ or n of A contains exactly two 1's. Hence the contraction of A either on $a_{n,n-1}$ or on a_{nn} is an elementary contraction. Thus the principal contraction on a_{nn} is pseudo-elementary and the conclusion holds.

Now suppose that each row and column of A has at least three 1's. We apply Lemma 6.5.2 and assume without loss of generality that $A(\{n\})$ is a fully indecomposable matrix. The same argument as given in the proof of Theorem 6.5.4 shows that row n of A and column n of A each contains exactly three 1's. Without loss of generality we assume that $a_{n,n-2} = a_{n,n-1} = 1$. Let u and v be integers with $1 \le u < v < n$ such that $a_{u,n} = a_{v,n} = 1$.

Let $M = A(\{n\})$. It follows as in the proof of Theorem 6.5.4 that M has the property that changing any 1 off the main diagonal to 0 results in a partly decomposable matrix. Hence $M - I_{n-1}$ is a nearly reducible matrix. Since A has at least three 1's in each row, $M - I_{n-1}$ contains at least $2n - 4$ 1's. Since a nearly reducible matrix of order $n - 1$ contains at most $2(n - 2)$ 1's with equality if and only if it is the adjacency matrix of a tree of order $n - 1$, we conclude that $M - I_{n-1}$ is the adjacency matrix of a tree T. Using again the assumption that A has at least three 1's in each row and column, we conclude that $u = n - 2$ and $v = n - 1$ and that T is a undirected path joining u and v. Hence $A - I_n$ is an adjacency matrix of a cycle of length n. Thus there exists a permutation matrix P such that $PAP^T = I_n + C_n + C_n^T$. Since $\mathcal{Z}(A)$ does not contain an SNS-matrix, Corollary 6.1.2 implies that n is odd. Hence the theorem follows by induction. $\qquad\square$

Theorem 6.5.8 *Let A be a $(0, 1)$-matrix of order n with $I_n \leq A$. Then the following are equivalent:*

   (i) *$\mathcal{Z}(A)$ does not contain an SNS-matrix.*
   (ii) *There exist a principal submatrix $A[\alpha]$ of order k and a $(0, 1)$-matrix B with $I_k \leq B \leq A[\alpha]$ such that for some odd integer $m \geq 3$ and some permutation matrix P of order m, B can be contracted to $P(I_m + C_m + C_m^T)P^T$ using only principal contractions.*
   (iii) *There exist a principal submatrix $A[\alpha']$ of order k' and a $(0, 1)$-matrix B' with $I_{k'} \leq B' \leq A[\alpha']$ such that for some odd integer $m' \geq 3$ and some permutation matrix P' of order m', B' can be contracted to $P'(I_{m'} + C_{m'} + C_{m'}T)P'^T$ using only pseudo-elementary, principal contractions.*

Proof Clearly, (iii) implies (ii). Now assume that (ii) holds. Then by Corollary 6.1.2 and Lemma 6.5.1, $\mathcal{Z}(B)$ does not contain an SNS-matrix, and hence (i) holds. Finally, assume that (i) holds. There exists a $(0,1)$-matrix Y with $I_n \leq Y \leq A$ such that $\mathcal{Z}(Y)$ does not contain an SNS-matrix but $\mathcal{Z}(Y')$ contains an SNS-matrix for each matrix Y' obtained from Y by changing a 1 off the main diagonal to a 0. Then there exists a permutation matrix U such that $UYU^T = C \oplus I_{n-k}$ where C is a fully indecomposable matrix with $I_k \leq C$ such that $\mathcal{Z}(C)$ does not contain an SNS-matrix but $\mathcal{Z}(C')$ contains an SNS-matrix for each matrix C' obtained from C by changing a 1 off the main diagonal to a 0. Hence, by Theorem 6.5.7, (iii) holds with $B' = C$.   □

Let $A = [a_{ij}]$ be a $(0,1)$-matrix of order n and let B be obtained from A by a pseudo-elementary, principal contraction. If the contraction is elementary, then the digraph of A can be obtained from that of B by splitting an arc with the insertion of a new vertex. Now suppose that the contraction is not elementary. Then $\mathcal{D}(A)$ can be obtained from $\mathcal{D}(B)$ by splitting a vertex v, that is, by inserting a new vertex v', the arc (v, v') from v to v', and replacing each arc of $\mathcal{D}(B)$ of the form (v, x) by the arc (v', x). A *subdivision of a digraph $\mathcal{D}$* is a digraph obtained from $\mathcal{D}$ by a sequence of arc splittings. A *splitting of a digraph $\mathcal{D}$* is a digraph obtained from $\mathcal{D}$ by a sequence of arc splittings and vertex splittings. A digraph is considered to be both a subdivision and a splitting of itself. A *doubly directed cycle* is a digraph obtained from the graph of a cycle of length $k \geq 3$ by replacing each edge by two oppositely directed arcs. If k is odd, then the doubly directed cycle is *odd*. If $k \geq 3$, then the digraph of $I_k + C_k + C_k^T$ is a doubly directed cycle. A direct translation of the equivalence of (i) and (iii) in Theorem 6.5.8 gives the following characterization of zero patterns of SNS-matrices in terms of digraphs [20].

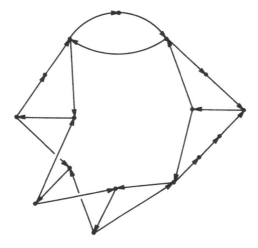

Fig. 6.5.

Corollary 6.5.9 *Let A be a $(0, 1)$-matrix of order n with $I_n \leq A$. Then $\mathcal{Z}(A)$ does not contain an SNS-matrix if and only if the digraph of A contains a splitting of an odd, doubly directed cycle.*

It is unknown whether there exists a polynomial-time algorithm for recognizing if a bipartite graph contains a spanning subgraph which is the vertex disjoint union of an even subdivision of $K_{3,3}$ and a matching, or for recognizing if a digraph contains a splitting of an odd, doubly directed cycle. Figure 6.5 illustrates a splitting of an odd, doubly directed cycle.

6.6 Combinatorially symmetric SNS-matrices

Let A be a symmetric $(0,1)$-matrix with $I_n \leq A$. Then there exists a permutation matrix P such that $PAP^T = A_1 \oplus \cdots \oplus A_k$ where $k \geq 1$ and each A_i is a fully indecomposable, symmetric matrix. Let B_1 and B_2 be SNS-matrices in $\mathcal{Z}(A)$. It follows from Theorem 6.1.3 that there exist strict signings D and E such that $B_1 = DB_2E$. Hence, up to multiplication of rows and columns by -1, there is at most one SNS-matrix in $\mathcal{Z}(A)$. In this section we give a characterization of symmetric $(0, 1)$-matrices A of order n with $I_n \leq A$ such that $\mathcal{Z}(A)$ contains an SNS-matrix.

A matrix is *combinatorially symmetric* provided that its zero pattern is a symmetric matrix. Let $B = [b_{ij}]$ be a combinatorially symmetric matrix of order n. Let $\mathcal{G}(B)$ be the graph of order n with vertices $1, 2, \ldots, n$ and an edge

$\{i, j\}$ joining vertices i and j if and only if $i \neq j$ and $b_{ij} \neq 0$. We call $\mathcal{G}(B)$ the *graph of the matrix* B. Two matrices with the same zero pattern have the same graph. If $b_{ii} \neq 0$ for each i, then the fully indecomposable components of B correspond to the connected components of $\mathcal{G}(B)$.

The matrices

$$\begin{bmatrix} -1 & 1 & 0 \\ -1 & -1 & 1 \\ 0 & -1 & -1 \end{bmatrix} \text{ and } \begin{bmatrix} -1 & 1 & 0 & 1 \\ -1 & -1 & 1 & 0 \\ 0 & -1 & -1 & 1 \\ -1 & 0 & -1 & -1 \end{bmatrix} \tag{6.22}$$

are combinatorially symmetric SNS-matrices whose graphs are a path of length 2 and a cycle of length 4, respectively.

A *cut-vertex*, respectively a *bridge*, of a connected graph is a vertex, respectively an edge, whose removal results in a graph which is not connected. A *2-connected graph* is a connected graph of order at least 3 without any cut-vertices. A standard fact from graph theory asserts that a graph with at least three vertices is 2-connected if and only if for each pair of distinct vertices u and v there exist two paths joining u and v whose only common vertices are u and v. Let $\mathcal{G}$ be a connected graph of order $n \geq 2$. The *blocks* of $\mathcal{G}$ are the connected subgraphs of $\mathcal{G}$ which have no cut-vertices and which are maximal with respect to these properties. The blocks of $\mathcal{G}$ partition its edges and two distinct blocks have at most one vertex in common. Each block of $\mathcal{G}$ is either 2-connected or is a bridge.

We regard the bipartition $\{X, Y\}$ of a bipartite graph as a coloring of its vertices with two colors. Thus two vertices of the same color are joined only by paths of even length, and two vertices of different colors are joined only by paths of odd length. Note that a bipartite graph contains an even subdivision of the complete bipartite graph $K_{2,3}$ if and only if there exist two distinct vertices u and v of the same color and three paths joining u and v no two of which have a common vertex different from u and v.

Lemma 6.6.1 *Let B be a combinatorially symmetric $(0, 1, -1)$ SNS-matrix of order n with a negative main diagonal. Then*

(i) *$\mathcal{G}(B)$ is a bipartite graph,*

(ii) *there exists a permutation matrix P and an integer k with $1 \leq k \leq n$ such that PBP^T has the form*

$$\begin{bmatrix} -I_k & B_1 \\ -B_1^T & -I_{n-k} \end{bmatrix},$$

(iii) *and $\mathcal{G}(B)$ does not contain an even subdivision of $K_{2,3}$.*

Proof If $\mathcal{G}(B)$ is not bipartite, then $\mathcal{G}(B)$ contains a cycle of odd length and hence the digraph $\mathcal{D}(B)$ contains a odd, doubly directed cycle. Thus (i) follows from Corollary 6.1.2. Let $\{X, Y\}$ be a bipartition of the vertices of $\mathcal{G}(B)$. Let P be the permutation matrix that reorders the vertices so that the vertices of X come first. Since B is an SNS-matrix with a negative main diagonal, PBP^T has the form given in (ii). Finally, let α, β, and γ be three paths joining distinct vertices u and v of the same color such that no two of these paths have a common vertex different from u and v. Then in $\mathcal{D}(B)$, following α from u to v, β from v to u, γ from u to v, and α from v to u we obtain a splitting of an odd, doubly directed cycle. Since B is an SNS-matrix, it follows from Corollary 6.5.9 that (iii) holds. □

Let $\mathcal{G}_1$ and $\mathcal{G}_2$ be graphs whose vertex sets are disjoint. A *vertex-join* of $\mathcal{G}_1$ and $\mathcal{G}_2$ is a graph obtained by identifying a vertex of $\mathcal{G}_1$ with a vertex of $\mathcal{G}_2$. An *edge-join* is a graph obtained by identifying an edge of $\mathcal{G}_1$ with an edge of $\mathcal{G}_2$. Let A_1 and A_2 be symmetric $(0,1)$-matrices each of whose diagonal entries equals 1. Then the graph of a 1-join, respectively 2-join, of A_1 and A_2 is a vertex-join, respectively edge-join, of $\mathcal{G}(A_1)$ and $\mathcal{G}(A_2)$.

The following corollary is a direct consequence of Theorem 6.2.7 and the observation that a connected graph of order $n \geq 2$ can be obtained from its blocks by a sequence of vertex-joins at cut-vertices.

Corollary 6.6.2 *Let $\mathcal{G}_1$ and $\mathcal{G}_2$ be connected graphs whose vertex sets are disjoint. Let $\mathcal{G}$ be either a vertex-join or an edge-join of $\mathcal{G}_1$ and $\mathcal{G}_2$. Then $\mathcal{G}$ is the graph of a combinatorially symmetric SNS-matrix with a negative main diagonal if and only if $\mathcal{G}_1$ and $\mathcal{G}_2$ are. A connected graph is the graph of a combinatorially symmetric SNS-matrix with a negative main diagonal if and only if each of its blocks is.*

A graph which is either an edge or can be obtained from an edge by a sequence of edge-joins with cycles of length 4 is called a 4-*cockade*. Any edge of a 4-cockade can be used as the first edge. A 4-cockade with 4 vertices is a cycle of length 4. A 4-cockade which is not an edge is a 2-connected bipartite graph and can be obtained by successively appending paths of length 3 on edges. A 4-cockade has an even number n of vertices and $(3n - 4)/2$ edges. The following theorem characterizes zero patterns of combinatorially symmetric SNS-matrices with a negative main diagonal [25]. The equivalence of (ii) and (iii) is also proved in [8, 17].

Theorem 6.6.3 *Let $\mathcal{G}$ be a connected graph. Then the following are equivalent:*

(i) *$\mathcal{G}$ is the graph of a combinatorially symmetric SNS-matrix with a negative main diagonal.*

(ii) $\mathcal{G}$ *is bipartite and does not contain an even subdivision of* $K_{2,3}$.

(iii) *Each block of* $\mathcal{G}$ *is a spanning subgraph of a* 4-*cockade*.

Proof First we prove that (iii) implies (i). By Corollary 6.6.2 we may assume that $\mathcal{G}$ is 2-connected. The matrix on the right in (6.22) shows that a cycle of length 4 is the graph of a combinatorially symmetric SNS-matrix with a negative main diagonal. Hence it follows from Corollary 6.6.2 that each 4-cockade is the graph of a combinatorially symmetric SNS-matrix with a negative main diagonal. Hence (iii) implies (i).

By Lemma 6.6.1, (i) implies (ii). We prove that (ii) implies (iii) by induction on the order of $\mathcal{G}$. Assume that (ii) holds. Without loss of generality we assume that $\mathcal{G}$ is 2-connected and that no graph obtained from $\mathcal{G}$ by inserting a new edge also satisfies (ii). If $\mathcal{G}$ is a cycle of even length, then $\mathcal{G}$ is a subgraph of a 4-cockade. Otherwise, $\mathcal{G}$ has a 2-connected subgraph of smaller order. Let $\mathcal{G}'$ be a 2-connected, proper subgraph of $\mathcal{G}$ with the largest number of vertices. Then there is a path $u, x_1, \ldots, x_k, w$ in $\mathcal{G}$ such that $k \geq 1$ and u and w are the only vertices of the path in $\mathcal{G}'$. By the choice of $\mathcal{G}'$ each of the vertices $x_2, \ldots, x_{k-1}$ has degree 2. The 2-connectivity of $\mathcal{G}'$ and (ii) imply that x_1 is not joined by an edge to a vertex $v \neq u$ of $\mathcal{G}'$. Hence $k \geq 2$ and the degree of x_1, and similarly of x_k, is 2. If $k \geq 3$, then it is easy to verify that the graph obtained from $\mathcal{G}$ by inserting the edge $\{u, x_3\}$ also satisfies (ii), contradicting our choice of $\mathcal{G}$. Hence $k = 2$. The graph obtained from $\mathcal{G}$ by inserting the edge $\{u, w\}$, if not already an edge of $\mathcal{G}$, also satisfies (ii). Thus, by our choice of $\mathcal{G}$, $\{u, w\}$ is an edge of $\mathcal{G}$. Therefore $\mathcal{G}$ is obtained from $\mathcal{G}'$ by the edge-join of a cycle of length 4, and (iii) holds by induction. $\qquad\Box$

Let $\mathcal{G}$ be a graph of order n for which there exists a sequence of subgraphs

$$\mathcal{G}_1, \mathcal{G}_2, \ldots, \mathcal{G}_k = \mathcal{G}$$

such that the vertices of $\mathcal{G}_i$ are $\{1, 2, \ldots, n_i\}$ where $\mathcal{G}_1$ is a graph with one vertex and $\mathcal{G}_{i+1}$ is either a vertex-join of $\mathcal{G}_i$ and an edge, or an edge-join of $\mathcal{G}_i$ and a cycle

$$p\text{---}q\text{---}n_i + 1\text{---}n_i + 2\text{---}p$$

of length 4 where $p < q$. Note that $n_1 = 1$, $n_k = n$, and $n_{i+1} - n_i = 1$ or 2. It follows inductively that the blocks of the graphs $\mathcal{G}_i$ are 4-cockades. Moreover, up to labeling of vertices, every connected graph whose blocks are 4-cockades can be obtained in this way. By Theorem 6.6.3, $\mathcal{G}$ is the graph of a combinatorially symmetric SNS-matrix with a negative main diagonal. Indeed, let A be the $(0, 1, -1)$-matrix of order n whose graph equals $\mathcal{G}$ and whose entries are 0 or 1 above the main diagonal, 0 or −1 below the main diagonal, and −1 on the main diagonal. Since each of the matrices

$$\begin{bmatrix} -1 & 1 \\ -1 & -1 \end{bmatrix}, \quad \begin{bmatrix} -1 & 1 & 0 & 1 \\ -1 & -1 & 1 & 0 \\ 0 & -1 & -1 & 1 \\ -1 & 0 & -1 & -1 \end{bmatrix}$$

is a SNS-matrix, it follows inductively using Theorem 6.2.7 (alternatively, using a simple inductive argument to show that all terms in the standard determinant expansion of A have the same sign) that A is an SNS-matrix.[6] We call a matrix A obtained in this way a *canonical combinatorially symmetric SNS-matrix with a negative main diagonal*. Theorems 6.6.3 and 6.1.3 imply the following characterization of combinatorially symmetric $(0, 1, -1)$ SNS-matrices with a negative main diagonal.

Corollary 6.6.4 *Up to simultaneous row and column permutations and multiplication of rows and columns by* −1, *the set of combinatorially symmetric* $(0, 1, -1)$ *SNS-matrices with a negative main diagonal is the set of matrices obtained from the canonical combinatorially symmetric SNS-matrices with a negative main diagonal by replacing symmetrically opposite entries off the main diagonal by zeros. In particular, if B is a combinatorially symmetric $(0, 1, -1)$ SNS-matrix with a negative main diagonal, then there exist a strict signing D and a permutation matrix P such that the entries of $PDBDP^T$ are 0 or* −1 *below the main diagonal,* −1 *on the main diagonal, and 0 or 1 above the main diagonal.*

We illustrate Corollary 6.6.4 for $n = 6$. It is easy to show that there are eleven connected graphs of order 6 whose blocks are 4-cockades. Of these, four of the graphs are not subgraphs of any of the others. These four graphs are the graphs of the canonical combinatorially symmetric SNS-matrices given on the following page:

---

[6]Note that the matrix

$$\begin{bmatrix} -1 & 1 & 1 & 0 \\ -1 & -1 & 0 & 1 \\ -1 & 0 & -1 & 1 \\ 0 & -1 & -1 & -1 \end{bmatrix}$$

is not an SNS-matrix and hence the requirement that we have $p < q$ in the cycle $p, q, n_i + 1, n_i + 2$ is essential.

$$
\begin{bmatrix}
-1 & 1 & 0 & 1 & 0 & 0 \\
-1 & -1 & 1 & 0 & 0 & 0 \\
0 & -1 & -1 & 1 & 0 & 1 \\
-1 & 0 & -1 & -1 & 1 & 0 \\
0 & 0 & 0 & -1 & -1 & 1 \\
0 & 0 & -1 & -1 & -1 & -1
\end{bmatrix},
\qquad
\begin{bmatrix}
-1 & 1 & 0 & 1 & 0 & 0 \\
-1 & -1 & 1 & 0 & 0 & 0 \\
0 & -1 & -1 & 1 & 1 & 1 \\
-1 & 0 & -1 & -1 & 0 & 0 \\
0 & 0 & 0 & -1 & -1 & 0 \\
0 & 0 & 0 & -1 & 0 & -1
\end{bmatrix},
$$

$$
\begin{bmatrix}
-1 & 1 & 0 & 1 & 1 & 0 \\
-1 & -1 & 1 & 0 & 0 & 0 \\
0 & -1 & -1 & 1 & 0 & 1 \\
-1 & 0 & -1 & -1 & 0 & 0 \\
-1 & 0 & 0 & 0 & -1 & 0 \\
0 & 0 & -1 & 0 & 0 & -1
\end{bmatrix},
\qquad
\begin{bmatrix}
-1 & 1 & 1 & 1 & 1 & 1 \\
-1 & -1 & 0 & 0 & 0 & 0 \\
-1 & 0 & -1 & 0 & 0 & 0 \\
-1 & 0 & 0 & -1 & 0 & 0 \\
-1 & 0 & 0 & 0 & -1 & 0 \\
-1 & 0 & 0 & 0 & 0 & -1
\end{bmatrix}.
$$

It follows from Corollary 6.6.4 that for each combinatorially symmetric $(0, 1, -1)$ SNS-matrix B of order 6 with a negative main diagonal there exist a permutation matrix P and a strict signing D such that $PDBDP^T$ can be obtained from one of these four canonical matrices by replacing certain symmetrically opposite entries by zero.

We now give a polynomial-time algorithm to decide whether or not a combinatorially symmetric matrix with a negative main diagonal is an SNS-matrix. A combinatorially symmetric matrix B with a negative main diagonal is an SNS-matrix if and only if each of the principal submatrices corresponding to the blocks of its graph $\mathcal{G}(B)$ is an SNS-matrix. Since there are polynomial-time algorithms to determine the connected components of a graph and their blocks, it suffices to assume that $\mathcal{G}(B)$ has only one block. The validity of this algorithm is a consequence of Theorem 6.6.3 and Corollary 3.2.2.

SNS-matrix recognition algorithm for combinatorially symmetric matrices with negative main diagonal

Let $B = [b_{ij}]$ be a combinatorially symmetric $(0, 1, -1)$-matrix of order $n \geq 2$ with a negative diagonal whose graph has only one block.

(0) If there exists $i \neq j$ such that $b_{ij} = b_{ji} \neq 0$, then B is not an SNS-matrix. Otherwise,

(1) let $W = [w_{ij}] = B$.

(2) If $\mathcal{G}(W)$ is an edge, then B is an SNS-matrix. Otherwise,

(3) if there are no vertices of degree 2 which are joined by an edge, then B is not an SNS-matrix. Otherwise,

(4) let u and v be vertices of degree 2 such that $\{u, v\}$ is an edge. Let $x \neq u$ and $y \neq v$ be vertices such that $\{x, u\}$ and $\{y, v\}$ are edges. If $x = y$, then B is not an SNS-matrix. Otherwise,

(5) if $\{x, y\}$ is an edge and $w_{xy} \neq w_{xu} w_{uv} w_{vy}$, then B is not an SNS-matrix. Otherwise,

(6) replace W by the matrix obtained from W by substituting $w_{xu}w_{uv}w_{yv}$ for w_{xy} and $-w_{xu}w_{uv}w_{yv}$ for w_{yx} and then deleting rows u and v and column u and v. Then go back to (2).

We conclude this section by investigating combinatorially symmetric SNS-matrices with a negative main diagonal whose graphs are not proper subgraphs of the graph of any combinatorially symmetric SNS-matrices with a negative main diagonal. Let B be a $(0, 1, -1)$-matrix with a negative main diagonal. Then B is a *maximal combinatorially symmetric SNS-matrix* provided it is a combinatorially symmetric SNS-matrix and no matrix obtained from B by replacing two symmetrically opposite zeros by $+1$ and -1 is an SNS-matrix. It follows from Corollary 6.1.5 that the graph of a maximal combinatorially symmetric SNS-matrix is connected. Theorem 6.1.3 implies that if B_1 and B_2 are two combinatorially symmetric SNS-matrices with the same graph, then B_1 is a maximal combinatorially symmetric SNS-matrix if and only if B_2 is. Thus the maximality property depends only on the graph. The matrices in (6.22), (6.23) are maximal combinatorially symmetric SNS-matrices. The first matrix in (6.22) shows that a maximal combinatorially symmetric SNS-matrix with a negative main diagonal need not be a maximal SNS-matrix.

Theorem 6.6.5 *Let B be a combinatorially symmetric $(0, 1, -1)$ SNS-matrix of order n with a negative main diagonal such that $\mathcal{G}(B)$ is 2-connected. Then the following are equivalent:*

(i) *B is a maximal combinatorially symmetric SNS-matrix.*

(ii) *$\mathcal{G}(B)$ is a 4-cockade.*

(iii) *n is even and the number of nonzero entries of B equals $4n - 4$.*

Proof Since the maximality property is determined by the graph of B, it follows from Theorem 6.2.7 and Theorem 6.6.3 that (i) and (ii) are equivalent. By Theorem 6.6.3, $\mathcal{G}(B)$ is a subgraph of a 4-cockade. Since the number of edges of a 4-cockade of order n is $(3n - 4)/2$, (ii) and (iii) are equivalent. □

We now characterize maximal combinatorially symmetric SNS-matrices with a negative main diagonal in terms of their graphs.

Theorem 6.6.6 *Let B be a combinatorially symmetric $(0, 1, -1)$ SNS-matrix of order $n \geq 2$ with a negative main diagonal. Then B is a maximal combinatorially symmetric SNS-matrix if and only if the graph of B satisfies the following three properties:*

(i) *$\mathcal{G}(B)$ is a connected graph,*

(ii) *each block is a 4-cockade, and*

(iii) *no two cut-vertices are joined by an edge.*

Proof First assume that B is a maximal combinatorially symmetric SNS-matrix. As already noted, (i) and (ii) hold. Suppose that u and v are cut-vertices which are joined by an edge $\{u, v\}$. Then there exist edges $\{u, x\}$ and $\{v, y\}$ which lie in different blocks $\mathcal{G}_1$ and $\mathcal{G}_2$. Let $\mathcal{H}$ be the graph obtained from $\mathcal{G}(B)$ by inserting the edge $\{x, y\}$. Then the blocks of $\mathcal{H}$ are the blocks of $\mathcal{G}(B)$ different from $\mathcal{G}_1, \mathcal{G}_2$, the edge $\{u, v\}$, and the 2-connected graph $\mathcal{H}_1$ obtained from $\mathcal{G}_1, \mathcal{G}_2$, and the edge $\{u, v\}$ by inserting the edge $\{x, y\}$. Since $\mathcal{G}_1$ and $\mathcal{G}_2$ are 4-cockades, it is easy to see that $\mathcal{H}_1$ is a 4-cockade. By Theorem 6.2.7, $\mathcal{H}$ is the graph of a combinatorially symmetric SNS-matrix with a negative main diagonal, contradicting the maximality assumption of B. Hence (iii) also holds.

Conversely, assume that (i), (ii), and (iii) hold. By (ii), $\mathcal{G}(B)$ is bipartite. Let u and v be two vertices of $\mathcal{G}(B)$ such that $\{u, v\}$ is not an edge, and let $\mathcal{H}$ be the graph obtained from $\mathcal{G}(B)$ by inserting the edge $\{u, v\}$. We show that $\mathcal{H}$ is not the graph of a combinatorially symmetric SNS-matrix with a negative main diagonal. If u and v have the same color, then $\mathcal{H}$ is not bipartite and by Lemma 6.6.1, the conclusion holds. If u and v are in the same block, then the conclusion holds by Corollary 6.6.5. Now suppose that u and v are not in the same block and that u and v have different colors. Let $u, x_1, \ldots, x_k, v$ be a path from u to v in $\mathcal{G}(B)$. Let $X = \{x_{i_1}, \ldots, x_{i_p}\}$ be the nonempty set of cut-vertices on this path different from u and v. First consider the case $p = 1$. Since $\mathcal{G}(B)$ is bipartite, either the block containing u and x_{i_1} or the block containing x_{i_1} and v is 2-connected. If only one of these blocks is 2-connected, say the one containing u, then u and x_{i_1} have the same color, and $\mathcal{G}(B)$ contains three paths joining u and x_{i_1} whose only common vertices are u and x_{i_1}, and the conclusion follows from Lemma 6.6.1. Suppose both of these blocks are 2-connected. Since u and v have different colors, we may assume without loss of generality that u and x_{i_1} have the same color. The 2-connectivity of the block of $\mathcal{G}(B)$ containing u and x_{i_1} implies that there exist two paths in $\mathcal{G}(B)$ whose common vertices are u and x_{i_1}. The connectivity of the block containing v and x_{i_1} implies that there exists a path in $\mathcal{G}(B)$ joining v and x_{i_1}. These three paths and the edge $\{u, v\}$ determine three paths joining u and x_{i_1} whose only common vertices are u and x_{i_1}. Hence the conclusion follows from (iii) of Lemma 6.6.1. Now consider the case $p > 1$. By (iii) the block $\mathcal{G}'$ containing x_{i_1} and x_{i_2} is 2-connected. By Corollary 6.6.5 and Theorem 6.6.3, the graph obtained from $\mathcal{G}'$ by inserting the edge $\{x_{i_1}, x_{i_2}\}$ contains an even subdivision of $K_{2,3}$. It is now easy to see that $\mathcal{H}$ contains an even subdivision of $K_{2,3}$. Hence the conclusion holds by Lemma 6.6.1. $\square$

An *end-block* of a graph is a block which does not contain more than one cut-vertex. It follows from (iii) of Theorem 6.6.6 that each bridge of the graph

Fig. 6.6.

of a maximally symmetric SNS-matrix with a negative main diagonal is an end-block.

We illustrate Theorem 6.6.6 with the matrices

$$B_1 = \begin{bmatrix} -1 & 1 & 1 & 0 & 1 & 0 \\ -1 & -1 & 0 & 1 & 0 & 1 \\ -1 & 0 & -1 & 1 & 0 & 0 \\ 0 & -1 & -1 & -1 & 0 & 0 \\ -1 & 0 & 0 & 0 & -1 & 0 \\ 0 & -1 & 0 & 0 & 0 & -1 \end{bmatrix}$$

and

$$B_2 = \begin{bmatrix} -1 & 1 & 1 & 0 & 1 & 0 \\ -1 & -1 & 0 & 1 & 0 & 0 \\ -1 & 0 & -1 & 1 & 0 & 0 \\ 0 & -1 & -1 & -1 & 0 & -1 \\ -1 & 0 & 0 & 0 & -1 & 0 \\ 0 & 0 & 0 & -1 & 0 & -1 \end{bmatrix}.$$

Each of these matrices is a combinatorially symmetric SNS-matrix. The graphs of B_1 and B_2 are drawn in Figure 6.6. It follows from Theorem 6.6.6 that B_2 but not B_1 is a maximal combinatorially symmetric SNS-matrix.

Bibliography

[1]   R. Aharoni, R. Manber, and B. Wajnryb. Special parity of perfect matchings in bipartite graphs, *Discrete Math.*, 79:221–8, 1990.

[2]   R.A. Brualdi. Counting permutations with restricted positions: Permanents of (0, 1)-matrices. A tale in four parts., in The 1987 Utah State University Department of Mathematics Conference Report by L. Beasley and E. E. Underwood, *Linear Alg. Appls.*, 104:173–83, 1988.

[3]   R.A. Brualdi and T. Foregger. Matrices with constant permanental minors, *Linear Multilin. Alg.*, 3:227–43, 1975.

[4]   R.A. Brualdi and H.J. Ryser. *Combinatorial Matrix Theory*, Cambridge University Press, New York, 1991.

[5]   R.A. Brualdi and B.L. Shader. On sign-nonsingular matrices and the conversion of the permanent into the determinant, in *Applied Geometry and Discrete Mathematics* (P. Gritzmann and B. Sturmfels, eds.), Amer. Math. Soc., Providence, 117–34, 1991.

[6]   R.A. Brualdi and B.L. Shader. Cutsets in bipartite graphs, *Linear Multilin. Alg.*, 34:51–4, 1993.

[7]   G. Engel and H. Schneider. Cyclic and diagonal products on a matrix, *Linear Alg. Appls.*, 7:301–35, 1973.

[8]   F. Harary, J.R. Lundgren, and J.S. Maybee. On signed digraphs with all cycles negative, *Discrete Appl. Math.*, 12:155–64, 1985.

[9]   P.W. Kasteleyn. The statistics of dimers on a lattice, *Physica*, 27:1209–25, 1961.

[10] P.W. Kasteleyn. Dimer statistics and phase transitions, *J. Math. Phys.*, 287–93, 1963.

[11] P. W. Kasteleyn. Graph theory and crystal physics, in *Graph Theory and Theoretical Physics* (Frank Harary, ed.), Academic Press, New York, 44–110, 1967.

[12] C.C. Lim. Nonsingular sign patterns and the orthogonal group, *Linear Alg. Appls.*, 184:1–12, 1993.

[13] C.H.C. Little. A characterization of convertible (0, 1)-matrices, *J. Combin. Theory, Ser. B*, 18:187–208, 1975.

[14] L. Lovász and M.D. Plummer. *Matching Theory*, Elsevier, Amsterdam, 1986.

[15] R. Lundgren and J.S. Maybee. A class of maximal *L*-matrices, *Congressus Numerantium*, 44:239–50, 1984.

[16] T. Lundy, J. Maybee, and J. van Buskirk. On maximal sign-nonsingular matrices, preprint.

[17] R. Manber and J. Shao. On digraphs with the odd-cycle property, *J. Graph Theory*, 10:155–65, 1986.

[18] G. Pólya. Aufgabe 424, *Arch. Math. Phys.*, 20:271, 1913.

[19] H.J. Ryser. Indeterminates and incidence matrices, *Linear Multilin. Alg.*, 1:149–57, 1973.

[20] P. Seymour and C. Thomassen. Characterization of even directed graphs, *J. Combin. Theory, Ser. B*, 42:36–45, 1987.

[21] B.L. Shader. Maximal convertible matrices, *Congressus Numerantium*, 81:161–72, 1991.

[22] B.L. Shader. Convertible, nearly decomposable and nearly reducible matrices, *Linear Alg. Appls.*, 184:37–53, 1993.

[23] R. Sinkhorn and P. Knopp. Problems involving diagonal products in nonnegative matrices, *Trans. Amer. Math. Soc.*, 136:67–75, 1969.

[24] G. Szegö. Lösung zu Aufgabe 424, *Arch. Math. Phys.*, 21:291, 1913.

[25] C. Thomassen. Sign-nonsingular matrices and even cycles in directed graphs, *Linear Alg. Appls.*, 75:27–41, 1986.

[26] L.G. Valiant. The complexity of computing the permanent, *Theoretical Computer Science*, 8:189–201, 1979.

[27] V. Vazirani and M. Yannakakis. Pfaffian orientations, 0–1 permanents and even cycles in directed graphs, *Discrete Appl. Math.*, 25:179–90, 1989.

7

S^2NS-matrices

7.1 Fully indecomposable S^2NS-matrices

We recall from section 1.2 that an S^2NS-matrix is an SNS-matrix A for which the sign pattern of A^{-1} is uniquely determined by the sign pattern of A. Clearly, if A is an S^2NS-matrix, then so is DAE for all strict signings D and E. By Corollary 1.2.8, an SNS-matrix is an S^2NS-matrix if and only if each submatrix of order $n-1$ either has an identically zero determinant or is an SNS-matrix. By Corollary 3.2.3, a matrix A with a negative main diagonal is an S^2NS-matrix if and only if every directed cycle of the digraph $\mathcal{D}(A)$ is negative and the sign of α equals the sign of β for every pair of paths α and β with the same initial vertex and the same terminal vertex. It follows that if A is an S^2NS-matrix of order n and γ and δ are subsets of $\{1, 2, \ldots, n\}$ with the same number of elements, then either one of $A[\gamma, \delta]$ and $A(\gamma, \delta)$ has an identically zero determinant or both are S^2NS-matrices. As with SNS-matrices, when we discuss S^2NS-matrices there is no loss of generality in considering only $(0, 1, -1)$-matrices.

Let A be a square matrix whose determinant is not identically zero. Then A is an SNS-matrix if and only if each of its fully indecomposable components is an SNS-matrix. However, there exist matrices A such that each fully indecomposable component of A is an S^2NS-matrix and yet A is not an S^2NS-matrix. For example, the two fully indecomposable components of the matrix

$$A = \begin{bmatrix} -1 & 0 & 0 \\ -1 & -1 & 1 \\ -1 & -1 & -1 \end{bmatrix} \tag{7.1}$$

are S^2NS-matrices, but A is not an S^2NS-matrix since $A[\{2, 3\}, \{1, 2\}]$ is not an SNS-matrix.

Let A be a $(0, 1)$-matrix of order n. It follows from Theorem 6.1.3 that if A is fully indecomposable, then up to multiplication of rows and columns by -1,

168

there is at most one S²NS-matrix in $\mathcal{Z}(A)$. This conclusion need not hold if A is partly decomposable. Indeed, let k and ℓ be positive integers, and let

$$A = \begin{bmatrix} I_\ell & O \\ J_{k,\ell} & I_k \end{bmatrix}$$

and

$$\widehat{A} = \begin{bmatrix} I_\ell & O \\ B & I_k \end{bmatrix}$$

where $\widehat{B}$ is a k by ℓ $(1, -1)$-matrix. Then $\widehat{A}$ is an S²NS-matrix whose inverse has sign pattern

$$\begin{bmatrix} I_\ell & O \\ -\widehat{B} & I_k \end{bmatrix}.$$

It follows that up to multiplication of rows and columns by -1, there are $2^{k\ell-k-\ell+1}$ S²NS-matrices in $\mathcal{Z}(A)$.

The inverse of an S²NS-matrix A contains no zeros if and only if A is fully indecomposable. We consider fully indecomposable matrices in this section and partly decomposable matrices in section 7.3. We first prove the following structure theorem for fully indecomposable matrices [1].

Lemma 7.1.1 *Let A be a fully indecomposable $(0, 1)$-matrix of order $n \geq 2$. Then there exist a positive integer $m < n$ and permutation matrices P and Q such that PAQ has the form*

$$\begin{bmatrix} A' & & & G & & \\ \hline & 1 & 1 & 0 & \cdots & 0 & 0 \\ & 0 & 1 & 1 & \cdots & 0 & 0 \\ & 0 & 0 & 1 & \cdots & 0 & 0 \\ F & \vdots & \vdots & \vdots & \ddots & \vdots & \vdots \\ & 0 & 0 & 0 & \cdots & 1 & 1 \\ & 0 & 0 & 0 & \cdots & 0 & 1 \end{bmatrix}, \qquad (7.2)$$

where A' is a fully indecomposable matrix of order m, the only nonzero entries of F are in its last row, and the only nonzero entries of G are in its first column.

Proof Let $A' = A[\alpha, \beta]$ be a fully indecomposable proper submatrix of A of maximal order $m < n$ such that $A(\alpha, \beta)$ does not have an identically zero determinant. Permutations of rows and columns of A allow us to assume that

$\alpha = \beta = \{1, 2, \ldots, m\}$ and $I_{n-m} \leq A(\alpha)$. The full indecomposability of A implies that there exists a nonzero entry in $G = A[\alpha, \overline{\alpha}]$ and this nonzero entry is contained in a nonzero term in the standard determinant expansion of A. This nonzero term and the maximal property of A' imply that after simultaneous permutations of the last $n - m$ rows and columns we may assume that

$$\begin{bmatrix} 1 & 1 & 0 & \cdots & 0 & 0 \\ 0 & 1 & 1 & \cdots & 0 & 0 \\ 0 & 0 & 1 & \cdots & 0 & 0 \\ \vdots & \vdots & \vdots & \ddots & \vdots & \vdots \\ 0 & 0 & 0 & \cdots & 1 & 1 \\ 0 & 0 & 0 & \cdots & 0 & 1 \end{bmatrix} \leq A(\alpha),$$

G has a nonzero entry in its first column, and $F = A[\overline{\alpha}, \alpha]$ has a nonzero entry in its last row. The maximal property of A' now immediately implies that the only nonzero entries of F and G are in their last row and first column, respectively, and that the entries of $A(\alpha)$ in positions (i, j) with $j \geq i + 2$ all equal zero. Suppose that $A(\alpha)$ has a nonzero entry in position (p, q) of A where $p > q$. Let

$$\gamma = \{1, 2, \ldots, q-1, p, p+1, \ldots, n\} \text{ and } \delta = \{1, 2, \ldots, q, p+1, p+2, \ldots, n\}.$$

Then $A[\gamma, \delta]$ is a fully indecomposable proper submatrix of A such that $A(\gamma, \delta)$ does not have an identically zero determinant, contradicting the choice of A'. Thus all of the entries of $A(\alpha)$ in positions (p, q) with $p > q$ equal zero. □

The following corollary is an immediate consequence of Lemma 7.1.1, and it shows that the first conclusion of Lemma 6.5.2 holds under a weaker hypothesis.

Corollary 7.1.2 *Let $A = [a_{ij}]$ be a fully indecomposable $(0, 1)$-matrix of order $n \geq 2$. If A has a row of all 1's or if each 1 of A is contained in a row or column with at least three 1's, then there exist integers k and ℓ such that $a_{k\ell} = 1$ and $A(\{k\}, \{\ell\})$ is fully indecomposable.*

Let A be a fully indecomposable $(0,1)$-matrix of order $n \geq 2$. In section 6.1 we took a permutation matrix $R \leq A$, replaced A by $R^T A$ to get all 1's on the main diagonal, and then used the digraph to obtain a recursive fully indecomposable construction (6.3) of A. For each matrix A_i in (6.3) there exists a subset α_i of $\{1, 2, \ldots, n\}$ such that $A_i \leq A[\alpha_i]$. Lemma 7.1.1 implies that there is a choice of R that gives another type of recursive fully indecomposable construction of A, namely,

$$I_1 = A'_1, A'_2, \ldots, A'_p = A,$$

where A_i' is a proper submatrix of A_{i+1}' and A_{i+1}' is obtained from A_i' as in Lemma 7.1.1 ($i = 1, 2, \ldots, p - 1$).

The next lemma relates fully indecomposable S^2NS-matrices to S^*-matrices and SNS*-matrices.

Lemma 7.1.3 *Let A be a fully indecomposable matrix of order n. Then the following are equivalent:*

(i) *A is an S^2NS-matrix.*

(ii) *A is an SNS-matrix, and each matrix obtained from A by deleting a row is an S^*-matrix.*

(iii) *For each $k = 1, 2, \ldots, n$ an SNS*-matrix can be obtained from A by replacing each nonzero entry in row k by 1 or -1.*

Proof The equivalence of (i) and (ii) follows from the full indecomposability of A. Clearly, (iii) implies (i). Now assume that A is an S^2NS-matrix, and let k be an integer with $1 \leq k \leq n$. For each j such that $a_{kj} = 0$, we replace a_{kj} by the sign of $(-1)^{i+j} \det A \det A(\{i\}, \{j\})$ and obtain an SNS*-matrix. Hence (i) implies (iii). □

The following corollary is contained in [5].

Corollary 7.1.4 *A fully indecomposable S^2NS-matrix of order $n \geq 2$ has at least two rows, each of which contains exactly two nonzero entries.*

Proof The corollary follows immediately from Lemma 7.1.3 and Theorem 4.1.1. □

Corollary 7.1.5 *A conformal contraction of an S^2NS-matrix is an S^2NS-matrix.*

Proof The corollary is a consequence of (i) and (iii) of Lemma 4.2.1 and Lemma 7.1.3. □

It follows from (iii) of Lemma 7.1.3 and Theorem 4.2.4 that if A is a fully indecomposable S^2NS-matrix, then there exists a strict signing D such that $\pm J_1$ can be obtained from AD by successive conformal contractions on rows. Thus the zero pattern of a fully indecomposable S^2NS-matrix can be contracted to J_1. Note that the lower Hessenberg matrices H_n defined in (1.1.11) can be conformally contracted on rows to $-J_1$ but are not S^2NS-matrices for $n \geq 3$.

We now give two different methods for constructing S^2NS-matrices from trees.

Theorem 7.1.6 *Let $A = [a_{ij}]$ be a combinatorially symmetric matrix with a negative main diagonal. Then A is a fully indecomposable S^2NS-matrix if and only if the graph $\mathcal{G}(A)$ is a tree and $a_{ij}a_{ji} \leq 0$ for all $i \neq j$.*

Proof First assume that $\mathcal{G}(A)$ is a tree and that $a_{ij}a_{ji} \leq 0$ for all $i \neq j$. Since $\mathcal{G}(A)$ is connected, A is fully indecomposable. The fact that $\mathcal{G}(A)$ is a tree implies that the only directed cycles of $\mathcal{D}(A)$ have length equal to 2 and that there is a unique path in $\mathcal{D}(A)$ from any given vertex to any other. Since $a_{ij}a_{ji} \leq 0$ for all $i \neq j$, it follows from Corollary 3.2.3 that A is an S^2NS-matrix.

Now assume that A is a fully indecomposable S^2NS-matrix. Then A is an SNS-matrix with a negative main diagonal and $a_{ij}a_{ji} \leq 0$ for all $i \neq j$. By Corollary 7.1.4, there is a row of A with exactly two nonzero entries and hence some vertex of $\mathcal{G}(A)$, say vertex 1, is a pendant vertex. The matrix $A(\{1\})$ is a fully indecomposable S^2NS-matrix, and the fact that $\mathcal{G}(A)$ is a tree now follows by induction. □

Theorem 7.1.7 *Let A be a $(0, 1, -1)$-matrix of order n each of whose elements in its last row equals 1. Then A is a fully indecomposable S^2NS-matrix if and only if the matrix obtained from A by deleting its last row is an oriented edge-vertex incidence matrix of a tree.*

Proof First assume that the matrix obtained from A by deleting its last row is an oriented edge-vertex incidence matrix of a tree. Then the full indecomposability of A is a consequence of the fact that the edge-vertex incidence matrix of a tree of order n does not have a p by q zero submatrix for any positive integers p and q with $p+q = n$. By (ii) of Lemma 7.1.3, it suffices to show that the matrix A_i obtained from A by deleting row i is an S^*-matrix for each $i = 1, 2, \ldots, n$. By (i) of Theorem 4.4.1, an oriented incidence matrix of a tree is an S-matrix, and hence A_n is an S-matrix. Now assume that $1 \leq i \leq n - 1$. Then by permuting rows and columns we may assume that

$$
A_i = \begin{bmatrix}
X_i & O \\
O & Y_i \\
1 \cdots 1 & 1 \cdots 1
\end{bmatrix}
$$

where X_i and Y_i are oriented edge-vertex incidence matrices of trees of orders k_i and $n - k_i$, respectively, and hence are S-matrices. Let $D = E \oplus F$ be a signing such that each row of $A_i D$ is balanced where E has order k_i. Since X_i and Y_i are S-matrices, E equals $I_{k_i}, -I_{k_i}$ or O, and F equals $I_{n-k_i}, -I_{n-k_i}$, or O. It follows that $D = \pm(I_{k_i} \oplus (-I_{n-k_i}))$, and hence by (iv) of Theorem 2.1.1, A_i is an S^*-matrix.

Now assume that $A = [a_{ij}]$ is a fully indecomposable S^2NS-matrix. We show that A_n is an oriented incidence matrix of a tree by induction on n. If $n \leq 2$, this is clear. Suppose that $n > 2$. By Corollary 7.1.2, there exist integers k and ℓ such that $a_{k\ell} = 1$ and $A(\{k\}, \{\ell\})$ is fully indecomposable. It follows from Lemma 7.1.3 and Theorem 4.1.2 that row k and column ℓ of A each contain exactly two nonzero entries, in particular $k \neq n$. By the induction hypothesis $A(\{k, n\}, \{\ell\})$ is an oriented edge-vertex incidence matrix of a tree. Since A is an SNS-matrix and each of the entries in its last row is positive, the two nonzero entries in row k have opposite sign, and it follows that A is an oriented edge-vertex incidence matrix of a tree. $\qquad\square$

If in Theorems 7.1.6 and 7.1.7 we take a path of length 5, then we obtain the S^2NS-matrices

$$
\begin{bmatrix}
-1 & 1 & 0 & 0 & 0 \\
-1 & -1 & 1 & 0 & 0 \\
0 & -1 & -1 & 1 & 0 \\
0 & 0 & -1 & -1 & 1 \\
0 & 0 & 0 & -1 & -1
\end{bmatrix}
\quad \text{and} \quad
\begin{bmatrix}
-1 & 1 & 0 & 0 & 0 \\
0 & -1 & 1 & 0 & 0 \\
0 & 0 & -1 & 1 & 0 \\
0 & 0 & 0 & -1 & 1 \\
1 & 1 & 1 & 1 & 1
\end{bmatrix},
$$

respectively.

By Lemma 7.1.3, every fully indecomposable S^2NS-matrix can be embedded in an SNS^*-matrix of the same order. Conversely, the next corollary shows that every SNS^*-matrix has a fully indecomposable S^2NS-matrix (indeed one which is also an SNS^*-matrix) of the same order embedded in it.

Corollary 7.1.8 *Let A be an $(0, 1, -1)$ SNS^*-matrix of order n. Then there exists a fully indecomposable S^2NS-matrix which can be obtained from A by replacing certain nonzero entries by zeros. Indeed, if the last row of A contains only 1's, then any matrix B obtained by replacing nonzero entries by zeros in rows $1, 2, \ldots, n - 1$ of A so as to leave exactly one 1 and one -1 in each of these rows is a fully indecomposable S^2NS-matrix.*

Proof By permuting rows and multiplying certain columns by -1, we may assume without loss of generality that each entry of the last row of A equals 1.

By Theorem 4.1.4, the matrix obtained from A by deleting the last row is an S-matrix. Hence each of rows $1, 2, \ldots, n-1$ of A contains both a 1 and a -1. Choosing both a 1 and -1 in each of these rows, we obtain a matrix B as described in the corollary. It follows from Theorem 4.4.4 that the matrix $B[\{1, 2, \ldots, n-1\}, :]$ is an oriented edge-vertex incidence matrix of a tree. By Theorem 7.1.6, B is a fully indecomposable S^2NS-matrix. $\qquad\square$

The following example shows that a fully indecomposable SNS-matrix need not have a fully indecomposable S^2NS-matrix of the same order embedded in it. Let

$$A = \begin{bmatrix} -1 & 0 & 0 & 0 & 0 & 0 & -1 & 0 \\ 1 & -1 & 0 & 0 & 1 & 0 & 0 & 0 \\ 0 & 0 & -1 & 0 & 0 & 0 & 0 & 1 \\ 0 & 0 & 0 & -1 & 0 & 0 & 1 & 0 \\ 0 & 0 & 0 & 0 & -1 & 1 & 0 & 0 \\ 0 & 0 & -1 & -1 & 0 & -1 & 0 & 0 \\ 0 & 0 & 0 & 0 & 0 & 0 & -1 & 1 \\ 0 & 1 & 0 & 0 & 0 & 0 & 0 & -1 \end{bmatrix}.$$

Then A is a fully indecomposable SNS-matrix. In the signed digraph $\mathcal{D}(A)$ there is a positive path and a negative path from vertex 6 to vertex 7, and hence A is not an S^2NS-matrix. Since each nonzero entry of A is contained in a row or a column with exactly two nonzero entries, A is nearly decomposable. Hence there is no fully indecomposable S^2NS-matrix of the same order embedded in A (although, of course, there is a partly decomposable S^2NS-matrix of the same order embedded in A).

Let $A = [a_{ij}]$ be a matrix of order n, and suppose that $a_{pq} \neq 0$. Let B be the matrix of order $n + 1$ obtained by bordering A as shown:

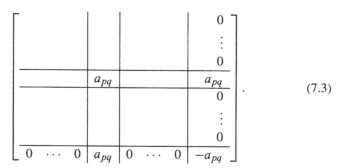 (7.3)

We say that B is the matrix obtained from A by *conformally copying the entry* a_{pq}. More generally, we say that a matrix B can be obtained from A by *conformally copying entries* provided that there is a sequence of matrices $A = A_0, A_1, \ldots, A_k = B$ $(k \geq 0)$ such that A_i can be obtained by conformally copying an entry of A_{i-1} $(i = 1, 2, \ldots, k)$.

Lemma 7.1.9 *Let B be a matrix obtained from a $(0, 1, -1)$-matrix A by conformally copying entries. Then B is fully indecomposable if and only if A is, and B is an S^2NS-matrix if and only if A is.*

Proof The assertion that B is fully indecomposable if and only if A is follows from (ii) of Lemma 4.2.1. By Corollary 7.1.5, if B is an S^2NS-matrix, so is A. Now suppose that $A = [a_{ij}]$ is an S^2NS-matrix of order n and B is the matrix of order $n + 1$ obtained from A by copying the entry a_{pq} as shown in (7.3). Without loss of generality we assume that A has a negative main diagonal. Let B' be the matrix obtained from B by multiplying its last column by a_{pq}. Then B' has a negative main diagonal. The signed digraph $\mathcal{D}(B')$ is obtained from the signed digraph $\mathcal{D}(A)$ by inserting a new vertex $n + 1$, a positive arc $(p, n + 1)$, and an arc $(n + 1, q)$ with sign a_{pq}. The assertion that B is an S^2NS-matrix now follows from Corollary 3.2.3. $\qquad\square$

It follows from Lemma 7.1.9 that each matrix of order n which can be obtained from $\pm J_1$ by conformally copying entries is a fully indecomposable S^2NS-matrix with exactly $3n - 2$ nonzero entries. For example, the matrix

$$\begin{bmatrix} 1 & 1 & 1 & 1 & 0 \\ 1 & -1 & 0 & 0 & 0 \\ 0 & 1 & -1 & 0 & -1 \\ 0 & 0 & 1 & -1 & 0 \\ 0 & 0 & -1 & 0 & 1 \end{bmatrix}$$

can be obtained from J_1 by conformally copying entries and hence is a fully indecomposable S^2NS-matrix.

A $(0, 1, -1)$-matrix A is a *maximal S^2NS-matrix* provided that it is an S^2NS-matrix and no matrix obtained from A by replacing a 0 by a 1 or a -1 is an S^2NS-matrix. We now characterize fully indecomposable, maximal S^2NS-matrices [1].

Theorem 7.1.10 *Up to row and column permutations and multiplication of rows and columns by -1, the fully indecomposable, maximal $(0, 1, -1)$ S^2NS-matrices are precisely the matrices that can be obtained from J_1 by conformally copying entries.*

Proof Let $A = [a_{ij}]$ be a matrix of order n which is obtained from J_1 by conformally copying entries. By Lemma 7.1.9, A is a fully indecomposable S^2NS-matrix, and we show by induction on n that A is maximal. If $n = 1$, this is clear. Now assume that $n > 1$. Without loss of generality we also assume that A is obtained from $A(\{n\})$ by conformally copying $a_{n-1,n-1}$. By the induction hypothesis, $A(\{n\})$ is a fully indecomposable, maximal S^2NS-matrix. Hence no matrix obtained from A by replacing a zero of $A(\{n\})$ by a 1 or a -1 is an S^2NS-matrix. By Lemma 7.1.3, the matrix $A[\{1, 2, \ldots, n - 1\}, :]$ is an S^*-matrix. Since $A(\{n\})$ is fully indecomposable, it follows from Theorem 4.1.2 and Lemma 7.1.3 that no matrix obtained from A by replacing a zero in its last column by 1 or -1 is an S^2NS-matrix. A similar argument shows that the same conclusion holds for a zero in the last row of A. Therefore A is a maximal S^2NS-matrix.

Now let $A = [a_{ij}]$ be a fully indecomposable, maximal S^2NS-matrix of order n. We prove by induction on n that up to row and column permutations and multiplication of rows and columns by -1, A can be obtained from J_1 by conformally copying entries. If $n = 1$, this is clear. Now assume that $n \geq 2$. We apply Lemma 7.1.1 to the zero pattern of A. By multiplying certain rows and columns by -1, we may assume that A has the form

$$
\begin{bmatrix}
A' & & & & G & & \\
\hline
 & -1 & 1 & 0 & \cdots & 0 & 0 \\
 & 0 & -1 & 1 & \cdots & 0 & 0 \\
 & 0 & 0 & -1 & \cdots & 0 & 0 \\
F & \vdots & \vdots & \vdots & \ddots & \vdots & \vdots \\
 & 0 & 0 & 0 & \cdots & -1 & 1 \\
 & 0 & 0 & 0 & \cdots & 0 & -1
\end{bmatrix}, \qquad (7.4)
$$

where A' is a fully indecomposable matrix of order $m < n$, the only nonzero entries of G are in its first column, and the only nonzero entries of F are in its last row. By Lemma 7.1.3, the matrix $A[\{1, 2, \ldots, n - 1\}, :]$, and hence the matrix $A[\{1, 2, \ldots, m\}, \{1, 2, \ldots, m + 1\}]$ is an S^*-matrix. Since A' is fully indecomposable, Theorem 4.1.2 implies that the first column of G and the last row of F each contain exactly one nonzero entry. Without loss of generality we assume that these nonzero entries occur in row m and column m, respectively. It follows from Lemma 7.1.9 and the maximality of A that $a_{mm} = a_{m,m+1}a_{nm}$. Thus A is the join of the fully indecomposable S^2NS-matrices A' and $A'' = A[\{m, m + 1, \ldots, n\}]$ at a matrix of order 1. Since A

is a maximal S^2NS-matrix, so is A'' and hence $n - m + 1 = 2$. Therefore A can be obtained from A' by conformally copying a_{mm}, and the conclusion now follows by induction. □

Corollary 7.1.11 *Every fully indecomposable, maximal S^2NS-matrix of order n has exactly $3n - 2$ nonzero entries.*

Corollary 7.1.12 *Up to row and column permutations and multiplication of rows and columns by -1, the fully indecomposable $(0, 1, -1)$ S^2NS-matrices are precisely the fully indecomposable matrices that can be obtained from J_1 by conformally copying entries and then replacing some of the nonzero entries by zeros.*

Corollary 7.1.13 *A fully indecomposable S^2NS-matrix of order $n \geq 2$ contains at least one nonzero entry whose row and column each contain exactly two nonzero entries.*

An inductive argument shows that there are at least two such entries satisfying the conclusions of Corollary 7.1.13.

The next corollary follows easily by induction using Theorem 7.1.10 and Lemma 4.2.1.

Corollary 7.1.14 *Every square submatrix of a fully indecomposable S^2NS-matrix is an SNS-matrix or has an identically zero determinant, that is, every fully indecomposable S^2NS-matrix of order n has a signed rth compound for each $r = 1, 2, \ldots, n$.*

The validity of the following polynomial-time algorithm for recognizing fully indecomposable S^2NS-matrices is a consequence of Lemma 7.1.9 and Theorem 7.1.10.

S^2NS-matrix recognition algorithm for fully indecomposable matrices

Let $B = [b_{ij}]$ be a fully indecomposable $(0, 1, -1)$-matrix.

(0) Let $W = [w_{ij}] = B$.
(1) If the order of W is 1, then B is an S^2NS-matrix. Otherwise,
(2) if there do not exist integers i and j such that $b_{ij} \neq 0$ and row i and column j each have exactly two nonzero entries, then B is not an S^2NS-matrix. Otherwise,
(3) let i, j, p and q be integers such that b_{ij} and b_{iq} are the only nonzero entries in row i, and b_{ij} and b_{pj} are the only nonzero entries in column j. If $b_{pq} \neq 0$ and $b_{pq} \neq -b_{ij}b_{iq}b_{pj}$, then B is not an SNS-matrix. Otherwise,

(4) replace W by the matrix obtained from W by substituting $-b_{ij}b_{iq}b_{pj}$ for b_{pq} and then deleting row i and column j. Then go back to (1).

7.2 Zero patterns of S^2NS-matrices

In this section we characterize zero patterns of fully indecomposable S^2NS-matrices in terms of bipartite graphs and of fully indecomposable S^2NS-matrices with a negative main diagonal in terms of digraphs. We first characterize zero patterns of fully indecomposable, maximal S^2NS-matrices [1].

Theorem 7.2.1 *Let A be a $(0, 1)$-matrix of order n. Then A is the zero pattern of a fully indecomposable, maximal S^2NS-matrix if and only if the bipartite graph of A is a 4-cockade.*

Proof Let B be obtained from a matrix C by conformally copying an entry. Then the bipartite graph of B is an edge-join of the bipartite graph of C with a cycle of length 4. Hence it follows from Theorem 7.1.10 and the definition of a 4-cockade that if A is the zero pattern of a fully indecomposable, maximal S^2NS-matrix, then its bipartite graph is a 4-cockade. Conversely, assume that the bipartite graph of A is a 4-cockade. Then there is a matrix $\widehat{A}$ in $\mathcal{Z}(A)$ which is obtained from J_1 by conformally copying entries. By Theorem 7.1.10, $\widehat{A}$ is a maximal S^2NS-matrix. ☐

We now characterize zero patterns of fully indecomposable S^2NS-matrices.

Theorem 7.2.2 *Let A be a $(0, 1)$-matrix of order n. Then the following are equivalent:*

 (i) *A is the zero pattern of a fully indecomposable S^2NS-matrix.*

 (ii) *The bipartite graph of A is a connected spanning subgraph of a 4-cockade, and each edge is contained in a perfect matching.*

 (iii) *The bipartite graph of A is connected, does not contain an even subdivision of $K_{2,3}$, and each edge is contained in a perfect matching.*

Proof It follows from the Frobenius–König theorem that A is fully indecomposable if and only if the bipartite graph of A is connected and each edge is contained in a perfect matching. The bipartite graph of a fully indecomposable matrix has no cut vertices and hence has only one block. The equivalence of (i) and (ii) is now a consequence of Theorem 7.2.1, and the equivalence of (ii) and (iii) is a consequence of Theorem 6.6.3. ☐

Corollary 7.2.3 *Let A be an m by n $(0, 1)$-matrix with $m \leq n$. Then the following are equivalent:*

(i) *A is the zero pattern of a matrix which for each $r = 1, 2, \ldots, m$ has a signed rth compound.*

(ii) *The blocks of the connected components of the bipartite graph of A are spanning subgraphs of 4-cockades.*

(iii) *The bipartite graph of A does not contain an even subdivision of $K_{2,3}$.*

(iv) *The bipartite graph of A is the graph of a combinatorially symmetric SNS-matrix of order $m + n$ with a negative main diagonal.*

Proof The equivalence of (ii), (iii), and (iv) is a consequence of Theorem 6.6.3. Consider a k by $k + 1$ or $k + 1$ by k $(0, 1)$-matrix B whose bipartite graph is an even subdivision of $K_{2,3}$. Then B can be contracted to $J_{2,3}$ or $J_{3,2}$. By Lemma 4.3.3 and Corollary 4.3.7, there does not exist a matrix in $\mathcal{Z}(B)$ with a signed kth compound, and hence (i) implies (iii).

A matrix $\widehat{A} = [\hat{a}_{ij}]$ in $\mathcal{Z}(A)$ has a signed rth compound for all $r = 1, 2, \ldots, m$ if and only if for each cycle $i_1, j_1, i_2, j_2, \ldots, i_p, j_p, i_1$ of the bipartite graph of A we have $\hat{a}_{i_1 j_1}\hat{a}_{j_1 i_2}\hat{a}_{i_2 j_2} \cdots \hat{a}_{i_p j_p}\hat{a}_{j_p i_1} = (-1)^{p-1}$. Since each cycle is contained in a block, the fact that (ii) implies (i) is a consequence of Theorem 7.2.1 and Corollary 7.1.14. □

The S^2NS-matrix recognition algorithm for fully indecomposable matrices given at the end of section 7.1 can be reversed to give a recursive algorithm for finding an S^2NS-matrix with a prescribed zero pattern.

Algorithm for the construction of an S^2NS-matrix with a prescribed fully indecomposable zero pattern

Let $A = [a_{ij}]$ be a fully indecomposable $(0, 1)$-matrix.

(1) If the order of A is 1, then A is an S^2NS-matrix. Otherwise,
(2) if there do not exist integers i and j such that $a_{ij} \neq 0$ and row i and column j each have exactly two nonzero entries, then A is not the zero pattern of an S^2NS-matrix. Otherwise,
(3) let k, ℓ, p, and q be integers such that $a_{k\ell}$ and a_{kq} are the only nonzero entries in row k, and $a_{k\ell}$ and $a_{p\ell}$ are the only nonzero entries in column ℓ. Let B be the matrix obtained from A by contracting on $a_{k\ell}$. Recursively apply the algorithm to B. If B is not the zero pattern of an S^2NS-matrix, then A is not the zero pattern of an S^2NS-matrix. Otherwise,
(4) let $\widehat{B} = [\hat{b}_{ij}]$ be an S^2NS-matrix whose zero pattern equals B with rows labeled $1, \ldots, k - 1, k + 1, \ldots, n$ and columns labeled $1, \ldots, \ell - 1, \ell + 1, \ldots, n$. Let $\widehat{A} = [\hat{a}_{ij}]$ be the matrix with zero pattern A such that $\hat{a}_{ij} = \hat{b}_{ij}$ if $i \neq k$

and $j \neq \ell$, $\hat{a}_{kl} = -\hat{b}_{pq}$, $\hat{a}_{kq} = \hat{a}_{p\ell} = 1$, and $\hat{a}_{pq} = a_{pq}\hat{b}_{pq}$. Then $\widehat{A}$ is an S^2NS-matrix.

We now investigate the digraphs of fully indecomposable S^2NS-matrices with a negative main diagonal. As we did for bipartite graphs, we obtain both a forbidden configuration characterization and a recursive description. First we prove the following lemma.

Lemma 7.2.4 *Let A be a fully indecomposable $(0, 1)$-matrix such that $\mathcal{Z}(A)$ contains an SNS-matrix. Then $\mathcal{Z}(A)$ contains an S^2NS-matrix if and only if $\mathcal{Z}(A')$ contains an SNS-matrix for each matrix A' obtained from A by changing a 0 to a 1.*

Proof Let $\widehat{A}$ be an SNS-matrix in $\mathcal{Z}(A)$ and let A' be a matrix obtained from A by changing a 0 in position (r, s) to a 1. By Corollary 6.1.4, $\widehat{A}(\{r\}, \{s\})$ is an SNS-matrix if and only if $\mathcal{Z}(A')$ contains an SNS-matrix. The lemma now follows from Corollary 1.2.8.   □

Let $\mathcal{D}^3$ be the digraph of order 3 with arcs $(1, 2)$, $(2, 1)$, $(1, 3)$, and $(2, 3)$. It is easy to see that there is no signing of $\mathcal{D}^3$ such that the directed cycle $1 \rightarrow 2 \rightarrow 1$ is negative, both of the paths from 1 to 3 have the same sign, and both of the paths from 2 to 3 have the same sign. Hence $\mathcal{D}^3$, indeed any subdivision of $\mathcal{D}^3$, is not a subdigraph of the digraph of an S^2NS-matrix with a negative main diagonal. Let $\mathcal{D}_1$ be a digraph and let $\mathcal{D}_2$ be obtained from $\mathcal{D}_1$ either by splitting an arc or by splitting a vertex. It is easy to see that $\mathcal{D}_1$ contains a subdivision of $\mathcal{D}^3$ if and only if $\mathcal{D}_2$ does. Similar conclusions hold for the digraph $\mathcal{D}'^3$ obtained from $\mathcal{D}^3$ by reversing the direction of each of its arcs. Either of the digraphs $\mathcal{D}^3$ and $\mathcal{D}'^3$ provides a forbidden configuration characterization of digraphs of fully indecomposable S^2NS-matrices with a negative main diagonal [5].

Theorem 7.2.5 *Let A be a fully indecomposable $(0, 1)$-matrix with $I_n \leq A$. Then the following are equivalent.*

  (i) *$\mathcal{Z}(A)$ contains an S^2NS-matrix.*
 (ii) *$\mathcal{D}(A)$ does not contain a subdivision of $\mathcal{D}^3$.*
(iii) *$\mathcal{D}(A)$ does not contain a subdivision of $\mathcal{D}'^3$.*

Proof As already noted, (i) implies (ii) and (iii). Now assume that (i) does not hold. By Lemma 7.2.4, either $\mathcal{Z}(A)$ does not contain an SNS-matrix or $\mathcal{Z}(A')$ does not contain an SNS-matrix for some matrix A' obtained from A by changing a 0 to a 1. Hence, by Corollary 6.5.9, $\mathcal{D}(A)$ contains all but at

most one arc of a splitting of an odd, doubly directed cycle. A digraph obtained
from a splitting of a doubly directed cycle of length at least 3 by removing an
arc contains both a subdivision of $\mathcal{D}^3$ and a subdivision of $\mathcal{D}'^3$. Thus $\mathcal{D}(A)$
contains a subdivision of both $\mathcal{D}^3$ and of $\mathcal{D}'^3$.   □

Corollary 7.2.6 *Let B be a fully indecomposable SNS-matrix with a negative
main diagonal. Then B is an S^2NS-matrix if and only if $\mathcal{D}(B)$ does not contain
a subdivision of $\mathcal{D}^3$.*

Proof If B is an S^2NS-matrix, then by Theorem 7.2.5, $\mathcal{D}(B)$ does not contain
a subdivision of $\mathcal{D}^3$. Now assume that $\mathcal{D}(B)$ does not contain a subdivision of
$\mathcal{D}^3$, and let A be the zero pattern of B. By Theorem 7.2.5, $\mathcal{Z}(A)$ contains an
S^2NS-matrix $\widehat{A}$. By Theorem 6.1.3, there exist strict signings D and E such
that $B = D\widehat{A}E$, and hence B is an S^2NS-matrix.   □

Let $A = [a_{ij}]$ be a fully indecomposable $(0, 1)$-matrix with $I_n \leq A$. Assume
that there exist integers p and q such that $a_{pq} = 1$ and each of row p and column
q of A contains exactly two 1's, say $a_{ps} = 1$ and $a_{rq} = 1$ where $r \neq p$ and
$s \neq q$. Let B be the matrix obtained from A by contracting on a_{pq}. It follows
from Lemma 7.1.9 that $\mathcal{Z}(A)$ contains an S^2NS-matrix if and only if $\mathcal{Z}(B)$
does. If $p = q$ and $r = s$, then $I_{n-1} \leq B$ and $\mathcal{D}(A)$ can be obtained from
$\mathcal{D}(B)$ by inserting a new vertex and a directed cycle of length 2 containing it.
If $p = q$ and $r \neq s$, then $I_{n-1} \leq B$, and $\mathcal{D}(A)$ can be obtained from $\mathcal{D}(B)$
by inserting a new vertex z and the arcs (x, z) and (z, y) where (x, y) is an arc
of $\mathcal{D}(B)$, and then possibly deleting the arc (x, y). Now assume that $p \neq q$.
Then $r = q, s = p$, and up to column permutations B is the same as the matrix
B' obtained from A by contracting on a_{pp}. We have $I_{n-1} \leq B'$, and $\mathcal{D}(A)$ can
be obtained from $\mathcal{D}(B')$ by splitting a vertex v and possibly inserting the arc
(v', v) from the new vertex v' to v.

Let $\mathcal{F}$ be the set of strongly connected digraphs defined recursively as follows:

   (i) The digraph of order 1 is in $\mathcal{F}$.
  (ii) If $\mathcal{D}$ is in $\mathcal{F}$ and v is a vertex of $\mathcal{D}$, then so are

       (a) the digraph obtained from $\mathcal{D}$ by inserting a new vertex v' and
           the arcs (v, v') and (v', v) and

       (b) the digraph obtained from $\mathcal{D}$ by inserting a new vertex v', the
           arc (v, v'), possibly the arc (v', v), and replacing each arc of the
           form (v, x) in $\mathcal{D}$ by the arc (v', x).

 (iii) If $\mathcal{D}$ is in $\mathcal{F}$ and (x, y) is an arc of $\mathcal{D}$, then so is the digraph obtained from $\mathcal{D}$ by inserting a new vertex z and the arcs (x, z) and (z, y), and then possibly deleting the arc (x, y).

If, in this recursive definition of $\mathcal{F}$, we require in (ii$b$) the insertion of the arc (v', v) and do not allow in (iii) the deletion of the arc (x, y), then we obtain a set of strongly connected digraphs $\mathcal{F}_0$ with $\mathcal{F}_0 \subseteq \mathcal{F}$. The following theorem is an immediate consequence of the earlier discussion and Corollaries 7.1.12 and 7.1.13.

Theorem 7.2.7 *The digraphs of fully indecomposable S^2NS-matrices with a negative main diagonal are precisely the digraphs in the set $\mathcal{F}$. The digraphs of fully indecomposable, maximal S^2NS-matrices with a negative main diagonal are precisely the digraphs in the set $\mathcal{F}_0$.*

We now give an alternative polynomial-time recognition algorithm for fully indecomposable S^2NS-matrices which we formulate in terms of signed digraphs. The validity of this algorithm is a consequence of Lemma 7.1.9 and Theorem 7.2.7.

S^2NS-matrix recognition algorithm for fully indecomposable matrices with a negative main diagonal

Let B be a fully indecomposable $(0, 1, -1)$-matrix with a negative main diagonal.

(0) Let $\mathcal{D}$ be the signed digraph of B.

(1) If the order of $\mathcal{D}$ is 1, then B is an S^2NS-matrix. Otherwise,

(2) if there are vertices v and v' of $\mathcal{D}$ such that the only arcs containing v are (v, v') and (v', v), then

    (a) if (v, v') and (v', v) have the same sign, B is not an S^2NS-matrix;

    (b) if (v, v') and (v', v) have the opposite sign, delete the vertex v from $\mathcal{D}$ and go back to (1).

(3) If there are vertices v and v' of $\mathcal{D}$ such that the only arc with initial vertex v is the arc $(v,.v')$ and the only arc with terminal vertex v' is the arc (v, v'), then

    (a) if (v', v) is an arc and its sign is the same as the sign of (v, v'), B is not an S^2NS-matrix;

    (b) if (v', v) is an arc and its sign is the negative of the sign of (v, v') or if (v', v) is not an arc, identify the vertices v and v' in $\mathcal{D}$ and multiply the sign of each arc of the form (v', x) by the sign of the arc (v, v') and go back to (1).

(4) If there exist distinct vertices x, y, and z such that the only arcs containing z are (x, z) and (z, y), then

    (a) if (x, y) is an arc and its sign is the negative of the product of the signs of the arcs (x, z) and (z, y), B is not an S^2NS-matrix;

(b) if (x, y) is an arc and its sign equals the product c of the signs of the arcs (x, z) and (z, y) or if (x, y) is not an arc, delete vertex z and if (x, y) is not an arc, insert the arc (x, y) with sign c, and go back to (1). Otherwise,

(5) B is not an S^2NS-matrix.

An alternative characterization of the set of digraphs of fully indecomposable, maximal S^2NS-matrices with a negative main diagonal is contained in [5]. The characterization is based on the following property of S^2NS-matrices.

Lemma 7.2.8 *Let $A = [a_{ij}]$ be a fully indecomposable $(0, 1)$-matrix of order $n \geq 2$ with $I_n \leq A$. Assume that $\mathcal{Z}(A)$ contains an S^2NS-matrix. Then there exists a permutation matrix P such that PAP^T has the form*

$$\begin{bmatrix} A_1 & U \\ V & A_2 \end{bmatrix} \tag{7.5}$$

where A_1 and A_2 are square matrices and the matrices U and V each contain exactly one 1.

Proof We prove the lemma by induction on n. If $n = 2$, the lemma is clear. Assume that $n > 2$. By Corollary 7.1.4, $\mathcal{D}(A)$ has a row 1, say row 1, with exactly one 1 off the main diagonal, say $a_{12} = 1$. If column 1 of A contains only one 1 off the main diagonal, then A has the form (7.5) where A_1 is of order 1. Now assume that column 1 of A contains at least two 1's off the main diagonal. Let B be the matrix obtained from A by contracting on a_{11}. By Lemma 4.2.1 and Corollary 7.1.5, B is fully indecomposable and $\mathcal{Z}(B)$ contains an S^2NS-matrix. Using the induction assumption we may assume that A has the form

$$\begin{bmatrix} 1 & 1 & 0 & \cdots & 0 & 0 & \cdots & 0 \\ x & & B_1 & & & U_1 & & \\ y & & V_1 & & & B_2 & & \end{bmatrix}$$

where B_1 is a square matrix of order p, U_1 contains exactly one 1, x contains at least one 1, each of y and V_1 contains at most one 1, and if both y and V_1 each contain a 1, then the 1 in V_1 is in its first column and in the same row as the 1 in y. If y and V_1 do not both contain a 1, then A has the form (7.5) where

A_1 is of order $p + 1$. Now assume that each of y and V_1 contains a 1. The strong connectivity of $\mathcal{D}(A)$ and the fact that y and V_1 each contain exactly one 1 imply that there are paths from vertex 2 to each vertex $2, \ldots, p + 1$ which only contain vertices in $\{2, \ldots, p + 1\}$. Since x contains at least one 1, $\mathcal{D}(A)$ contains a subdivision of $\mathcal{D}'^3$, and hence $\mathcal{Z}(A)$ does not contain an S^2NS-matrix. This contradiction completes the proof of the lemma.   □

Let $\mathcal{D}$ be the digraph of a fully indecomposable, maximal S^2NS-matrix of order $n \geq 2$ with a negative main diagonal. By Theorem 7.2.5, $\mathcal{D}$ does not contain a subdivision of $\mathcal{D}^3$. By Lemma 7.2.8, the vertices of $\mathcal{D}$ can be partitioned into two sets X and Y such that there is exactly one arc (x_1, y_1) from a vertex in X to a vertex in Y and exactly one arc (y_2, x_2) from a vertex in Y to a vertex in X. Let $\mathcal{D}_X$ be the subdigraph of $\mathcal{D}$ induced on X, and let $\mathcal{D}_Y$ be the subdigraph induced on Y. If $x_1 \neq x_2$, then the insertion of the arc (x_1, x_2) does not create a subdivision of $\mathcal{D}^3$. Hence, by Theorem 7.2.5 and the maximality assumption, if $x_1 \neq x_2$, then (x_1, x_2) is an arc of $\mathcal{D}$ and similarly, if $y_1 \neq y_2$, then (y_1, y_2) is an arc of $\mathcal{D}$. This implies that $\mathcal{D}_X$ and $\mathcal{D}_Y$ are both strongly connected.

Now let $\mathcal{D}_X$ and $\mathcal{D}_Y$ be two arbitrary strongly connected digraphs with vertex sets X and Y such that $X \cap Y = \emptyset$. Let x_1 and x_2 be vertices in X such that if $x_1 \neq x_2$, then (x_1, x_2) is an arc of $\mathcal{D}_X$. Similarly, let y_1 and y_2 be vertices in Y such that if $y_1 \neq y_2$, then (y_1, y_2) is an arc of $\mathcal{D}_Y$. Then $\mathcal{D}_X \diamond \mathcal{D}_Y$ denotes the strongly connected digraph with vertex set $X \cup Y$ obtained from $\mathcal{D}_X$ and $\mathcal{D}_Y$ by inserting the arcs (x_1, y_1) and (y_2, x_2). It is easy to verify that $\mathcal{D}_X \diamond \mathcal{D}_Y$ contains a subdivision of $\mathcal{D}^3$ if and only if at least one of $\mathcal{D}_X$ and $\mathcal{D}_Y$ does. Hence, by Theorem 7.2.5, $\mathcal{D}_X \diamond \mathcal{D}_Y$ is the digraph of a fully indecomposable S^2NS-matrix with a negative main diagonal if and only if both $\mathcal{D}_X$ and $\mathcal{D}_Y$ are. Since the insertion in $\mathcal{D}_X \diamond \mathcal{D}_Y$ of any arc containing a vertex in X and a vertex in Y creates a subdivision of $\mathcal{D}^3$, it follows that $\mathcal{D}_X \diamond \mathcal{D}_Y$ is the digraph of a fully indecomposable, maximal S^2NS-matrix with a negative main diagonal if and only if both $\mathcal{D}_X$ and $\mathcal{D}_Y$ are.

Let $\mathcal{F}_0'$ be the set of digraphs defined recursively as follows: $\mathcal{F}_0'$ contains the digraph of order 1, and if $\mathcal{D}_X$ and $\mathcal{D}_Y$ are in $\mathcal{F}_0'$, then so are all the digraphs $\mathcal{D}_X \diamond \mathcal{D}_Y$. The preceding discussion now implies that $\mathcal{F}_0'$ is the set of digraphs of fully indecomposable, maximal S^2NS-matrices with a negative main diagonal. Hence, by Theorem 7.2.7, $\mathcal{F}_0' = \mathcal{F}_0$. This is used in [5] to obtain a different polynomial-time recognition algorithm for fully indecomposable S^2NS-matrices.

An inductive argument shows that a digraph in $\mathcal{F}_0'$ of order n has exactly $2n - 2$ arcs. This gives an alternative proof of Corollary 7.1.11 [5].

7.3 Partly decomposable S²NS-matrices

Let B be a $(0, 1, -1)$-matrix of order n with a negative main diagonal. Without loss of generality we assume that B has the form

$$\begin{bmatrix} B_1 & O & \cdots & O \\ B_{21} & B_2 & \cdots & O \\ \vdots & \vdots & \ddots & \vdots \\ B_{k1} & B_{k2} & \cdots & B_k \end{bmatrix} \qquad (B_1, B_2, \ldots, B_k \text{ fully indecomposable}) \quad (7.6)$$

where $k \geq 1$.

If B is an S²NS-matrix, then each of its fully indecomposable components B_i is an S²NS-matrix. As already noted, the converse need not hold. Also if B is an S²NS-matrix, then B is an SNS-matrix and $\mathcal{D}(B)$ does not contain a subdivision of $\mathcal{D}^3$ (nor a subdivision of $\mathcal{D}'^3$). Assume that B is an SNS-matrix and $\mathcal{D}(B)$ does not contain a subdivision of $\mathcal{D}^3$ nor of $\mathcal{D}'^3$. By Corollary 7.2.6, if B is fully indecomposable ($k = 1$), then B is an S²NS-matrix. The matrix

$$\begin{bmatrix} -1 & 0 & 0 \\ 1 & -1 & 0 \\ -1 & 1 & -1 \end{bmatrix}$$

shows that this conclusion need not hold if B is partly decomposable ($k > 1$), although each fully indecomposable component is an S²NS-matrix.

The next theorem shows that the study of partly decomposable S²NS-matrices can be reduced to the study of fully indecomposable S²NS-matrices and partly decomposable S²NS-matrices, each of whose fully indecomposable components has order 1 [5].

Theorem 7.3.1 *Let $B = [b_{ij}]$ be a $(0, 1, -1)$-matrix of order n of the form (7.6) where each of the fully indecomposable components $B_1, B_2, \ldots, B_k$ is an S^2NS-matrix. Let $\mathcal{D}^*$ be the signed digraph of order $2n$ with vertices $1, 2, \ldots, n$ and $1', 2', 3', \ldots, n'$ and with signed arcs determined as follows. There is a positive arc (i, i') in $\mathcal{D}^*$ for each $i = 1, 2, \ldots, n$. If i and j are in the same strong component of $\mathcal{D}(B)$, then there is an arc (i, j') in $\mathcal{D}^*$ whose sign equals the common sign of the paths from i to j in $\mathcal{D}(B)$. If i and j are in different strong components of $\mathcal{D}(B)$ and there is an arc (i, j) in $\mathcal{D}(B)$, then there is an arc (i', j) in $\mathcal{D}^*$ with the same sign. Then B is an S^2NS-matrix if and only if the sign of α^* equals the sign of β^* for every pair of paths α^* and β^* in $\mathcal{D}^*$ with the same initial vertex in $\{1, 2, \ldots, n\}$ and the same terminal vertex in $\{1', 2', \ldots, n'\}$.*

Proof Since the fully indecomposable components of B are S^2NS-matrices, B is an S^2NS-matrix if and only if the sign of α equals the sign of β for each pair of paths α and β with the same initial vertex and the same terminal vertex where the initial vertex and terminal vertex belong to different strong components of $\mathcal{D}(B)$. Let α be a path in $\mathcal{D}(B)$ from a vertex i to a vertex j where i and j belong to different strong components. Then there is a path α^* in $\mathcal{D}^*$ from i to j' of the same sign. Conversely, a path α^* in $\mathcal{D}^*$ from i to j' corresponds to a path α in $\mathcal{D}(B)$ from i to j of the same sign. $\square$

The digraph $\mathcal{D}^*$ defined in Theorem 7.3.1 is an acyclic digraph. Let B^* be a $(0, 1, -1)$-matrix with negative main diagonal whose signed digraph is $\mathcal{D}^*$. Since $\mathcal{D}^*$ is an acyclic digraph, there exists a permutation matrix P such that PB^*P^T has only zeros above its main diagonal. It follows from Theorem 7.3.1 that B is an S^2NS-matrix if and only if B^* is. Let $\mathcal{D}^{**}$ be the digraph obtained from $\mathcal{D}^*$ by splitting each negative arc with the insertion of a new vertex. Let the new vertices obtained this way be $\{z_1, z_2, \ldots, z_p\}$. Note that $p \leq n^2$. Then B^* is an S^2NS-matrix if and only if for each i and j with $1 \leq i, j \leq n$, the lengths of the paths from i to j' in $\mathcal{D}^{**}$ have the same parity. Let B^{**} be the adjacency matrix of $\mathcal{D}^{**}$ obtained by listing the vertices in the order $1, 2, \ldots, n, 1', 2', \ldots, n', z_1, z_2, \ldots, z_p$.

The validity of the following polynomial-time algorithm [5] for recognizing partly decomposable S^2NS-matrices is a consequence of Theorem 7.3.1 and the preceding discussion.

S²NS-matrix recognition algorithm for partly decomposable matrices with a negative main diagonal

Let B be a partly decomposable $(0, 1, -1)$-matrix with a negative main diagonal.

(0) Determine the fully indecomposable components of B by finding the strong components of $\mathcal{D}(B)$.

(1) Use one of the two S^2NS-matrix recognition algorithms for fully indecomposable matrices to check whether the fully indecomposable components of B are S^2NS-matrices. If some component is not an S^2NS-matrix, then B is not an S^2NS-matrix. Otherwise,

(2) for each pair of vertices i and j of $\mathcal{D}(B)$ which are in the same strong component, determine the sign of one path (and hence every path) from i to j, and then form the matrix B^{**}.

(3) Calculate the matrices $B^{**}, (B^{**})^2, \ldots, (B^{**})^{(2n+p)}$. If there exist i and j with $1 \leq i, j \leq n$, and an even integer r and an odd integer s with $1 \leq r, s \leq 2n + p$ such that the (i, j') entries of $(B^{**})^r$ and $(B^{**})^s$ are nonzero, then B is not an S^2NS-matrix. Otherwise,

(4) B is an S^2NS-matrix.

We now investigate the zero patterns of partly decomposable S²NS-matrices.

Lemma 7.3.2 *Let $B = [b_{ij}]$ be an S^2NS-matrix of the form (7.6). Then*

(i) *B_{ij} contains at most one nonzero entry for $j < i$, and*

(ii) *if r, s, and t are integers such that $B_{rs} \neq O$ and $B_{st} \neq O$, and the nonzero entries of B_{rs} and B_{st} are in positions (p_{rs}, q_{rs}) and (p_{st}, q_{st}) of A, respectively, then either $B_{rt} = O$ or the unique nonzero entry of B_{rt} is in position (p_{rs}, q_{st}), and $b_{p_{rs}q_{st}} = \epsilon b_{p_{rs}q_{rs}} b_{p_{st}q_{st}}$ where ϵ equals the common sign of the paths from vertex q_{rs} to vertex p_{st}.*

Proof If some B_{ij} contains more than one nonzero entry, then $\mathcal{D}(B)$ contains a subdivision of $\mathcal{D}^3$. Now let r, s, and t be integers such that $B_{rs} \neq O$ and $B_{st} \neq O$. Assume that $B_{rt} \neq O$ and that its unique nonzero entry is in position (p_{rt}, q_{rt}) of B. Suppose, for instance, that $p_{rt} \neq p_{rs}$. Then taking the path from the vertex p_{rs} to the vertex q_{st}, a path from q_{st} to q_{rt}, a path in $\mathcal{D}(B_r)$ from p_{rs} to p_{rt}, a path in $\mathcal{D}(B_r)$ from p_{rt} to p_{rs}, and the arc (p_{rt}, q_{rt}), we obtain a digraph which contains a subdivision of $\mathcal{D}^3$. Hence $p_{rt} = p_{rs}$. A similar argument shows that $q_{rt} = q_{st}$. The formula for $b_{p_{rs}q_{st}}$ is a consequence of the fact that all paths from p_{rs} to q_{st} have the same sign. □

If in Lemma 7.3.2, B is a maximal S²NS-matrix, $B_{rs} \neq 0$, and $B_{st} \neq 0$, then it follows easily that $B_{rt} \neq O$.

Let A be a $(0, 1)$-matrix of order n whose determinant is not identically zero. It follows from Corollary 7.2.3 that if the bipartite graph of A does not contain an even subdivision of $K_{2,3}$, then A is the zero pattern of an S²NS-matrix, indeed A is the zero pattern of a matrix which for each $r = 1, 2, \ldots, n$ has a signed rth compound. The S²NS-matrix

$$\begin{bmatrix} -1 & 0 & 0 & 0 \\ 1 & -1 & 0 & 0 \\ 1 & 1 & -1 & 0 \\ 1 & 1 & 1 & -1 \end{bmatrix}$$

shows that the bipartite graph of an S²NS-matrix can contain a $K_{2,3}$. Now assume that $I_n \leq A$. If A is the zero pattern of an S²NS-matrix, then the digraph of A cannot contain a subdivision of $\mathcal{D}^3$ or of $\mathcal{D}'^3$. The following example shows that not containing a subdivision of $\mathcal{D}^3$ or of $\mathcal{D}'^3$ is not sufficient

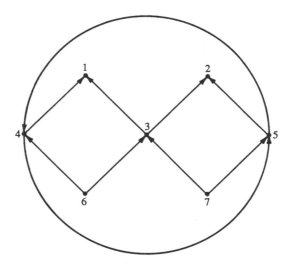

Fig. 7.1.

for a matrix to be the zero pattern of an S^2NS-matrix. Let

$$A = \begin{bmatrix} 1 & 0 & 0 & 0 & 0 & 0 & 0 \\ 0 & 1 & 0 & 0 & 0 & 0 & 0 \\ 1 & 1 & 1 & 0 & 0 & 0 & 0 \\ 1 & 0 & 0 & 1 & 1 & 0 & 0 \\ 0 & 1 & 0 & 1 & 1 & 0 & 0 \\ 0 & 0 & 1 & 1 & 0 & 1 & 0 \\ 0 & 0 & 1 & 0 & 1 & 0 & 1 \end{bmatrix}.$$

The digraph of A is illustrated in Figure 7.1.

The digraph $\mathcal{D}(A)$ has a unique directed cycle γ. There are two paths from each of 6 and 7 to each of 1 and 2, where the arcs of γ are each contained in one of these four paths and the other arcs are each contained in two. It follows that in any signing of $\mathcal{D}(A)$ either γ is positive or there are two paths with the same initial vertex and the same terminal vertex of opposite sign. Hence A is not the zero pattern of an S^2NS-matrix. More generally, let Γ_t be the digraph of order $t+6$ obtained from the digraph in Figure 7.1 by replacing the directed cycle by a doubly directed path of length t. Then, if t is odd, neither Γ_t nor any splitting of Γ_t is contained in the digraph of an S^2NS-matrix with a negative main diagonal. Thus $\mathcal{D}_3$, $\mathcal{D}^{\prime 3}$, and splittings of Γ_t with t odd are forbidden configurations for S^2NS-matrices. Indeed, these digraphs are minimal forbidden configurations. It is unknown whether there are other minimal forbidden configurations.

We now describe a polynomial-time algorithm for determining whether a partly decomposable matrix is the zero pattern of an S^2NS-matrix.

Algorithm for the construction of an S^2NS-matrix with a prescribed partly decomposable zero pattern

Let A be a partly decomposable $(0, 1)$-matrix with $I_n \leq A$.

(0) Determine the fully indecomposable components $A_1, A_2, \ldots, A_k$ of A by finding the strong components of $\mathcal{D}(A)$.

(1) Apply the algorithm for the construction of an S^2NS-matrix with a prescribed fully indecomposable zero pattern to each of the matrices $A_1, A_2, \ldots, A_k$. If at least one of these matrices is not the zero pattern of an S^2NS-matrix, then A is not the zero pattern of an S^2NS-matrix. Otherwise, let $B_1, B_2, \ldots, B_k$ be S^2NS-matrices with negative main diagonals whose zero patterns are A_1, $A_2, \ldots, A_k$, respectively.

(2) Let $A' = A$. While there exist distinct strong components $\mathcal{D}_1, \mathcal{D}_2$, and $\mathcal{D}_3$ of $\mathcal{D}(A)$ and vertices u_1, u_2, v_2, and v_3 such that u_i and v_i are vertices of $\mathcal{D}_i$, (u_1, u_2), and (v_2, v_3) are arcs of $\mathcal{D}(A')$, and (u_1, v_3) is not an arc of $\mathcal{D}(A')$, replace the (u_1, v_3) entry of A' by 1.

(3) If there exist two arcs whose initial vertices are in the same strong component of $\mathcal{D}(A')$ and whose terminal vertices are in the same strong component of $\mathcal{D}(A')$ and if these strong components are distinct, then $\mathcal{D}(A')$ [and hence $\mathcal{D}(A)$] contains a subdivision of $\mathcal{D}_3$ or $\mathcal{D}'_3$, and A is not the zero pattern of an S^2NS-matrix. Otherwise,

(4) for each pair of vertices x and y in the same strong component A_i, determine the (common) sign $(-1)^{\epsilon(x, y)}$ of the paths from x to y in the signed digraph $\mathcal{D}(B_i)$ where $\epsilon(x, y)$ equals 0 or 1.

(5) Form a system of linear equations over the integers modulo 2 as follows: For each triple of strong components $\mathcal{D}_1, \mathcal{D}_2$, and $\mathcal{D}_3$ for which there exist vertices u_1, u_2, v_2, and v_3 such that u_i and v_i are vertices of $\mathcal{D}_i$ and (u_1, u_2) and (v_2, v_3) are arcs of $\mathcal{D}(A')$, we have the equation

$$z_{(u_1, u_2)} + z_{(v_2, v_3)} + z_{(u_1, v_3)} = \epsilon(u_2, v_2).$$

If this system of equations does not have a solution, then A is not the zero pattern of an S^2NS-matrix. Otherwise,

(6) let $\widehat{A}$ be the matrix obtained from A by replacing each A_i by B_i and each 1 corresponding to an arc (u_1, u_2) between distinct strong components by $(-1)^{z_{(u_1, u_2)}}$. Then $\widehat{A}$ is an S^2NS-matrix whose zero pattern equals A.

The validity of the algorithm is a consequence of Theorem 6.1.3, Lemma 7.3.2, and the observation that if $\widehat{A'}$ is a matrix with zero pattern A' and a negative main diagonal, then $\widehat{A'}$ is an S^2NS-matrix if and only if the sign of α equals the sign of β for every pair of paths α and β in $\mathcal{D}(\widehat{A'})$ whose initial vertices are the same, whose terminal vertices are the same, and whose vertices are contained in the union of at most three different strong components.

We now consider partly decomposable, maximal S^2NS-matrices. Let

$$
T_n = \begin{bmatrix}
-1 & 0 & \cdots & 0 & 0 \\
1 & -1 & \cdots & 0 & 0 \\
\vdots & \vdots & \ddots & \vdots & \vdots \\
1 & 1 & \cdots & -1 & 0 \\
1 & 1 & \cdots & 1 & -1
\end{bmatrix}.
$$

Then $\mathcal{D}(T_n)$ is the acyclic digraph of order n with arcs (i, j) for all i and j with $i > j$. Since every path in $\mathcal{D}(T_n)$ is positive, T_n is an S^2NS-matrix. The S^2NS-matrix

$$
T_3' = \begin{bmatrix}
-1 & 0 & -1 \\
1 & -1 & 0 \\
1 & 1 & -1
\end{bmatrix}
$$

shows that T_3 is not a maximal S^2NS-matrix. It is easy to verify that if $n \geq 4$, then any digraph obtained from $\mathcal{D}(T_n)$ by the insertion of any new arc contains either a $\mathcal{D}^3$ or a $\mathcal{D}'^3$. This implies that for $n \geq 4$, T_n is a maximal S^2NS-matrix. In fact, every S^2NS-matrix whose zero pattern equals the zero pattern of T_n ($n \geq 4$) is a maximal S^2NS-matrix. The matrices

$$
\begin{bmatrix}
-1 & 0 & 0 & 0 & 0 & 0 \\
0 & -1 & 0 & 0 & 0 & 0 \\
-1 & 1 & -1 & 0 & 0 & 0 \\
1 & 1 & 0 & -1 & 0 & 0 \\
1 & -1 & -1 & 0 & -1 & 0 \\
1 & 1 & 0 & 1 & 0 & -1
\end{bmatrix}
\quad \text{and} \quad
\begin{bmatrix}
-1 & 0 & 0 & 0 & 0 \\
0 & -1 & 0 & 0 & 0 \\
1 & 0 & -1 & 1 & 0 \\
0 & 1 & -1 & -1 & 0 \\
1 & 1 & 1 & 0 & -1
\end{bmatrix}
$$

are also maximal S^2NS-matrices. The matrix on the left and T_6 imply that in contrast to fully indecomposable matrices, the maximality of a S^2NS-matrix which is partly decomposable is not determined by its zero pattern. Every S^2NS-matrix whose zero pattern equals the zero pattern of the matrix on the right is a maximal S^2NS-matrix.

We now show how the join operation can be used to obtain partly decomposable, maximal S^2NS-matrices.

Theorem 7.3.3 *Let A and B be $(0, 1, -1)$ S^2NS-matrices with a negative main diagonal. Then their join $A \star B$ at $-J_1$ is an S^2NS-matrix. If A and B are maximal S^2NS-matrices and at least one of them is fully indecomposable, then $A \star B$ is a maximal S^2NS-matrix.*

Proof The digraph of $A \star B$ is obtained from the digraphs $\mathcal{D}(A)$ and $\mathcal{D}(B)$ by identifying a vertex of one with a vertex of the other. Let v be this vertex of $\mathcal{D}(A \star B)$. The assertion that $A \star B$ is an S^2NS-matrix follows from Corollary 3.2.3. Now assume that A and B are maximal S^2NS-matrices and that, say, A is a fully indecomposable matrix. The maximality property of A and B implies that no matrix obtained from $A \star B$ by changing a 0 of A or of B to ± 1 is an S^2NS-matrix. Suppose that there is an entry in some position (p, q) of $A \star B$ which is not an entry of A or B which can be changed to $a = \pm 1$ in order to obtain an S^2NS-matrix C. First suppose that p is a vertex of $\mathcal{D}(A)$ and q is a vertex of $\mathcal{D}(B)$. If (v, q) is an arc of $\mathcal{D}(A \star B)$, then since $\mathcal{D}(A)$ is strongly connected, it follows that by taking paths from v to p and p to v and the arcs (p, q) and (v, q) we obtain a digraph which contains a subdivision of $\mathcal{D}^3$. Hence (v, q) is not an arc of $\mathcal{D}(A \star B)$. It now follows that the zero in the entry of B corresponding to the position (v, q) of $A \star B$ can be replaced by the product of a and the sign of a path from v to p in $\mathcal{D}(A \star B)$, contradicting the maximal property of B. A similar contradiction results if p is a vertex of $\mathcal{D}(B)$ and q is a vertex of $\mathcal{D}(A)$. Hence $A \star B$ is a maximal S^2NS-matrix. $\square$

For example, the matrix $T_3' \star T_4$ obtained by joining T_3' and T_4 at a $-J_1$ is a partly decomposable, maximal S^2NS-matrix of order 6. The join of T_4 with itself at $-J_1$ shows that the join of two maximal S^2NS-matrices is not necessarily a maximal S^2NS-matrix.

7.4 S^2NS-matrices whose inverses are S^2NS-matrices

Let A be an S^2NS-matrix. Then A^{-1} need not be an S^2NS-matrix. Indeed, A^{-1} need not even be an SNS-matrix. In this section we consider the very special subclass of S^2NS-matrices consisting of those matrices whose inverses are also S^2NS-matrices. This subclass of S^2NS-matrices was first studied in [4]. If A is a direct sum of matrices, then both A and A^{-1} are S^2NS-matrices if and only if each direct summand and its inverse are S^2NS-matrices.

Let $\mathcal{D}$ be a digraph of order n. An arc (u, v) is a *transitive arc* of $\mathcal{D}$ provided there is a path from u to v which does not contain the arc (u, v). Thus an arc (u, v) is a *nontransitive arc* of $\mathcal{D}$ if and only if there is no path from u to v in the digraph obtained from $\mathcal{D}$ by removing the arc (u, v). The *transitive closure* of $\mathcal{D}$ is the digraph of order n such that (x, y) is an arc if and only if there is a path in $\mathcal{D}$ from x to y. The digraph $\mathcal{D}$ is *transitively closed* if the transitive closure of $\mathcal{D}$ is $\mathcal{D}$.

Lemma 7.4.1 *Let A be a square $(0, 1, -1)$-matrix with a negative main diagonal such that A and A^{-1} are S^2NS-matrices. Then there exists a permutation* *matrix P such that PAP^T is a direct sum of matrices of the form*

$$[-1], \quad \begin{bmatrix} -1 & 1 \\ -1 & -1 \end{bmatrix}, \quad \begin{bmatrix} -I_\ell & O \\ X & -I_k \end{bmatrix} \tag{7.7}$$

where k and ℓ are positive integers and X is a k by ℓ matrix with at least one nonzero entry in each row and column.

Proof Without loss of generality we may assume that no matrix obtained from A by simultaneous row and column permutations is a direct sum of matrices and that

$$A = \begin{bmatrix} A_1 & O & \cdots & O \\ A_{21} & A_2 & \cdots & O \\ \vdots & \vdots & \ddots & \vdots \\ A_{m1} & A_{m2} & \cdots & A_m \end{bmatrix}$$

where m is a positive integer and $A_1, A_2 \ldots, A_m$ are the fully indecomposable components of A. Since $(A^{-1})^{-1} = A$, the inverse of each matrix in $\mathcal{Q}(A^{-1})$ is in $\mathcal{Q}(A)$. It follows from Theorem 3.2.5 that $\mathcal{D}(A^{-1})$ is the transitive closure of $\mathcal{D}(A)$ and that $\mathcal{D}(A) = \mathcal{D}((A^{-1})^{-1})$ is the transitive closure of $\mathcal{D}(A^{-1})$. We conclude that $\mathcal{D}(A)$ is transitively closed and that the digraphs $\mathcal{D}(A)$ and $\mathcal{D}(A^{-1})$ are the same. Theorem 3.2.5 now implies that the signed digraph $\mathcal{D}(A^{-1})$ is obtained from the signed digraph $\mathcal{D}(A)$ by negating the sign of each arc.

Suppose that $i \to j \to k$ is a path of length 2 in $\mathcal{D}(A)$. Then (i, k) is an arc in $\mathcal{D}(A)$. By Corollary 3.2.3 the sign of the path $i \to j \to k$ and the sign of the arc (i, k) in $\mathcal{D}(A)$ are equal. Similarly, the signs of these two paths in $\mathcal{D}(A^{-1})$ are equal. Since the sign of each arc of $\mathcal{D}(A^{-1})$ is opposite to that of the corresponding arc of $\mathcal{D}(A)$, this is impossible. Thus $\mathcal{D}(A)$ has no paths of length 2. This implies that each fully indecomposable component of A has order 1 or 2 and, up to simultaneous row and column permutations, is either the first or second matrix in (7.7).

Suppose A_i has order 2 for some integer i. Since $\mathcal{D}(A)$ has no paths of length 2, $A_{ij} = O$ and $A_{ji} = O$ for all j. Since there does not exist a permutation matrix P such that PAP^T is a direct sum, we conclude that $m = 1$ and A is the second matrix in (7.7).

Now suppose that each A_i has order 1 and that $m \geq 2$. Then $\mathcal{D}(A)$ is an acyclic graph. Since $\mathcal{D}(A)$ has no paths of length 2, each vertex has indegree 0 or outdegree 0. Since no matrix obtained from A by simultaneous row and column permutations is a direct sum of matrices and since $m \geq 2$, the indegree and outdegree of a vertex cannot both be zero. If the number of vertices of indegree zero equals k and the number of vertices of outdegree zero equals ℓ, then there exists a permutation matrix P such that PAP^T has the form of the third matrix in (7.7).   □

It is easy to verify that if A has the form of the third matrix in (7.7), then no matrix obtained from A by simultaneous row and column permutations is a direct sum of matrices if and only if the bipartite graph of X is connected.

Since row and column permutations and multiplication by strict signings preserve the property of a matrix and its inverse being S²NS-matrices, we have the following characterization of S²NS-matrices whose inverses are S²NS-matrices.

Theorem 7.4.2 *Let A be a square $(0, 1, -1)$-matrix. Then both A and A^{-1} are S²NS-matrices if and only if there exist permutation matrices P and Q and strict signings D and E such that $DPAQE$ is a direct sum of matrices of the forms given in (7.7).*

By definition, a $(0, 1, -1)$ SNS-matrix B is an S²NS-matrix provided there exists a $(0, 1, -1)$-matrix C such that C is the sign pattern of the inverse of each matrix in $\mathcal{Q}(B)$. The S²NS-matrix B has a *self-inverse sign pattern* provided that $C = B$. Any strict signing D is an S²NS-matrix with a self-inverse sign pattern. The class of S²NS-matrices which have a self-inverse sign pattern is a subclass of the S²NS-matrices whose inverses are also S²NS-matrices. Since the property of having a self-inverse sign pattern is not invariant under arbitrary row and column permutations, but is invariant under simultaneous row and column permutations, we consider irreducible matrices rather than fully indecomposable matrices. The irreducible S²NS-matrices with a self-inverse sign pattern were determined in [2].

Lemma 7.4.3 *Let A be an irreducible $(0, 1, -1)$-matrix. Then A is an S²NS-matrix with a self-inverse sign pattern if and only if there exist a permutation matrix P and a strict signing D such that $PDADP^T$ is one of the following five matrices:*

$$\begin{bmatrix} 1 \end{bmatrix}, \begin{bmatrix} -1 \end{bmatrix}, \begin{bmatrix} 0 & 1 \\ 1 & 0 \end{bmatrix}, \begin{bmatrix} 1 & 1 \\ 1 & -1 \end{bmatrix}, \text{ and } \begin{bmatrix} 0 & 0 & 1 & 1 \\ 0 & 0 & 1 & -1 \\ 1 & 1 & 0 & 0 \\ 1 & -1 & 0 & 0 \end{bmatrix}.$$

$$(7.8)$$

Proof It is easy to verify that each of the five matrices in (7.8) has a self-inverse sign pattern. Now assume that A is an S^2NS-matrix of order n with a self-inverse sign pattern. First suppose that A is fully indecomposable. Then A^{-1} has no zero entries and hence $n < 3$. It is now easy to see that there exist a permutation matrix P and a strict signing D such that $PDADP^T$ is either the first, second, or fourth matrix in (7.8). Now suppose that A is partly decomposable. Consider a k by ℓ zero submatrix $A[\alpha, \beta]$ with $k + \ell = n$. It follows that $A^{-1}[\overline{\beta}, \overline{\alpha}] = O$, and hence $A[\overline{\beta}, \overline{\alpha}] = O$. This implies that $A[\alpha \setminus \beta, \overline{\alpha \setminus \beta}] = O$ and that $A[\overline{\beta \setminus \alpha}, \beta \setminus \alpha] = O$. Since A is irreducible, we conclude that $\alpha = \beta$ and $k = \ell$. If $A[\alpha, \overline{\beta}]$ is partly decomposable, then there exists a k' by ℓ' zero submatrix of A with $k' + \ell' = n$ but $k' \neq \ell'$ in contradiction to what has already been proved. Hence $A[\alpha, \overline{\beta}]$ and similarly $A[\beta, \overline{\alpha}]$ are fully indecomposable, and thus there exists a permutation matrix Q such that

$$QAQ^T = \begin{bmatrix} O & A_1 \\ A_2 & O \end{bmatrix}$$

where A_1 and A_2 are fully indecomposable SNS-matrices of order k. We have

$$(QAQ^T)^{-1} = \begin{bmatrix} O & A_2^{-1} \\ A_1^{-1} & O \end{bmatrix}$$

where A_1^{-1} and A_2^{-1} have no zero entries. Since A has a self-inverse sign pattern, we conclude that A_1 and A_2 are SNS-matrices with no zero entries and hence that $k \leq 2$. It is now easy to see that there exist a permutation matrix P and a strict signing D such that $PDADP^T$ is either the third or fifth matrix in (7.8). $\qquad\square$

Lemma 7.4.4 *Let $A = [a_{ij}]$ be an S^2NS-matrix with a self-inverse sign pattern such that $A[\alpha]$ is an irreducible component of A.*

  (i) *If $A[\alpha]$ is the fourth or fifth matrix in (7.8), then $A[\alpha, \overline{\alpha}] = O$ and $A[\overline{\alpha}, \alpha] = O$.*

  (ii) *If $A[\alpha]$ is the third matrix in (7.8), then either both rows of $A[\alpha, \overline{\alpha}]$ are zero or both rows are nonzero, and either both columns of $A[\alpha, \overline{\alpha}]$ are zero or both columns are nonzero.*

Proof First suppose that $A[\alpha]$ is either the fourth or fifth matrix in (7.8). Then each fully indecomposable component of $A[\alpha]$ is a fully indecomposable component of A. Theorem 7.4.2 now implies that (i) holds.

Now suppose that $A[\alpha]$ is the third matrix in (7.8). If A is irreducible, then certainly (ii) holds. Assume A is reducible. Without loss of generality we may assume that $\alpha = \{1, 2\}$, A has exactly two irreducible components, and $A(\alpha)$ is one of the matrices in (7.8). If $A(\alpha)$ is either the fourth or fifth matrix in (7.8), then it follows from (i) that (ii) holds.

Suppose that $A(\alpha)$ is the third matrix in (7.8) and that, for example, $a_{31} \neq 0$. Then $A(\{2\}, \{4\})$ does not have an identically zero determinant and hence is an SNS-matrix. Thus, since A is an S^2NS-matrix with a self-inverse sign pattern, $a_{24} \neq 0$. Similarly, if $a_{41} \neq 0$, then $a_{32} \neq 0$. Hence (ii) holds. In the case that $A(\alpha)$ is either the first or second matrix similar arguments show that (ii) holds. $\square$

Lemma 7.4.5 *Let k, ℓ, m, n, p, and q be nonnegative integers and let*

$$N = \begin{bmatrix} R & S & T \\ U & V & W \\ X & Y & Z \end{bmatrix}$$

be a $(0, 1, -1)$-matrix where R is k by ℓ, V is m by n, and Z is $2p$ by $2q$. Then the matrix

$$A = \left[\begin{array}{c|c} I_\ell \oplus -I_n \oplus B & O \\ \hline N & I_k \oplus -I_m \oplus C \end{array} \right] \tag{7.9}$$

where

$$B = \overbrace{\begin{bmatrix} 0 & 1 \\ 1 & 0 \end{bmatrix} \oplus \cdots \oplus \begin{bmatrix} 0 & 1 \\ 1 & 0 \end{bmatrix}}^{q} \quad and \quad C = \overbrace{\begin{bmatrix} 0 & 1 \\ 1 & 0 \end{bmatrix} \oplus \cdots \oplus \begin{bmatrix} 0 & 1 \\ 1 & 0 \end{bmatrix}}^{p}$$

is an S^2NS-matrix with a self-inverse sign pattern if and only if $R = O$, $V = O$, columns $2j - 1$ and $2j$ of T are negatives of each other, columns $2j - 1$ and $2j$ of W are equal, rows $2i - 1$ and $2i$ of X are negatives of each other, rows $2i - 1$ and $2i$ of Y are equal, the $(2i - 1, 2j - 1)$ and $(2i, 2j)$ entries of Z are negatives of each other, and the $(2i - 1, 2j)$ and $(2i, 2j - 1)$ entries of Z are negatives of each other $(i = 1, 2, \ldots, p$ and $j = 1, 2, \ldots, q)$.

Proof Let

$$M = (-I_\ell) \oplus I_n \oplus (-B) \oplus (-I_k) \oplus I_m \oplus (-C).$$

Then MA has a negative main diagonal, and $\mathcal{D}(MA)$ is an acyclic digraph with no paths of length 2. It follows from Corollary 3.2.3 that MA and $(MA)^{-1}$ are S^2NS-matrices. Since MA has a negative diagonal and $\mathcal{D}(MA)$ is an acyclic digraph, Theorem 3.2.5 implies that the matrix obtained from MA by negating the off diagonal entries is the sign pattern of $(MA)^{-1}$. Hence the sign pattern of $(MA)^{-1}$ is

$$
\begin{bmatrix}
-I_\ell & O & O & O & O & O \\
O & -I_n & O & O & O & O \\
O & O & -I_{2q} & O & O & O \\
R & S & T & -I_k & O & O \\
-U & -V & -W & O & -I_m & O \\
CX & CY & CZ & O & O & -I_{2p}
\end{bmatrix}. \tag{7.10}
$$

Since $(MA)^{-1} = A^{-1}M^{-1}$ and M is the product of a permutation matrix and a strict signing, it follows from (7.10) that the sign pattern of A^{-1} is

$$
\begin{bmatrix}
I_\ell & O & O & O & O & O \\
O & -I_n & O & O & O & O \\
O & O & B & O & O & O \\
-R & S & -TB & I_k & O & O \\
U & -V & WB & O & -I_m & O \\
-CX & CY & -CZB & O & O & C
\end{bmatrix}. \tag{7.11}
$$

Hence A has a self-inverse sign pattern if and only if the matrix in (7.10) equals A, and the lemma now follows.   □

We now characterize S^2NS-matrices which have a self-inverse sign pattern.

Theorem 7.4.6 *Let A be a $(0, 1, -1)$-matrix. Then A is an S^2NS-matrix with a self-inverse sign pattern if and only if there exist a permutation matrix P and a strict signing D such that $DPAP^T D$ is a direct sum of matrices in (7.8) and (7.9).*

Proof It follows from Lemmas 7.4.3 and 7.4.5 that each matrix which is a direct sum of the matrices in (7.8) and matrices of the form described in (7.9) is an S^2NS-matrix with a self-inverse sign pattern.

Suppose that A is an S^2NS-matrix with a self-inverse sign pattern, and that no matrix obtained from A by simultaneous row and column permutations is a direct sum of matrices. We may assume that

$$
A = \begin{bmatrix}
A_1 & O & \cdots & O \\
A_{21} & A_2 & \cdots & O \\
\vdots & \vdots & \ddots & \vdots \\
A_{m1} & A_{m2} & \cdots & A_m
\end{bmatrix}
$$

where $m \geq 1$ and $A_1, \ldots, A_m$ are the irreducible components of A. Further, by Lemma 7.4.3 we may assume that each irreducible component is one of the matrices given in (7.8). If either the fourth or fifth matrix in (7.8) is an irreducible component of A, then it follows from Lemma 7.4.4 and the assumption that A is not a direct sum of matrices that $m = 1$.

Assume that each irreducible component is either the first, second, or third matrix in (7.8) and that $m \geq 2$. Let $A' = -(A_1 \oplus A_2 \oplus \cdots A_m)A$. Then A' has a negative main diagonal, and $\mathcal{D}(A)$ is an acyclic digraph with no paths of length 2. It follows from Lemma 7.4.4 that the vertices of $\mathcal{D}(A')$ corresponding to the columns of A_i both have indegree one or both have outdegree one ($i = 1, 2, \ldots, m$), and hence that there exists a permutation matrix P such that PAP^T has the form

$$
\left[
\begin{array}{c|c}
I_\ell \oplus -I_n \oplus \overbrace{\left[\begin{smallmatrix} 0 & 1 \\ 1 & 0 \end{smallmatrix}\right] \oplus \cdots \oplus \left[\begin{smallmatrix} 0 & 1 \\ 1 & 0 \end{smallmatrix}\right]}^{q} & O \\
\hline
N & I_k \oplus -I_m \oplus \overbrace{\left[\begin{smallmatrix} 0 & 1 \\ 1 & 0 \end{smallmatrix}\right] \oplus \cdots \oplus \left[\begin{smallmatrix} 0 & 1 \\ 1 & 0 \end{smallmatrix}\right]}^{p}
\end{array}
\right].
$$

The assumption that no matrix obtained from A by simultaneous row and column permutation is a direct sum of matrices implies that each row and column of N contains at least one nonzero entry. The theorem now follows from Lemma 7.4.5.   □

7.5 Inverses of fully indecomposable S^2NS-matrices

Let A be a $(0, 1, -1)$ S^2NS-matrix of order n. Then the inverse of each matrix in $\mathcal{Q}(A)$ has the same sign pattern B. If A is fully indecomposable, then B is a $(1, -1)$-matrix. We note that in general A is not uniquely determined by B, that is, a $(1, -1)$-matrix may be the sign pattern of the inverses of fully indecomposable S^2NS-matrices with different sign patterns. For example, let

$$
A = \begin{bmatrix} 1 & 1 & -1 \\ 1 & 0 & 1 \\ 0 & 1 & 1 \end{bmatrix}.
$$

Then A is an S^2NS-matrix, and the sign pattern of the inverses of the matrices in $\mathcal{Q}(A)$ is

$$
B = \begin{bmatrix} 1 & 1 & -1 \\ 1 & -1 & 1 \\ -1 & 1 & 1 \end{bmatrix}.
$$

But B is also the sign pattern of the inverses of the matrices in $Q(A')$ where A' is the S^2NS-matrix obtained from A by changing the entry in position (1,3) to zero. More generally, if A is a fully indecomposable S^2NS-matrix and if a fully indecomposable matrix A' can be obtained from A by changing a nonzero entry to zero, then A' is an S^2NS-matrix and the sign patterns of the inverses of matrices in $Q(A)$ and $Q(A')$ are the same. This implies that the sign pattern of the inverse of a fully indecomposable S^2NS-matrix A is determined by any nearly decomposable matrix obtained from A by replacing certain nonzero entries with zeros.

In the next lemma we characterize S^2NS-matrices in terms of signings.

Lemma 7.5.1 *Let $A = [a_{ij}]$ be an SNS-matrix of order n.*

   (i) *A is an S^2NS-matrix if and only if for each $i = 1, 2, \ldots, n$ there exists a unique signing D_i such that row i is the only unsigned row of $A_i D_i$ and the nonzero entries in row i are positive. Moreover, if A is a fully indecomposable S^2NS-matrix, then each of the signings D_i is a strict signing.*

   (ii) *If A is an S^2NS-matrix and $D_1, D_2, \ldots, D_n$ are the signings described in (i), then the sign pattern of A^{-1} is the matrix whose j column equals the diagonal vector of D_j $(1 \le j \le n)$.*

   (iii) *If A is a fully indecomposable S^2NS-matrix and $a_{pq} \ne 0$, then the sign of the entry in position (q, p) of A^{-1} equals the sign of a_{pq}.*

Proof Assertion (i) is a consequence of the fact that a matrix A of order n is an S^2NS-matrix if and only if $AX = I_n$ is sign-solvable (see section 1.2). Assertion (ii) is a consequence of (i). Now assume that A is a fully indecomposable S^2NS-matrix. Then the entry in position (q, p) of A^{-1} equals

$$(-1)^{p+q} \frac{\det A(\{p\}, \{q\})}{\det A}.$$

Since A is fully indecomposable and A is an SNS-matrix, the sign of $\det A$ equals the sign of $(-1)^{p+q} a_{pq} \det A(\{p\}, \{q\})$ and assertion (iii) follows. $\square$

We next describe the relationship between the sign patterns of the inverse of an S^2NS-matrix A and the sign pattern of an S^2NS-matrix obtained from A by copying an entry [1].

Theorem 7.5.2 *Let $A = [a_{ij}]$ be a fully indecomposable S^2NS-matrix of order n and assume that p and q are integers such that $a_{pq} \ne 0$. Let $B = [b_{ij}]$ be the* *fully indecomposable S^2NS-matrix obtained from A by conformally copying the entry a_{pq}. Then the sign pattern of B^{-1} equals the sign pattern of the matrix*

$$\left[\begin{array}{c|c} A^{-1} & v \\ \hline u & -a_{pq} \end{array}\right], \tag{7.12}$$

where $u = (u_1, \ldots, u_n)$ is the qth row of A^{-1} and $v = (v_1, \ldots, v_n)^T$ is the pth column of A^{-1}.

Proof For each $j = 1, 2, \ldots, n$ let D_j be the strict signing of order n whose diagonal vector is the sign pattern of the j column vector of A^{-1}, and let $E_j = D_j \oplus (u_j)$. By (i) and (ii) of Lemma 7.5.1, row j is the only unsigned row of AD_j and the nonzero entries in row j of A_jD are positive. It follows from the definitions of B and E_j that row j is the only unsigned row of BE_j and that its nonzero entries are positive. Thus the sign patterns of the first n columns of B^{-1} are equal to the sign patterns of the corresponding columns of (7.12). A similar argument applied to A^T shows that the sign patterns of the first n rows of B^{-1} are equal to the sign patterns of the corresponding rows of (7.12). Since $b_{n+1,n+1} = -a_{pq}$, it follows from (iii) of Lemma 7.5.1 that the entry in position $(n+1, n+1)$ of B^{-1} equals $-a_{pq}$. $\square$

Let $B = [b_{ij}]$ be the sign pattern of the inverse of a fully indecomposable S^2NS-matrix. By (i) and (ii) of Lemma 7.5.1, no two columns of B are equal. Corollary 7.1.12 and Theorem 7.5.2, however, imply that B has two columns that are almost equal or are almost opposite. More precisely, there exists an integer r such that the matrix B_r obtained from B by deleting row r contains two columns which either are equal or are negatives of one another. The following lemma implies that if columns p and q of B_r satisfy this property, then for every S^2NS-matrix whose inverse has sign pattern equal to B, the signs of the entries in positions (p, r) and (q, r) are equal to b_{rp} and b_{rq}, respectively.

Lemma 7.5.3 *Let $A = [a_{ij}]$ be a fully indecomposable S^2NS-matrix of order n and let $B = [b_{ij}]$ be the sign pattern of A^{-1}. Suppose that p, q, and r are integers with $1 \le p < q < n$ and $1 \le r \le n$ such that the vectors u and v obtained from columns p and q of B by deleting their rth entries satisfy $u = \pm v$. Then*

$$\text{sign } a_{pr} = b_{rp} \text{ and sign } a_{qr} = b_{rq}.$$

In addition, the matrix $\widehat{A}$ obtained from A by replacing each entry a_{sr} with $s \ne p, q$ by zero is a fully indecomposable S^2NS-matrix.

Proof Let D and E be the (strict) signings of order n whose diagonal vectors equal columns p and q, respectively, of B. The assumptions on columns p and q of B imply that $D(r, r) = \pm E(r, r)$. Since row p of AD is unsigned and row q of AD is balanced, it now follows that $a_{pr} \neq 0$. Since the nonzero entries of row p of AD are positive, the sign of a_{pr} equals b_{rp}. A similar argument shows that $a_{qr} \neq 0$ and that the sign of a_{qr} equals b_{rq}.

Since A is a fully indecomposable S^2NS-matrix, $\widehat{A}$ is an S^2NS-matrix. Since columns p and q of B are neither equal nor opposite, and since u and v are equal or opposite, we have

$$v = -\frac{b_{rp}}{b_{rq}}u. \tag{7.13}$$

For $i = 1, 2, \ldots, n$ let F_i be the strict signing whose diagonal vector equals the transpose of row i of B. Since A is an S^2NS-matrix, it follows from (iii) of Lemma 7.5.1 that column i is the only unsigned column of $F_i A$, and the nonzero entries in column i are positive. Using (7.13) it is now easy to verify that column i is the only unsigned column of $F_i \widehat{A}$, and its nonzero entries are positive ($1 \leq i \leq n$). Now (iii) of Lemma 7.5.1 implies that the sign pattern of $\widehat{A}^{-1}$ equals B and hence that $\widehat{A}$ is fully indecomposable.   □

We now characterize the $(1, -1)$-matrices B for which there exists a fully indecomposable S^2NS-matrix whose inverse has sign pattern equal to B [1]. The assumption that B has no zeros implies that any S^2NS-matrix whose inverse has sign pattern equal to B is fully indecomposable. The recursive nature of the characterization is such that it is necessary to prove a somewhat more general theorem in which the signs of some of the nonzero entries are prescribed. By (iii) of Lemma 7.5.1, a prescribed sign for an entry of A must equal the entry in the symmetrically opposite position of B. In order to state the theorem more concisely we make the following definition. If $X = [x_{ij}]$ is a $(0, 1, -1)$-matrix of order n, then A *conforms to* X provided sign $a_{ij} = x_{ij}$ for all i and j with $x_{ij} \neq 0$. By (iii) of Lemma 7.5.1, if A conforms to X and the sign pattern of the inverse of A equals B, then B conforms to X^T.

Theorem 7.5.4 *Let $B = [b_{ij}]$ be a $(1, -1)$-matrix of order $n \geq 2$, and let $X = [x_{ij}]$ be a $(0, 1, -1)$-matrix of order n such that $b_{\ell k} = x_{k\ell}$ for each k and ℓ with $x_{k\ell} \neq 0$.*

    (a) *If B is the sign pattern of the inverse of an S^2NS-matrix A which conforms to X, then there exist integers p, q, r, s such that*

        (i) *the vectors $u^{(p)}$ and $u^{(q)}$ obtained from rows p and q of B by deleting their rth entry satisfy $u^{(p)} = \pm u^{(q)}$,*

(ii) *the vectors $v^{(r)}$ and $v^{(s)}$ obtained from columns r and s of B by deleting their pth entry satisfy $v^{(r)} = \pm v^{(s)}$,*
(iii) *$x_{rj} \neq 0$ only if $j = p$ or $j = q$, and*
(iv) *$x_{ip} \neq 0$ only if $i = r$ or $i = s$.*

(b) *The matrix B is the sign pattern of the inverse of an S^2NS-matrix which conforms to X if and only if there exist integers p, q, r, s satisfying (i) to (iv) such that $B(\{p\}, \{r\})$ is the sign pattern of the inverse of an S^2NS-matrix whose sign pattern conforms to $\widehat{X}(\{r\}, \{p\})$ where $\widehat{X}$ is obtained from X by replacing the entry in position (s, q) with $-x_{rp}x_{sp}x_{rq}$.*

Proof First assume that there exists an S^2NS-matrix A whose sign pattern conforms to X and whose inverse has sign pattern B. Then A is fully indecomposable, and without loss of generality we may assume that A is a maximal S^2NS-matrix. By Theorem 7.1.10, there exist integers p, q, r, s such that row p of A has exactly two nonzero entries and they occur in columns r and s; and column r has exactly two nonzero entries and they occur in rows p and q. Thus the integers p, q, r, s satisfy (iii) and (iv), and by Theorem 7.5.2 they also satisfy (i) and (ii).

Now assume that p, q, r, s are integers satisfying (i) to (iv). Lemma 7.5.3 implies that B is the sign pattern of an S^2NS-matrix if and only if B is the sign pattern of an S^2NS-matrix whose nonzero entries in row r are in columns p and q, whose nonzero entries in column p are in rows r and s, and whose entry in position (s, q) is nonzero. The assertion in (b) now follows from Theorem 7.5.2. □

Theorem 7.5.4 implies the validity of the following polynomial-time algorithm for determining whether a $(1, -1)$-matrix is the sign pattern of the inverse of a fully indecomposable S^2NS-matrix [1].

Algorithm for determining if a matrix is the sign pattern of the inverse of a fully indecomposable S^2NS-matrix

Let B be a $(1, -1)$-matrix of order n.

(0) Let $E = \emptyset$, $\alpha = \{1, 2, \ldots, n\}$, and $\beta = \{1, 2, \ldots, n\}$.
(1) If α contains only one element, then B is the sign pattern of an inverse of an S^2NS-matrix. Otherwise,
(2) if there exist integers p, q, r, and s with $p, q \in \alpha$ and $r, s \in \beta$ such that

(i) rows p and q of $B[\alpha, \beta \setminus \{r\}]$ are equal or negatives of one another,
(ii) columns r and s of $B[\alpha \setminus \{p\}, \beta]$ are equal or negatives of one another,
(iii) if $(x, r) \in E$ and $x \in \alpha$, then $x = p$ or $x = q$, and
(iv) if $(p, y) \in E$ and $y \in \beta$, then $y = r$ or $y = s$,

then replace E by $E \cup \{(q, s)\}$, α by $\alpha \setminus \{p\}$, and β by $\beta \setminus \{r\}$, and then go to (1). Otherwise,

(3) B is not the sign pattern of the inverse of an S^2NS-matrix.

We conclude this section by stating without proof the following theorem [1], which gives a characterization of fully indecomposable, maximal S^2NS-matrices which are uniquely determined by the sign pattern of their inverse.

Theorem 7.5.5 *Let $A = [a_{ij}]$ be a fully indecomposable, $(0, 1, -1)$ maximal S^2NS-matrix of order n. Then A is the unique fully indecomposable, $(0, 1, -1)$ maximal S^2NS-matrix whose inverse has the same sign pattern as that of A^{-1} if and only if no edge of the bipartite graph of A is contained in exactly two cycles of length 4.*

Bibliography

[1] R. A. Brualdi, K. L. Chavey, and B. L. Shader. Bipartite graphs and inverse sign patterns of strong sign-nonsingular matrices. *J. Combin. Th. Ser B*, 62:133–52, 1994.

[2] C. Eschenbach, F. Hall, and C.R. Johnson. Self-inverse sign patterns, in *Combinatorial and Graph-Theoretical Problems in Linear Algebra* (R. A. Brualdi, S. Friedland and V. Klee, eds.), Springer-Verlag, 245–57, 1993.

[3] C.C. Lim. Nonsingular sign patterns and the orthogonal group, *Linear Alg. Appls.*, 184:1–12, 1993.

[4] C.C. Lim and D.A. Schmidt. Full sign-invertibility and symplectic matrices, *Linear Alg. Appls.*, to appear.

[5] C. Thomassen. When the sign pattern of a square matrix determines uniquely the sign pattern of its inverse, *Linear Alg. Appls.*, 119:27–34, 1989.

8

Extremal properties of L-matrices

8.1 Maximal SNS-matrices

In this section we determine the maximum number of nonzero entries in a SNS-matrix of order n and obtain a lower bound for the number of nonzero entries of a maximal SNS-matrix of order n. We also investigate the number of nonzero entries in a doubly indecomposable SNS-matrix.

The following theorem is contained in [8].

Theorem 8.1.1 *Let A be an SNS-matrix of order n. Then the number of nonzero entries of A satisfies*

$$\#(A) \le \frac{n^2 + 3n - 2}{2} \tag{8.1}$$

with equality if and only if there exist permutation matrices P and Q and strict signings D and E such that $DPAQE$ equals the lower Hessenberg matrix H_n in (1.12).

Proof The matrix H_n is an SNS-matrix with $\#(H_n) = (n^2 + 3n - 2)/2$. We prove by induction on n that (8.1) holds with equality only if the conditions in the theorem are satisfied. This is clear for $n = 1$, and we now assume that $n > 1$. If A is an SNS*-matrix, that is, a fully indecomposable SNS-matrix which has a row with no zero entries, then the conclusion is an immediate consequence of Theorems 4.1.4 and 4.4.1. First suppose that $A = [a_{ij}]$ is fully indecomposable. Let τ_{ij} equal the number of nonzero entries in the union of row i and column j ($i, j = 1, 2, \ldots, n$). Let p and q be integers such that $a_{pq} \ne 0$. Then by the induction hypothesis

$$\#(A(\{p\}, \{q\})) \le \frac{n^2 + n - 4}{2} \tag{8.2}$$

and hence

$$\#(A) = \#(A(\{p\}, \{q\})) + \tau_{pq} \leq \frac{n^2 + n - 4}{2} + \tau_{pq}. \tag{8.3}$$

If $\tau_{pq} < n + 1$, then strict inequality holds in (8.1). We now assume that $\tau_{pq} \geq n + 1$ for all integers p and q such that $a_{pq} \neq 0$.

Without loss of generality, assume that $a_{ii} \neq 0$ for $i = 1, 2, \ldots, n$. Then

$$\sum_{i=1}^{n} \tau_{ii} = 2\#(A) - n,$$

and using (8.3), we get

$$n\#(A) \leq n \left(\frac{n^2 + n - 4}{2} \right) + \sum_{i=1}^{n} \tau_{ii} = n \left(\frac{n^2 + n - 4}{2} \right) + (2\#(A) - n).$$

This implies that

$$\#(A) \leq \frac{n^2 + 3n - 2}{2} + 1$$

and that either (8.1) is a strict inequality or there is an integer i such that $\#(A(\{i\})) = (n^2 + n - 4)/2$. Assume the latter possibility holds with, say, $i = 1$. By the induction hypothesis, we may assume that $A(\{1\}) = H_{n-1}$. If $a_{n-1,1} \neq 0$ or $a_{n1} \neq 0$, then A is an SNS*-matrix and we are done. Otherwise, since $\tau_{nn} \geq n + 1$, we have $a_{1n} \neq 0$. If $a_{k1} \neq 0$ for some k with $2 \leq k \leq n - 2$, then it follows from Lemma 4.2.1 that one of the matrices obtained from $A(\{1\})$ by replacing the zero in its $(k - 1, n - 1)$ position by 1 or -1 is an SNS-matrix, contradicting the induction hypothesis. Therefore $a_{k1} = 0$ for $k \geq 1$, contradicting $\tau_{11} \geq n + 1$.

If A is partly decomposable, then it follows from the induction hypothesis that strict inequality holds in (8.1).   □

The digraphs of SNS-matrices with a negative diagonal for which equality holds in (8.1) are characterized in [13].

The maximum number of nonzero entries in a fully indecomposable SNS-matrix with only nonnegative entries is about half of the maximum number in a fully indecomposable SNS-matrix with negative entries allowed. As shown in section 3.2, the matrix

$$
A_n = \begin{bmatrix}
1 & 1 & \cdots & 1 & 0 & \cdots & 0 \\
0 & & & & & & \\
\vdots & & I_{\left\lfloor \frac{n-1}{2} \right\rfloor} & & & J_{\left\lfloor \frac{n-1}{2} \right\rfloor, \left\lceil \frac{n-1}{2} \right\rceil} & \\
0 & & & & & & \\
1 & & & & & & \\
\vdots & & O & & & I_{\left\lceil \frac{n-1}{2} \right\rceil} & \\
1 & & & & & &
\end{bmatrix}
$$

is a fully indecomposable SNS-matrix of order n with

$$
\#(A_n) = \left\lfloor \frac{n^2 + 6n - 3}{4} \right\rfloor. \tag{8.4}
$$

It was conjectured in [5] and proved in [6] that A_n has the largest number of nonzero entries among all fully indecomposable SNS-matrices of order n with only nonnegative entries.

Since the lower Hessenberg matrix H_n is not doubly indecomposable for $n \geq 3$, Theorem 8.1.1 implies that a doubly indecomposable SNS-matrix of order $n \geq 3$ has at most $(n^2 - 3n - 4)/2$ nonzero entries. In the next theorem we obtain a much better bounds for doubly indecomposable SNS-matrices and for $(0, 1)$ SNS-matrices.

Theorem 8.1.2 *Let A be a doubly indecomposable SNS-matrix of order n. Then*

$$
\#(A) \leq \frac{n + n\sqrt{8n - 7}}{2}.
$$

If A has only nonnegative entries, then

$$
\#(A) \leq \frac{n + n\sqrt{4n - 3}}{2}.
$$

Proof Let B be the matrix obtained from A by replacing each nonzero entry by a 1. Let $r = (r_1, r_2, \ldots, r_n)^T$ be the row sum vector of B. Then $\#(B) = r_1 + r_2 + \cdots + r_n$, and the sum of the entries of $B^T B$ equals $r^T r$. By the Cauchy–Schwarz inequality,

$$
r^T r \geq \frac{\left(\sum_{k=1}^{n} r_k \right)^2}{n}.
$$

By Lemma 6.2.10, every 3 by 2 submatrix of A has a zero entry, and hence entries off the main diagonal of $B^T B$ are either 0, 1, or 2. Thus the sum $r^T r$ of the entries of $B^T B$ is bounded above by

$$2n(n-1) + \sum_{k=1}^{n} r_k.$$

The number of nonzero entries of B now satisfies

$$\#(B)^2 - n\#(B) - 2n^2(n-1) \le 0,$$

and the first part of the theorem follows.

If A has only nonnegative entries, then every submatrix of order 2 has a zero entry, and hence each entry off the main diagonal of $B^T B$ is either 0 or 1. Arguing as before we now have $\#(B)^2 - n\#(B) - n^2(n-1) \le 0$ and the theorem follows. $\qquad\Box$

Theorem 8.1.2 implies that a strongly 2-connected digraph with no loops having more than

$$n\sqrt{4n-3} - n/2$$

arcs contains a directed cycle of even length.

Theorem 8.1.2 furnishes an alternative proof of Theorem 8.1.1, and hence of Theorem 4.4.1, which we now outline. The proof is by induction on n. If $n \le 2$, then clearly (8.1) holds. Assume that $n \ge 3$. Then

$$\frac{n + n\sqrt{8n-7}}{2} < \frac{n^2 + 3n - 2}{2},$$

and hence if A is doubly indecomposable, then strict inequality holds in (8.1). Now assume that A is fully indecomposable but not doubly indecomposable. It follows from Theorem 6.2.4 that there exist fully indecomposable SNS-matrices

$$\left[\begin{array}{c} A' \\ \hline u^T \end{array}\right] \text{ and } \left[\ v\ |\ B'\ \right] \tag{8.5}$$

of orders p and q, respectively, such that $p + q = n + 1$ and A has the form

$$\left[\begin{array}{cc} A' & O \\ W & B' \end{array}\right]$$

where W is a matrix obtained from vu^T by replacing certain nonzero entries by zeros. By induction,

$$\begin{aligned} \#(A) &\le \frac{p^2 + 3p - 2}{2} + \frac{q^2 + 3q - 2}{2} + \#(W) \\ &\le \frac{p^2 + 3p - 2}{2} + \frac{q^2 + 3q - 2}{2} + (p-1)(q-1). \end{aligned}$$

Hence (8.1) holds with equality if and only if u and v have no nonzero entries, $W = uv^T$, and both of the matrices in (8.5) have $(p^2+3p-2)/2$ and $(q^2+3q-2)/2$ nonzero entries, respectively. The conclusion for fully indecomposable matrices now follows by induction. As before, strict inequality in (8.5) holds for partly decomposable matrices.

We note that the first proof of Theorem 8.1.1 relies on the characterization of S^*- and SNS^*-matrices, whereas the second proof relies on the characterization of SNS-matrices which are not doubly indecomposable.

The maximum number of nonzero entries in a doubly indecomposable SNS-matrix of order $n \geq 4$ has not been determined. All known matrices of this type have at most $4n-4$ nonzero entries. The doubly indecomposable SNS-matrices described after Theorem 6.2.11 have exactly this number of nonzero entries.

We now show that a fully indecomposable SNS-matrix of order $n \geq 4$ can be embedded in an SNS-matrix of order n with at least $3n$ nonzero entries [5].

Theorem 8.1.3 *Let A be a fully indecomposable, maximal $(0, 1, -1)$ SNS-matrix of order $n \geq 4$. Then $\#(A) \geq 3n$, and if $n \geq 5$, then equality holds only if A has exactly three nonzero entries in each row and column.*

Proof We prove the theorem by induction on n. It is not difficult to show that up to row and column permutations and transposition, and multiplication of rows and columns by -1, the fully indecomposable, maximal SNS-matrices of order 4 are

$$\begin{bmatrix} -1 & 1 & 0 & 0 \\ -1 & -1 & 1 & 0 \\ -1 & -1 & -1 & 1 \\ -1 & -1 & -1 & -1 \end{bmatrix}, \begin{bmatrix} -1 & 0 & 1 & 0 \\ 0 & -1 & 0 & 1 \\ -1 & 1 & -1 & 1 \\ -1 & -1 & -1 & -1 \end{bmatrix}, \text{ and} \qquad (8.6)$$

$$\begin{bmatrix} -1 & 1 & 0 & 1 \\ -1 & -1 & 1 & 0 \\ 0 & -1 & -1 & 1 \\ -1 & 0 & -1 & -1 \end{bmatrix}. \qquad (8.7)$$

Each of these matrices has at least 12 nonzero entries. Now assume that $n \geq 5$. If each row and column of A contains at least three nonzero entries, then the theorem holds. Otherwise, we assume without loss of generality that row 1 of A contains exactly two nonzero entries $a_{11} = 1$ and $a_{12} = -1$. By Lemma 4.2.1, the matrix B obtained from A by the conformal contraction on a_{11} is a fully indecomposable SNS-matrix. By Lemma 6.2.1, the matrix A' obtained from B by copying column 1 is an SNS-matrix, and it follows from the maximal property of A that we have $A' = A$ and that B is a maximal SNS-matrix. The induction hypothesis now implies that A has at least $3(n - 1) + 4 = 3n + 1$ nonzero entries. $\square$

It has been noted in section 6.2 that the incidence matrix $I_7 + C_7 + C_7^3$ of the projective plane of order 2 is a fully indecomposable, maximal SNS-matrix of order 7. It has been conjectured [5] that there is no fully indecomposable, maximal SNS-matrix of order $n \geq 8$ with exactly $3n$ nonzero entries. The following theorem [5] shows that such matrices are highly restricted.

Theorem 8.1.4 *Let A be a fully indecomposable, maximal SNS-matrix of order $n \geq 5$ such that $\#(A) = 3n$. Then A is doubly indecomposable and each submatrix of A of order 2 contains a zero.*

Proof By Theorem 8.1.3, A has exactly three nonzero entries in each row and column. First suppose that A is not doubly indecomposable. By Theorems 6.2.2 and 6.2.4, there exist integers p and q with $p + q = n + 1$, and there exist fully indecomposable, maximal SNS-matrices

$$\left[\frac{A'}{u^T} \right] \text{ and } \left[\; v \; | \; B' \; \right] \tag{8.8}$$

such that after row and column permutations,

$$A = \left[\begin{array}{cc} A' & O \\ vu^T & B' \end{array} \right].$$

Since each row and column of A has exactly three nonzero entries, the matrices in (8.8) have at least $3p - 1$ and $3q - 1$ nonzero entries, respectively. Thus $\#(A) \geq (3p - 1) + (3q - 1) = 3n + 1 > 3n$, and we obtain a contradiction. Hence we conclude that A is doubly indecomposable.

By Lemma 6.2.10, every 2 by 3 submatrix and every 3 by 2 submatrix of A contains a zero entry. Suppose that $A[\{1, 2\}]$ has no zero entries. Since A is a doubly indecomposable SNS-matrix, $A[\{1, 2\}]$ is an SNS-matrix. Without loss of generality we may now assume that

$$A[\{1, 2, 3, 4\}] = \left[\begin{array}{cccc} 1 & -1 & 1 & 0 \\ 1 & 1 & 0 & 1 \\ 1 & 0 & a & b \\ 0 & 1 & c & d \end{array} \right].$$

By Lemma 6.2.12, it follows that each of a, b, c, and d is nonzero. Since A has exactly three nonzeros in each row and column, $n = 4$. $\square$

Let

$$\left[\, v \mid B' \,\right]$$

be a fully indecomposable, maximal SNS-matrix of order m where B' is an m by $m-1$ matrix and v is a column vector with exactly 3 nonzero entries. Let A be the maximal SNS-matrix obtained from the maximal SNS-matrix $I_7 + C_7 + C_7^3$ by conformally copying its last column. Let A' be the 7 by 8 matrix obtained from A by deleting its last row u. By Theorem 6.2.2,

$$\left[\begin{array}{cc} A' & O \\ vu^T & B \end{array}\right]$$

is a maximal SNS-matrix of order $m+7$ with $\#(B)+27$ nonzero entries. Starting with $B = I_7 + C_7 + C_7^3$ and recursively applying the preceding construction, we can obtain maximal SNS-matrices of order $7k$ with $27k - 6$ nonzero entries for each positive integer k. By conformally copying rows, it is thus possible to construct for each $n \geq 7$ a maximal SNS-matrix with

$$\frac{27}{7}n - \frac{34}{7}$$

or fewer nonzero entries. It is not known if there exists a maximal SNS-matrix B of order $n \geq 5$ with $\#(B) < 27n/7 - 6$.

It is easy to construct (nonmaximal) fully indecomposable SNS-matrices with exactly three nonzero entries in each row and column. Let $A = [a_{ij}]$ and $B = [b_{ij}]$ be SNS-matrices with exactly three nonzero entries in each row and column, for instance, $I_7 + C_7 + C_7^3$ or the matrix in (8.7). Suppose that $a_{11} \neq 0$ and $b_{11} \neq 0$. Let

$$X = \left[\begin{array}{cc} A' & E \\ F & B' \end{array}\right]$$

where A' is obtained from A by changing a_{11} to 0, B' is obtained from B by changing b_{11} to 0, the only nonzero entry in E is a 1 in its (1,1) position, and the only nonzero entry in F is the number $-a_{11}b_{11}$ in its (1,1)-position. Then X is a fully indecomposable SNS-matrix with exactly three nonzero entries in each row and column.

It has been proved in [15] that if P and Q are permutation matrices of order $n \geq 8$ such that $PQ = QP$ and $I_n + P + Q$ is fully indecomposable with exactly three 1's in each row and column, then $I_n + P + Q$ is not an SNS-matrix.

8.2 SNS-matrices

In this section we consider the number of nonzero entries in rows and columns of SNS-matrices. It follows from Theorem 8.1.1 that if A is an SNS-matrix of order n, then at least one row (and hence at least one column) contains at most $\lfloor (n+2)/2 \rfloor$ nonzero entries. The following theorem [12] implies that if each row of a matrix of order n contains at least $\lfloor \log_2 n \rfloor + 2$ nonzero entries, then the matrix is not an SNS-matrix.

Theorem 8.2.1 *Every SNS-matrix of order n has a row with at most $\lfloor \log_2 n \rfloor + 1$ nonzero entries. Moreover, for each positive integer n there exists an SNS-matrix of order n each of whose rows contains at least $\lfloor \log_2 n \rfloor + 1$ nonzero entries.*

Proof Let A be an SNS-matrix of order n and let the number of nonzero entries in row i be s_i ($i = 1, 2, \ldots, n$). Let s be the minimum of the numbers $s_1, s_2, \ldots, s_n$. For each strict signing D, at least one row of AD is unsigned. Since row i contains s_i nonzero entries, there are exactly 2^{n-s_i+1} strict signings D such that row i of AD is unsigned. Hence

$$2^n \leq \sum_{i=1}^{n} 2^{n-s_i+1} \leq n2^{n-s+1},$$

which implies that $s \leq \lfloor \log_2 n \rfloor + 1$.

Let B be an SNS-matrix of order n which is obtained from $-J_1$ by successively copying a row with the fewest number of nonzero entries. Then it follows from Theorem 4.4.6 that each row of B contains at least $\lfloor \log_2 n \rfloor + 1$ nonzero entries. $\square$

It is natural to expect that the upper bound on the minimum number of nonzero entries in a row given in Theorem 8.2.1 can be improved if we consider only $(0, 1)$ SNS-matrices. It was conjectured in [9] that there exists a constant c such that every digraph with no directed cycle of even length has a vertex v with at most c arcs leaving v, that is, there exists a constant $c + 1$ such that every $(0, 1)$ SNS-matrix has a row containing at most $c+1$ ones. The following construction [12] disproves this conjecture and shows that there are $(0, 1)$-matrices of order n each of whose row sums has the same order of magnitude as the upper bound in Theorem 8.2.1. Let B be an SNS-matrix of order n which is obtained from $-J_1$ by successively copying a row with the fewest number of nonzero entries and multiplying the new row and new column by -1 if the row which is copied has the same number of 1's and -1's. Then in each row of B, there are at least $\lfloor \log_2 n \rfloor + 1$ nonzero entries and the number of -1's is at least as large as the number of 1's. Let C be the matrix obtained from B by changing each of the 1's to 0 and each of the -1's to 1. Then $I_n \le C$, and C is a $(0, 1)$ SNS-matrix of order n each of whose rows contains at least

$$\frac{1}{2} \lfloor \log_2 n \rfloor + 1$$

1's. The incidence matrix of the projective plane shows that there are $(0,1)$ SNS-matrices each of whose rows contain more than this number of 1's. It has been proved [1] that there is a constant d such that every $(0,1)$ SNS-matrix of order $n \ge 2$ has a row with at most

$$\log_2 n - \frac{1}{3} \log_2 (\log_2 n) - d$$

1's.

The remainder of this section is devoted to proving two theorems of Thomassen [14]. We have included some new ideas in reworking his difficult proofs. The first of these two theorems implies that a doubly indecomposable SNS-matrix of order $n \ge 4$ has at least four rows, each of which contains exactly three nonzero entries. Each matrix obtained by consecutive 2-joins of the matrix

$$\begin{bmatrix} -1 & 1 & 0 & 1 \\ -1 & 1 & 1 & 0 \\ 0 & -1 & -1 & 1 \\ -1 & 0 & -1 & -1 \end{bmatrix}$$

is a doubly indecomposable SNS-matrix with exactly four rows with three nonzero entries. The second theorem asserts that a fully indecomposable SNS-matrix has a row or column which contains at most three nonzero entries. Recall by Lemma 6.2.10, each 2 by 3 submatrix and each 3 by 2 submatrix of a doubly indecomposable SNS-matrix contains a zero entry. Hence, if A is a doubly indecomposable $(0, 1)$-matrix such that $J_{2,3}$ or $J_{3,2}$ is a submatrix of A, then $\mathcal{Z}(A)$ does not contain an SNS-matrix.

Let A be a $(0, 1)$-matrix of order $n \ge 2$ such that $X = A[\alpha, \beta]$ is a p by $p + 1$ matrix and $A[\alpha, \bar{\beta}] = O$ for some positive integer p. Then the X-*reduction* of A is the matrix of order $p + 1$ obtained from X by appending on the bottom a new row whose ith entry is 0 if column i of $A[\bar{\alpha}, \beta]$ is a zero column and is 1, otherwise. For example, if

$$A = \begin{bmatrix} 1 & 1 & 0 & 0 & 0 \\ 1 & 1 & 1 & 0 & 0 \\ 0 & 0 & 1 & 1 & 0 \\ 1 & 0 & 1 & 1 & 1 \\ 1 & 0 & 0 & 1 & 1 \end{bmatrix},$$

then the $A[\{1, 2\}, \{1, 2, 3\}]$-reduction of A is the matrix

$$\begin{bmatrix} 1 & 1 & 0 \\ 1 & 1 & 1 \\ 1 & 0 & 1 \end{bmatrix}.$$

The following is an immediate consequence of Theorem 6.2.6.

Lemma 8.2.2 *Let A be a fully indecomposable $(0, 1)$-matrix of order $n \geq 2$ and let B be the X-reduction of A for some p by $p + 1$ submatrix X of A. If $\mathcal{Z}(A)$ contains an SNS-matrix, then so does $\mathcal{Z}(B)$.*

Theorem 8.2.3 *Let $A = [a_{ij}]$ be a fully indecomposable $(0, 1)$-matrix such that every row contains at least four 1's except for three rows, say rows 1, 2, and 3, which contain at least three 1's. Assume that each p by q zero submatrix of A satisfying $p + q = n - 1$ does not intersect row 1. Then $\mathcal{Z}(A)$ does not contain an SNS-matrix.*

Proof We assume that A is a counterexample to the theorem with the smallest order and, subject to that condition, has the fewest number of 1's. The proof consists of the verification of a number of assertions which eventually lead to a contradiction. These assertions are numbered from (1) to (5), and the end of the verification of an assertion is indicated with a $\diamond$.

(1) A is doubly indecomposable and has a row with exactly three 1's.

Suppose that $A[\alpha, \overline{\beta}]$ is a p by $n - 1 - p$ zero matrix with the minimum number of rows. Since each row of A contains at least three 1's, we have $p \geq 2$. The hypotheses of the theorem imply that 1 is not contained in α. Let $X = A[\alpha, \beta]$ and consider the X-reduction B of A. Each of the first p rows of B contains the same number of 1's as the corresponding row of A. If there is an r by s zero submatrix of B which intersects its last row, then there is an $r + n - 1 - p$ by s zero submatrix of A which intersects row 1. It follows that no r by s zero submatrix of B with $r + s \geq p$ intersects the last row of B. In particular, the last row of B contains at least three 1's. By the minimum property of p, B is fully indecomposable. Therefore, B satisfies the hypotheses of the theorem but has smaller order than A. We conclude that A is a doubly indecomposable matrix. Suppose that each row of A has at least four 1's. Let $i \neq 1$ and j be integers such that $a_{1j} = a_{ij} = 1$. Let A' be the matrix obtained from A by changing a_{ij} to 0. Then A' satisfies the hypotheses of the theorem, and since $\mathcal{Z}(A)$ contains an SNS-matrix, so does A'. This contradicts the choice of A. $\diamond$

Since A is now a doubly indecomposable matrix, each of rows 2 and 3 can play the role of row 1. Hence we assume that row 1 has exactly three 1's and that $a_{11} = 1$.

(2) At least one of the matrices obtained from A by changing a 1 in the first row to 0, and then contracting on that row, has the property that each of its rows contains at least three 1's.

Suppose the 1's in the first row are in columns 1, 2, and 3, and suppose that each of the three matrices obtained by changing a 1 in the first row to 0 and contracting on that row contains a row with two 1's. By Lemma 6.2.10, A does not have a submatrix equal to $J_{2,3}$ and it follows that A has a submatrix whose columns can be permuted to give

$$C = \begin{bmatrix} 1 & 1 & 1 \\ 1 & 1 & 0 \\ 1 & 0 & 1 \\ 0 & 1 & 1 \end{bmatrix}.$$

The first row of C corresponds to the first row of A. The other rows of C correspond to rows of A with exactly three 1's. Hence we contradict the fact that at most three rows of A have only three 1's. ◇

Let A_1 be a matrix satisfying (2). Since A does not contain a submatrix equal to $J_{2,3}$, at most three rows of A_1 contain exactly three 1's. By Lemma 4.2.1, A_1 is fully indecomposable and $\mathcal{Z}(A_1)$ contains an SNS-matrix. Thus, by the choice of A, the matrix A_1, and hence A, contains a p by $n - 2 - p$ zero submatrix $A[\alpha, \beta]$ for some integer $p \geq 2$. We assume that p is a minimal such integer. Let $\widetilde{\alpha}$ and $\widetilde{\beta}$ denote the complements of α and β, respectively, in $\{2, 3, \ldots, n\}$. The matrix A now has the form

$$\begin{bmatrix} 1 & u^1 & v^1 \\ \hline w^1 & A[\alpha, \widetilde{\beta}] & O \\ \hline x^1 & A[\widetilde{\alpha}, \widetilde{\beta}] & A[\widetilde{\alpha}, \beta] \end{bmatrix}$$

where $A[\alpha, \widetilde{\beta}]$ is a p by $p + 1$ matrix, u^1 and v^1 each contain exactly one 1 (if v^1 does not contain a 1, then A is clearly not a doubly indecomposable matrix, contradicting (1); if u^1 does not contain a 1, then $w^1 = 0$ and A is again not a doubly indecomposable matrix), and A_1 is the matrix obtained from A by changing the 1 in v^1 to 0 and then contracting on a_{11}. Changing the 1 in v^1 to 0 and taking the X-reduction, where $X = [\{1\}\cup\alpha, \{1\}\cup\widetilde{\beta}]$, and then contracting on a_{11}, we obtain a matrix

$$A' = \left[\begin{array}{c} A[\alpha, \widetilde{\beta}]' \\ \hline z' \end{array}\right]$$

where $A[\alpha, \widetilde{\beta}]'$ differs from $A[\alpha, \widetilde{\beta}]$ only in the column corresponding to the 1 in u^1.

(3) A' is a doubly indecomposable matrix of order $p+1$, and $\mathcal{Z}(A')$ contains an SNS-matrix.

Consider an r by s zero submatrix M of A'. If M intersects the last row of A' but not the column corresponding to the 1 in u^1, then A has an $r+(n-1-p)$ by s zero submatrix. If M intersects the last row of A' and the column corresponding to the 1 in u^1, then A has an $r+(n-2-p)$ by $s+1$ zero submatrix. Otherwise, A has an r by $s+(n-2-p)$ zero submatrix. It follows from (1) and the minimal property of p that A' does not have an r by s zero submatrix with $r+s = p$. Hence A' is a doubly indecomposable matrix. By Lemmas 8.2.2 and 4.2.1, $\mathcal{Z}(A')$ contains an SNS-matrix. $\diamond$

It now follows from the choice of A that A' has at least four rows, each of which contains exactly three 1's. By Lemma 6.2.10, at most one row of $A[\alpha, \widetilde{\beta}]'$ has fewer 1's than the corresponding row of A. Since A has at most three rows (one of which is row 1) which contain exactly three 1's, we conclude that A' has exactly four rows with exactly three 1's, and that these rows are: a row of $A[\alpha, \widetilde{\beta}]'$ that corresponds to a row of A with exactly four 1's (thus this row of A has a 1 in column 1 and the column corresponding to the 1 in u^1), two rows of $A[\alpha, \widetilde{\beta}]'$ that correspond to rows of A which contain exactly three 1's, and the row z'. This implies the following assertion.

(4) A has three rows with exactly three 1's; both 2 and 3 are in α; there exists an integer ℓ_1 in α such that row ℓ_1 of A contains a 1 in the first column, a 1 in the column corresponding to the 1 in u^1 and two other 1's; and $A[\widetilde{\alpha}, \widetilde{\beta}]$ contains at most three nonzero columns. $\diamond$

Since both 2 and 3 are in α, it now follows that each of the rows of the matrix A_2 obtained from A by changing the 1 in u^1 to 0 and then contracting on a_{11}

contains at least three 1's. Hence the matrix A also has the form

$$\begin{bmatrix} 1 & u^2 & v^2 \\ \hline w^2 & A[\gamma, \tilde{\delta}] & O \\ \hline x^2 & A[\tilde{\gamma}, \tilde{\delta}] & A[\tilde{\gamma}, \delta] \end{bmatrix}$$

where $A[\gamma, \tilde{\delta}]$ is a q by $q+1$ matrix (as with A_1 we assume that q is a minimal integer such that A_2 contains a q by $n-2-q$ zero submatrix) and u^2 and v^2 each contain exactly one 1 (the one in u^2 corresponds to the 1 in v^1 and the 1 in v^2 corresponds to the 1 in u^1). Arguing with A_2 as we did with A_1, we obtain the following assertion.

(5) Both 2 and 3 are in γ; there exists an integer ℓ_2 in γ such that row ℓ_2 of A contains a 1 in the first column, a 1 in the column corresponding to the 1 in u^2, and two other 1's; and $A[\tilde{\gamma}, \tilde{\delta}]$ contains at most three nonzero columns. ◇

The existence of ℓ_2 implies that $x^1 \neq 0$ and hence $A[\tilde{\alpha}, \tilde{\beta}]$ contains exactly two nonzero columns other than the column corresponding to the 1 in u^1. Similarly, $A[\tilde{\gamma}, \tilde{\delta}]$ contains exactly two nonzero columns other than the column corresponding to the 1 in u^2. We now write A in the form

$$\begin{array}{cccccc} & \{1\} & \tilde{\beta}\cap\tilde{\delta} & \tilde{\beta}\cap\delta & \beta\cap\tilde{\delta} & \beta\cap\delta \\ \{1\} & \begin{bmatrix} 1 & 0\cdots0 & 10\cdots0 & 10\cdots0 & 0\cdots0 \\ & & O & O & O \\ & & & O & O \\ & & O & & O \\ & & & & \end{bmatrix} \\ \alpha\cap\gamma \\ \alpha\cap\tilde{\gamma} \\ \tilde{\alpha}\cap\gamma \\ \tilde{\alpha}\cap\tilde{\gamma} \end{array}.$$

Let e be the column in $\tilde{\beta}\cap\delta$ containing a 1 in the first row, and let f be the column in $\beta\cap\tilde{\delta}$ containing a 1 in the first row. The minimality of p implies that $A[\alpha, \tilde{\beta}\setminus\{e\}]$ is fully indecomposable. Hence at least two columns of $A[\alpha\cap\tilde{\gamma}, \tilde{\beta}\cap\tilde{\delta}]$ are nonzero. Since $A[\tilde{\gamma}, \tilde{\delta}\setminus\{f\}]$ contains exactly two nonzero columns, $A[\alpha\cap\tilde{\gamma}, \tilde{\beta}\cap\tilde{\delta}]$ contains exactly two nonzero columns, say columns c and d, and $A[\alpha\cap\tilde{\gamma}, \tilde{\beta}\cap\delta]$ is a square matrix of some order k_1. In addition, $A[\tilde{\gamma}, \tilde{\delta}\setminus\{c,d,f\}] = O$. Similarly, $A[\tilde{\alpha}\cap\gamma, \tilde{\beta}\cap\tilde{\delta}]$ contains exactly two nonzero columns, say columns a and b, $A[\tilde{\alpha}\cap\gamma, \beta\cap\tilde{\delta}]$ is a square matrix of some order k_2, and $A[\tilde{\alpha}, \tilde{\beta}\setminus\{a,b,e\}] = O$. The matrix $A[\alpha\cap\gamma, \tilde{\beta}\cap\tilde{\delta}]$ is a $p-k_1$ by $p-k_1+1$ matrix (since $\{2,3\}\subseteq\alpha\cap\gamma$, $p-k_1\geq 2$ and hence

$p - k_1 + 1 \geq 3$), and the matrix $A[\tilde{\alpha} \cap \tilde{\gamma}, \beta \cap \delta]$ is a $n - 1 - p - k_2$ by $n - 2 - p - k_2$ matrix. It follows from (1) that $\beta \cap \delta = \emptyset$ and that $\tilde{\alpha} \cap \tilde{\gamma}$ contains a unique element t. If $\{a, b\} = \{c, d\}$, then $A[:, \overline{\beta} \cap \tilde{\delta}]$ contains a $(n - p + k_1)$ by $(p - 1 - k_1)$ nonvacuous zero submatrix, contradicting (1). Hence $\{a, b\} \neq \{c, d\}$. The 1's in row t can only be contained in columns $1, e,$ f, and those columns in $\{a, b\} \cap \{c, d\}$. Therefore, since row t of A contains at least four 1's, row t contains a 1 in each of columns $1, e,$ and f. But then $J_{2,3}$ is a submatrix of A, contradicting Lemma 6.2.10. $\square$

The next theorem implies that if A is a (0,1)-matrix with exactly $k \geq 4$ 1's in each row and column, then $\mathcal{Z}(A)$ does not contain an SNS-matrix.

Theorem 8.2.4 *A fully indecomposable SNS-matrix has a row or column which contains at most three nonzero entries.*

Proof Let A be a fully indecomposable (0,1)-matrix with at least four 1's in each row and column. If A is a doubly indecomposable matrix, then it follows from Theorem 8.2.3 that $\mathcal{Z}(A)$ does not contain an SNS-matrix. Now assume that A has a p by q zero submatrix $A[\alpha, \overline{\beta}]$ where p and q are positive integers such that $p + q = n - 1$. Clearly, p and q are at least 3. We assume that no r by s zero submatrix of A with $r + s = n - 1$ satisfies $\min\{r, s\} < \min\{p, q\}$. Since $\mathcal{Z}(A)$ contains an SNS-matrix if and only if $\mathcal{Z}(A^T)$ does, we may assume without loss of generality that $p \leq q$. Let B be the $A[\alpha, \beta]$-reduction of A. Suppose M is a c by d zero submatrix of B with $c + d = p$. If M intersects the last row of B, then A contains a $c + n - 1 - p$ by d zero submatrix with $d < p$, a contradiction. If M does not intersect the last row of B, then A contains a c by $d + n - 1 - p$ zero submatrix with $c < p$, again a contradiction. Hence B is a doubly indecomposable matrix. The last row of B contains at least three 1's, and each of the other rows contains at least four. By Theorem 8.2.3, $\mathcal{Z}(B)$ does not contain an SNS-matrix. By Lemma 8.2.2, $\mathcal{Z}(A)$ does not contain an SNS-matrix. The theorem now follows. $\square$

It is proved in [1] using probabilistic techniques that if A is a (0,1)-matrix with exactly nine 1's in each row and column, then $\mathcal{Z}(A)$ does not contain an SNS-matrix. In [7] the validity of the van der Waerden conjecture for permanents (see [11]) is used to show that a (0,1)-matrix B with exactly eight 1's in each row and column is not an SNS-matrix. Both of these results are immediate consequences of Theorem 8.2.4. However, because of the elegant mathematics involved we give separate proofs of these two results.

We use the following special case of the Lovász Local Lemma from probability, which generalizes the fact that given a finite number of mutually independent events each with a positive probability, then the probability that all the events occur simultaneously is positive (see [2]).

Lemma 8.2.5 *Let E_1, E_2, ..., E_n be events in an arbitrary probability space. Assume that p and t are numbers such that for each i, the probability of event E_i is at most p and that the event E_i is mutually independent of a set of at least $n - 1 - t$ other events E_j. If*

$$ep(t + 1) < 1,$$

then the probability that none of the events occur is positive.[1]

Theorem 8.2.6 *Let $A = [a_{ij}]$ be a $(0, 1)$-matrix of order n with exactly k ones in each row and column. If $k \geq 9$, then $\mathcal{Z}(A)$ does not contain an SNS-matrix.*

Proof Suppose that $\widehat{A}$ is an SNS-matrix in $\mathcal{Z}(A)$. Consider the probability space consisting of the 2^n strict column signings of $\widehat{A}$ each with probability $1/2^n$. For $i = 1, 2, \ldots, n$, let E_i be the event consisting of all strict signings of $\widehat{A}$ which unisign row i. Since A is an SNS-matrix, each column signing of A has at least one unsigned row, and hence the probability that none of the events E_i occur is 0. For each $i = 1, 2, \ldots, n$, row i of $\widehat{A}$ has k nonzero entries and hence there are exactly 2^{n-k+1} strict signings D such that row i of $\widehat{A}D$ is unsigned. Thus the probability of E_i is $1/2^{k-1}$. Let

$$S_i = \{E_\ell : \text{there does not exist } j \text{ such that } a_{ij} = a_{\ell j} = 1\}.$$

Then E_i is mutually independent of the set of events S_i. Since each row and column of A has exactly k ones, it follows that $|S_i| \geq n - 1 - k(k - 1)$. It follows from Lemma 8.2.5 with $p = 1/2^{k-1}$ and $t = k(k - 1)$ that

$$ep(t + 1) = \frac{e(k^2 - k + 1)}{2^{k-1}} \geq 1,$$

implying that $k \leq 8$. □

Theorem 8.2.7 *Let A be a $(0, 1)$-matrix of order n with exactly k ones in each row and column. If $k \geq 8$, then $\mathcal{Z}(A)$ does not contain an SNS-matrix.*

Proof Suppose that $\widehat{A}$ is an SNS-matrix in $\mathcal{Z}(A)$. Without loss of generality we may assume that $\det \widehat{A} > 0$. Since $\widehat{A}$ is a $(0, 1, -1)$ SNS-matrix with a positive determinant,

$$\det \widehat{A} = \text{per } A.$$

---

[1] In the inequality, e is the base of the natural logarithm.

The matrix $(1/k)A$ is a doubly stochastic matrix of order n. Since the minimum permanent of a doubly stochastic matrix of order n is

$$\mathrm{per}\left(\tfrac{1}{n}J_n\right) = \frac{n!}{n^n},$$

we have

$$\mathrm{per}\, A \geq \frac{k^n n!}{n^n}.$$

By the classical Hadamard inequality for determinants,

$$\det \widehat{A} \leq k^{n/2}.$$

Therefore,

$$\frac{k^n n!}{n^n} \leq k^{n/2},$$

and hence $k \leq 7$. $\square$

8.3 Barely L-matrices

Let A be an m by n $(0, 1, -1)$ L-matrix. Recall that A is a barely L-matrix provided each of the matrices obtained from A by deleting a column is an L-matrix. By (ii) of Theorem 2.1.1, A is a barely L-matrix if and only if for each $i = 1, 2, \ldots, n$ the set $\mathcal{A}_i$ of signings D for which column i is the only unsigned column of DA is nonempty. It follows that the sign pattern of a column of a barely L-matrix cannot equal or be the negative of the sign pattern of another column. Hence a barely L-matrix with m rows has at most $(3^m - 1)/2$ columns. In section 2.1 we showed that the m by 2^{m-1} matrix Λ_m whose columns are all the m by 1 $(1, -1)$-column vectors with leading entry 1 is a barely L-matrix. In this section we show that Λ_m has the maximum number of columns of a barely L-matrix with m rows, and we also determine the maximum number of columns of a $(0, 1)$ barely L-matrix [3].

Theorem 8.3.1 *Let A be an m by n $(0, 1, -1)$ barely L-matrix. Then $n \leq 2^{m-1}$. If $m \geq 3$, then equality holds if and only if there exists a strict signing D and a permutation matrix P such that $A = \Lambda_m DP$.*

Proof First assume that $A = [a_{ij}]$ is L-indecomposable. By Theorem 2.2.5, $\mathcal{A}_i$ contains a strict signing D_i for each i. Since $-D_i$ is also in $\mathcal{A}_i$ and since the number of strict signings of order m is 2^m, it follows that $n \leq 2^{m-1}$. Assume that $n = 2^{m-1}$ and $m \geq 3$. Then each strict signing of order m belongs to exactly one $\mathcal{A}_i$, and the only strict signings in $\mathcal{A}_i$ are D_i and $-D_i$. Suppose that $a_{pq} = 0$ for some integers p and q. Let E be the strict signing obtained by replacing the pth diagonal entry of D_q by its negative. Then column q of EA is unsigned and hence E is in $\mathcal{A}_q$. Since $E \neq D_q, -D_q$, we conclude that no entry of A equals 0. Thus A is a $(1, -1)$-matrix such that each m by 1 $(1, -1)$-column vector, or its negative, is a column of A. Hence, up to multiplication of columns by -1 and column permutations, A equals Λ_m.

Now assume that A is L-decomposable and let $A_1, A_2, \ldots, A_k$ be the L-indecomposable components of A. Let m_i be the number of rows of A_i ($i = 1, 2, \ldots, k$). By Theorem 2.2.11, each A_i is a barely L-matrix. Using what we have proved above, we now obtain

$$n \leq 2^{m_1-1} + 2^{m_2-1} + \cdots + 2^{m_k-1}.$$

Hence $n \leq 2^{m-1}$ with equality only if $m \leq 2$. The theorem now follows. $\quad\square$

Corollary 8.3.2 *Let m be a positive integer. Then there exists an m by n barely L-matrix if and only if $m \leq n \leq 2^{m-1}$.*

Proof It follows from Theorem 8.3.1 that if A is an m by n barely matrix, then $m \leq n \leq 2^{n-1}$. We show by induction on m that if n is an integer with $m \leq n \leq 2^{m-1}$, then there exists an m by n barely L-matrix. This is clear if $m = 1$. Assume that $m > 1$ and that $m \leq n \leq 2^{m-1}$. If $n \leq 2^{m-2} + 1$, then by the induction hypothesis there exists an $m - 1$ by $n - 1$ barely L-matrix B and hence $B \oplus I_1$ is an m by n barely L-matrix. Now suppose that $n = 2^{m-2} + t$ where $t \geq 2$. Let $\alpha = \{1, 2, \ldots, t\}$ and let C be the m by n matrix

$$\left[\begin{array}{c|c|c} \Lambda_{m-1}[:, \alpha] & \Lambda_{m-1}[:, \alpha] & \Lambda_{m-1}[:, \overline{\alpha}] \\ \hline 1 \quad \cdots \quad 1 & -1 \quad \cdots \quad -1 & 0 \quad \cdots \quad 0 \end{array} \right].$$

It is now easy to verify that every row signing of C has a unsigned column and that, for each j with $1 \leq j \leq n$, there is a signing D_j such that only column j of $D_j C$ is unsigned. Hence C is a barely L-matrix. $\quad\square$

Let $A = [a_{ij}]$ be an m by n $(0,1)$-matrix. Then A is the incidence matrix of the family $\mathcal{X} = \{X_1, X_2, \ldots, X_n\}$ of subsets of $\{1, 2, \ldots, m\}$ where $X_j = \{i : a_{ij} = 1\}$ ($j = 1, 2, \ldots, n$). The family $\mathcal{X}$ is an *antichain* provided $X_i \not\subseteq X_j$ for all $i \neq j$. It follows from Theorem 2.2.5 that if A is an L-indecomposable, barely L-matrix, then the family $\mathcal{X}$ is an antichain. A well-known theorem of

Sperner asserts that if $\mathcal{X}$ is an antichain, then

$$n \leq \left(\begin{array}{c} m \\ \lfloor \frac{m+1}{2} \rfloor \end{array} \right), \tag{8.9}$$

with equality if and only if m is even and $\mathcal{X}$ consists of all subsets of $\{1, 2, \ldots, m\}$ of cardinality $m/2$, or m is odd and $\mathcal{X}$ consists of either all subsets of $\{1, 2, \ldots, m\}$ of cardinality $(m-1)/2$ or all subsets of cardinality $(m+1)/2$. It follows that if A is an L-indecomposable, barely L-matrix, then (8.9) holds. If m is odd, say $m = 2k + 1$, the matrix Γ_k defined in section 2.1 is the incidence matrix of the family of all subsets of $\{1, 2, \ldots, m\}$ of cardinality $k + 1$ and, by Corollary 2.2.6, is an L-indecomposable, barely L-matrix. Hence, if m is odd, Sperner's theorem implies that the maximum number of columns of a $(0,1)$ L-indecomposable, barely L-matrix with m rows is

$$\left(\begin{array}{c} m \\ \lfloor \frac{m+1}{2} \rfloor \end{array} \right).$$

If $m \geq 4$ is even, then the following lemma implies that the bound (8.9) on the number n of columns of an L-indecomposable, barely L-$(0, 1)$-matrix with m rows cannot be attained.

Lemma 8.3.3 *Let A be the incidence matrix of the family of all subsets of $\{1, 2, \ldots, m\}$ of a fixed cardinality $k > 0$. Then A is an L-matrix if and only if $1 \leq k \leq \lfloor (m + 1)/2 \rfloor$. The matrix A is a barely L-matrix if and only if $k = 1$, or m is odd and $k = (m + 1)/2$.*

Proof First assume that $k > \lfloor (m + 1)/2 \rfloor$. Let D be the strict signing of order m whose first $\lfloor (m + 1)/2 \rfloor$ diagonal entries are 1 and whose remaining diagonal entries are -1. Then each column of DA is balanced and hence A is not an L-matrix. Now assume that $k \leq \lfloor (m + 1)/2 \rfloor$. Each signing of order m contains either $\lfloor (m + 1)/2 \rfloor$ diagonal entries which are nonnegative and not all zero, or the same number of diagonal entries which are nonpositive and not all zero. Hence each row signing of A contains a unisigned column, and A is an L-matrix.

If $k = 1$, then A is a permutation matrix of order m and hence is a barely L-matrix. If m is odd and $k = (m + 1)/2$, then up to column permutations A equals Γ_k and hence is a barely L-matrix. Finally, assume that $2 \leq k \leq \lfloor m/2 \rfloor$. Let D be a signing such that the first column of DA is unisigned. Without loss of generality we assume that the first column of DA contains a positive entry. If there is an integer p such that the pth entry of the first column of A equals 0 and the pth diagonal entry of D is 0 or 1, then since $k \geq 2$, some column of DA other than column 1 is also unsigned. Otherwise, since $k \leq \lfloor m/2 \rfloor$, D has at least $\lceil m/2 \rceil$ diagonal entries equal to -1, and again we conclude that some column of DA other than column 1 is unsigned. Therefore the matrix obtained by deleting column 1 of A is an L-matrix, and hence A is not a barely L-matrix.   □

In order to obtain the best bound for the number of columns of a barely L-matrix with an even number of rows, we use the following extremal result [10], which we state without proof. An antichain $\mathcal{X} = \{X_1, X_2, \ldots, X_n\}$ is *intersecting* provided that $X_i \cap X_j \neq \emptyset$ for all $i \neq j$.

Lemma 8.3.4 *If $\mathcal{X} = \{X_1, X_2, \ldots, X_n\}$ is an intersecting antichain of subsets of $\{1, 2, \ldots, m\}$, then*

$$n \leq \left(\begin{array}{c} m \\ \lceil \frac{m+1}{2} \rceil \end{array} \right)$$

with equality if and only if one of the following holds:

(i) *$\mathcal{X}$ is the family of all subsets of $\{1, 2, \ldots, m\}$ of cardinality $\lceil (m+1)/2 \rceil$;*
(ii) *m is even and there exists an integer p in $\{1, 2, \ldots, m\}$ such that $\mathcal{X}$ consists of all subsets of $\{1, 2, \ldots, m\}$ of cardinality $m/2$ containing p and all subsets of cardinality $m/2 + 1$ not containing p.*

Let $m = 2k$ be a positive even integer. We define Ω_k to be the incidence matrix of the family $\mathcal{X}^{(k)}$ of subsets of $\{1, 2, \ldots, 2k\}$ consisting of all subsets of $\{1, 2, \ldots, 2k\}$ of cardinality k containing 1 and all subsets of cardinality $k + 1$ not containing 1. The

$$\binom{2k-1}{k-1} + \binom{2k-1}{k+1} = \binom{2k}{k+1}$$

columns of Ω_k are taken to be in lexicographic order. For example,

$$\Omega_3 = \left[\begin{array}{cccccccccccccccc}
0 & 0 & 0 & 0 & 0 & 1 & 1 & 1 & 1 & 1 & 1 & 1 & 1 & 1 & 1 \\
0 & 1 & 1 & 1 & 1 & 0 & 0 & 0 & 0 & 0 & 0 & 1 & 1 & 1 & 1 \\
1 & 0 & 1 & 1 & 1 & 0 & 0 & 0 & 1 & 1 & 1 & 0 & 0 & 0 & 1 \\
1 & 1 & 0 & 1 & 1 & 0 & 1 & 1 & 0 & 0 & 1 & 0 & 0 & 1 & 0 \\
1 & 1 & 1 & 0 & 1 & 1 & 0 & 1 & 0 & 1 & 0 & 0 & 1 & 0 & 0 \\
1 & 1 & 1 & 1 & 0 & 1 & 1 & 0 & 1 & 0 & 0 & 1 & 0 & 0 & 0
\end{array} \right].$$

Lemma 8.3.5 *Let A be an m by n incidence matrix of an intersecting antichain and assume that A is an L-matrix. Then A is an L-indecomposable, barely L-matrix. The matrix Ω_k is an L-indecomposable, barely L-matrix for each integer $k \geq 2$.*

Proof For each $j = 1, 2, \ldots, n$ let D_j be the strict signing of order m whose ith diagonal entry equals 1 if and only if the ith entry in column j of A equals 1. Since A is the incidence matrix of an intersecting antichain, the only unsigned column of $D_j A$ is column j. Hence, by Theorem 2.2.5, A is an L-indecomposable, barely L-matrix.

Let $k \geq 2$. To complete the proof of the lemma we show that Ω_k is an L-matrix. Let D be a signing of order $2k$. If the first diagonal entry of D is 0, then clearly some column of $D\Omega_k$ corresponding to a set of $\mathcal{X}^{(k)}$ containing 1 is unsigned. Now assume that the first diagonal entry of D is nonzero, and without loss of generality assume that it equals 1. If D contains at most k negative 1's, then again some column of $D\Omega_k$ corresponding to a set containing 1 is unsigned. Otherwise, some column of $D\Omega_k$ corresponding to a set not containing 1 is unsigned. Hence Ω_k is an L-matrix. $\qquad\square$

Theorem 8.3.6 *Let A be an m by n $(0, 1)$ barely L-matrix with $m \geq 3$. Then*

$$n \leq \binom{m}{\lceil \frac{m+1}{2} \rceil}. \tag{8.10}$$

If $m \geq 5$, then equality holds if and only if either m is odd, say $m = 2k + 1$, and there exists a permutation matrix P such that $A = \Lambda_k P$; or m is even, say $m = 2k$, and there exist permutation matrices Q and R such that $A = Q\Omega_k R$.

Proof Matrix A is the incidence matrix of a family $\mathcal{X} = \{X_1, X_2, \ldots, X_n\}$ of subsets of $\{1, 2, \ldots, m\}$. First, assume that A is L-indecomposable. By Theorem 2.2.5, $\mathcal{A}_j$ contains a strict signing D_j for each integer j with $1 \leq j \leq n$. Without loss of generality we assume that each entry of column j of $D_j A$ is 0 or 1. Let Y_j be the subset of $\{1, 2, \ldots, m\}$ consisting of those integers i such that the ith diagonal entry of D_j equals 1, and let $\mathcal{Y} = \{Y_1, Y_2, \ldots, Y_n\}$. Let ℓ be an integer with $1 \leq \ell \leq n$ and $\ell \neq j$. Since D_j is a strict signing and column ℓ of $D_j A$ is balanced, $X_\ell \cap Y_j \neq \emptyset$ and $X_\ell \cap \overline{Y}_j \neq \emptyset$. Since column ℓ of $D_\ell A$ contains only 0's and 1's, we have $X_\ell \subseteq Y_\ell$. Hence $Y_\ell \cap Y_j \neq \emptyset$ and $Y_\ell \cap \overline{Y}_j \neq \emptyset$, and $\mathcal{Y}$ is an intersecting antichain. Applying Theorem 8.3.4, we conclude that (8.10) holds. Assume that equality holds in (8.10). Then $\mathcal{Y}$ satisfies either (i) or (ii) of Lemma 8.3.4. Suppose that X_j is a proper subset of Y_j for some j. Then since $\mathcal{Y}$ satisfies (i) or (ii) of Lemma 8.3.4, it follows that there exists an integer $i \neq j$ such that X_j is also a subset of Y_i. This implies that column j of $D_i A$ is unsigned, contradicting the fact that only column i of $D_i A$ is unsigned. We conclude that $X_j = Y_j$ for all j, and hence $\mathcal{X}$ satisfies (i) or (ii) of Lemma 8.3.4. If $\mathcal{X}$ satisfies (i), then by Lemma 8.3.3, m is odd, say $m = 2k + 1$, and there exists a permutation matrix P such that $A = \Gamma_k P$. If $\mathcal{X}$

satisfies (ii), then m is even, say $m = 2k$, and there exist permutation matrices Q and R such that $A = Q\Omega_k R$.

Now assume that A is L-decomposable, and let $A_1, A_2, \ldots, A_\ell$ be the L-indecomposable components of A. Since A is a barely L-matrix, each A_i is also a barely L-matrix. Let m_i be the number of rows of A_i ($i = 1, 2, \ldots, \ell$). Since there is no (0,1) L-indecomposable, barely L-matrix with two rows and since (8.10) holds for $m = 1$, we obtain

$$n \leq \binom{m_1}{\lceil \frac{m_1+1}{2} \rceil} + \cdots + \binom{m_\ell}{\lceil \frac{m_\ell+1}{2} \rceil} \leq \binom{m}{\lceil \frac{m+1}{2} \rceil},$$

with strict inequality if $m \geq 5$.

Applying Corollary 2.2.6 and Lemma 8.3.5, we complete the proof of the theorem. $\qquad\square$

Bibliography

[1] N. Alon and N. Lineal. Cycles of length 0 modulo k in directed graphs, *J. Combin. Theory, Ser. B*, 47:114–19, 1989.

[2] N. Alon and J.H. Spencer. *The Probabilistic Method*, Wiley, New York, 1992.

[3] R.A. Brualdi, K.L. Chavey, and B.L. Shader. Rectangular L-matrices, *Linear Alg. Appls.*, 196:37–61, 1994.

[4] R.A. Brualdi and H.J. Ryser. *Combinatorial Matrix Theory*, Cambridge University Press, New York, 1991.

[5] R.A. Brualdi and B.L. Shader. On sign-nonsingular matrices and the conversion of the permanent into the determinant, in *Applied Geometry and Discrete Mathematics* (P. Gritzmann and B. Sturmfels, eds.), Amer. Math. Soc., Providence, 117–34, 1991.

[6] F.R.K. Chung, W. Goddard, and D.J. Kleitman. Even cycles in directed graphs, *SIAM J. Disc. Math.*, 7:474–83, 1994.

[7] S. Friedland. Every 7-regular digraph contains an even cycle, *J. Combin. Theory, Ser. B*, 46:249–52, 1989.

[8] P.M. Gibson. Conversion of the permanent into the determinant, *Proc. Amer. Math. Soc.*, 27:471–6, 1971.

[9] C.H.C. Little. A characterization of convertible (0, 1)–matrices, *J. Combin. Theory, Ser. B*, 18:187–208, 1975.

[10] E.C. Milnor. A combinatorial theorem on systems of sets, *J. London Math. Soc.*, 43:204–206, 1968.

[11] H. Minc. *Nonnegative Matrices*, Wiley, New York, 1988.

[12] C. Thomassen. Even cycles in directed graphs, *European J. Combin.*, 6:85–90, 1985.

[13] C. Thomassen. Sign-nonsingular matrices and even cycles in directed graphs, *Linear Alg. Appls.*, 75:27–41, 1986.

[14] C. Thomassen. The even cycle problem for directed graphs, *J. Amer. Math. Soc.* 5:217–30, 1992.

[15] M.F. Tinsley. Permanents of cyclic matrices, *Pacific J. Math.*, 10:1067–82, 1960.

9

The inverse sign pattern graph

9.1 General properties of the inverse sign pattern graph

Let n be a positive integer. The *inverse sign pattern graph* $\mathcal{I}_n$ is the graph whose vertices are the $(0, 1, -1)$-matrices of order n which do not have an identically zero determinant and whose edges are the pairs $\{A, B\}$ of vertices such that there exist matrices $\widetilde{A}$ in $\mathcal{Q}(A)$ and $\widetilde{B}$ in $\mathcal{Q}(B)$ which are inverses of one another. Since there are matrices A such that A and A^{-1} have the same sign pattern, the graph $\mathcal{I}_n$ has loops, that is, edges joining a vertex to itself. Note also that $\{A, -A^T\}$ is not an edge of $\mathcal{I}_n$ for any A. If P and Q are permutation matrices, then $\{A, B\}$ is an edge if and only if $\{PAQ, Q^T B P^T\}$ is an edge. Also, if D and E are strict signings, then $\{A, B\}$ is an edge if and only if $\{DAE, EBD\}$ is an edge. Since the inverse of a partly decomposable matrix is also partly decomposable, no edge of $\mathcal{I}_n$ joins a partly decomposable matrix and a fully indecomposable matrix. In particular, $\mathcal{I}_n$ is not connected if $n \geq 2$. We denote the induced subgraph of $\mathcal{I}_n$ whose vertices are the fully indecomposable $(0, 1, -1)$-matrices of order n by $\mathcal{F}_n$. Most of the results in this section are contained in [1], but we have simplified the exposition.

Lemma 9.1.1 *Let A be a fully indecomposable $(0, 1, -1)$-matrix. Then there exists a matrix $\widetilde{A}$ in $\mathcal{Q}(A)$ such that $\widetilde{A}$ is nonsingular. For each such matrix $\widetilde{A}$, there exists a nonsingular matrix $\widetilde{A}$ in $\mathcal{Q}(A)$ such that $\widetilde{A}^{-1}$ has no zero entries and the sign patterns of $\widetilde{A}^{-1}$ and $\widetilde{A}^{-1}$ agree on the nonzero positions of $\widetilde{A}^{-1}$.*

Proof Since A is fully indecomposable, there is a nonzero term in the standard determinant expansion of A, and hence there is a matrix $\widetilde{A} = [\tilde{a}_{ij}]$ in $\mathcal{Q}(A)$ with a nonzero determinant. Suppose that the (p, q)-entry of $\widetilde{A}^{-1}$ is zero. Since $\widetilde{A}$ is fully indecomposable, $\widetilde{A}(\{q\}, \{p\})$ does not have an identically zero determinant. By perturbing the nonzero entries of $\widetilde{A}(\{q\}, \{p\})$, we see that there is a matrix $\tilde{A}$ in $\mathcal{Q}(A)$ such that $\det \tilde{A}(\{q\}, \{p\}) \neq 0$, sign $\det \tilde{A} =$ sign $\det \tilde{A}$, and sign $\det \tilde{A}(\{i\}, \{j\}) =$ sign $\det \tilde{A}(\{i\}, \{j\})$ for each i and j such that $\det \tilde{A}(\{i\}, \{j\}) \neq 0$. The lemma now follows. $\qquad \square$

Lemma 9.1.1 implies that each fully indecomposable $(0, 1, -1)$-matrix of order n is joined in $\mathcal{F}_n$ to a $(1, -1)$-matrix. We denote the induced subgraph of $\mathcal{F}_n$ whose vertices are the $(1, -1)$-matrices of order n by $\mathcal{B}_n$.

Let X be a nonsingular matrix of order n such that X^{-1} has no zero entries. There exists a small neighborhood of X such that each matrix Y in this neighborhood is nonsingular and Y^{-1} has the same sign pattern as X^{-1}. This observation implies the following.

Lemma 9.1.2 *Let* $\{A, B\}$ *be an edge of* $\mathcal{F}_n$ *where* B *is a* $(1, -1)$*-matrix. Then* $\{A', B\}$ *is an edge of* $\mathcal{F}_n$ *for each* $(0, 1, -1)$*-matrix* A' *of order* n *such that* A' *and* A *agree on the nonzero positions of* A.

For example, let C_n be the permutation matrix of order $n \geq 2$ whose 1's are in positions $(1, 2), \ldots, (n - 1, n), (n, 1)$. Since

$$\left(I_n - \frac{1}{2} C_n \right)^{-1} = \frac{2^n}{2^n - 1} \left(I_n + \frac{1}{2} C_n + \frac{1}{2^2} C_n^2 + \cdots + \frac{1}{2^{n-1}} C_n^{n-1} \right),$$

it follows from Lemma 9.1.2 that J_n is contained in at least $3^{n^2 - 2n}$ edges of $\mathcal{F}_n$ and at least $2^{n^2 - 2n}$ edges of $\mathcal{B}_n$.

Lemma 9.1.3 *Let* R *be a* $(0, 1, -1)$*-matrix such that each of the main diagonal entries of* R *is zero. Then the following hold:*

   (i) *There exists a matrix* B *such that* $\{I_n + R, B\}$ *is an edge of* $\mathcal{I}_n$, *and* B *and* $I_n - R$ *agree on the nonzero positions of* $I_n - R$.

  (ii) *If* $I_n + R$ *is fully indecomposable, then the matrix* B *in* (i) *can be chosen to be a* $(1, -1)$*-matrix.*

 (iii) *If* $I_n + R$ *is a* $(1, -1)$*-matrix, then* $\{I_n + R, I_n - R\}$ *is an edge of* $\mathcal{B}_n$.

Proof If $\tilde{R}$ is a matrix in $\mathcal{Q}(R)$, each of whose nonzero entries is sufficiently small, then

$$(I_n + \tilde{R})^{-1} = I_n - \tilde{R} + \tilde{R}^2 - \tilde{R}^3 + \cdots$$

and the sign patterns of $(I_n + \tilde{R})^{-1}$ and $I_n - \tilde{R}$ agree on the nonzero positions of $I_n - \tilde{R}$. Thus (i) holds. Assertion (ii) follows from (i) and Lemma 9.1.1, and assertion (iii) follows from (ii). $\qquad \square$

Corollary 9.1.4 *Let $A = [a_{ij}]$ be a $(0, 1, -1)$-matrix, and let $(k_1, k_2, \ldots, k_n)$ be a permutation of $\{1, 2, \ldots, n\}$ such that $a_{1k_1} a_{2k_2} \cdots a_{nk_n} \neq 0$. Let $U = [u_{ij}]$ be the $(0, 1, -1)$-matrix such that $u_{ik_i} = a_{ik_i}$ for $i = 1, 2, \ldots, n$ and $u_{ij} = 0$ otherwise. Then the following hold:*

    (i) *There exists a matrix B such that $\{A, B\}$ is an edge of $\mathcal{I}_n$, and B and $2U^T - U^T A U^T$ agree on the nonzero positions of $2U^T - U^T A U^T$.*

    (ii) *If A is fully indecomposable, then the matrix B in (i) can be chosen to be a $(1, -1)$-matrix.*

    (iii) *If A is a $(1, -1)$-matrix, then $\{A, 2U^T - U^T A U^T\}$ is an edge of $\mathcal{B}_n$.*

Proof The matrix $U^T A$ has a positive main diagonal. Let $R = U^T A - I_n$. Then each of the main diagonal entries of R is zero, and there exists a matrix B satisfying (i) of Lemma 9.1.3. Thus $\{U^T A, B\}$ is an edge of $\mathcal{I}_n$, and hence so is $\{A, BU^T\}$. Since $I_n - R = 2I_n - U^T A$, the matrices BU^T and $2U^T - U^T A U^T$ agree on the nonzero positions of $2U^T - U^T A U^T$, and hence (i) holds. Assertions (ii) and (iii) follow in a similar way from (ii) and (iii) of Lemma 9.1.3. $\qquad\qquad\square$

The next corollary implies that two $(0, 1, -1)$-matrices of order n whose common nonzero entries determine a fully indecomposable matrix of order n are both joined in $\mathcal{F}_n$ to some matrix E.

Corollary 9.1.5 *Let $A = [a_{ij}]$ and $B = [b_{ij}]$ be $(0, 1, -1)$-matrices of order n. Let $M = [m_{ij}]$ be the $(0, 1, -1)$-matrix such that $m_{ij} = a_{ij}$ if $a_{ij} = b_{ij}$ and $m_{ij} = 0$ otherwise. Assume that M is a fully indecomposable matrix, and let $(k_1, k_2, \ldots, k_n)$ be a permutation of $\{1, 2, \ldots, n\}$ such that $m_{1k_1} m_{2k_2} \cdots m_{nk_n} \neq 0$. Let $U = [u_{ij}]$ be the $(0, 1, -1)$-matrix such that $u_{ik_i} = m_{ik_i}$ for $i = 1, 2, \ldots, n$ and $u_{ij} = 0$ otherwise. Then there exists a $(1, -1)$-matrix E such that $\{A, E\}$ and $\{B, E\}$ are edges of $\mathcal{F}_n$, and E and $2U^T - U^T M U^T$ agree on the nonzero positions of $2U^T - U^T M U^T$.*

Proof By (ii) of Corollary 9.1.4, there exists a $(1, -1)$-matrix E such that $\{M, E\}$ is an edge of $\mathcal{F}_n$, and E and $2U^T - U^T M U^T$ agree on the nonzero positions of $2U^T - U^T M U^T$. By Lemma 9.1.2, E is joined in $\mathcal{F}_n$ to each $(0, 1, -1)$-matrix which agrees with M on the nonzero entries of C. In particular, $\{A, E\}$ and $\{B, E\}$ are edges. $\qquad\qquad\square$

Since

$$\begin{bmatrix} 1 & 1 \\ 1 & -1 \end{bmatrix}$$

is an S^2NS-matrix with a self-inverse sign pattern, neither $\mathcal{F}_2$ nor $\mathcal{B}_2$ is connected. For $n \geq 3$ we have the following.

Theorem 9.1.6 *The graphs $\mathcal{F}_n$ and $\mathcal{B}_n$ are connected for each integer $n \geq 3$.*

Proof Let $n \geq 3$. It follows from Corollary 9.1.5 that given any two $(1, -1)$-matrices A and B which differ in only one entry there is a $(1, -1)$-matrix E such that $\{A, E\}$ and $\{B, E\}$ are edges of $\mathcal{B}_n$. This implies that $\mathcal{B}_n$ is connected. The connectedness of $\mathcal{F}_n$ now follows from Lemma 9.1.1.   □

Let A be a $(0, 1, -1)$-matrix whose determinant is not identically zero. We now consider the number of distinct sign patterns in

$$\{\widetilde{A}^{-1} : \widetilde{A} \in Q(A)\}, \tag{9.1}$$

that is, the degree of A in the graph $\mathcal{I}_n$. If $A = I_n$, then the degree of I_n in $\mathcal{I}_n$ equals one. If A is a fully indecomposable S^2NS-matrix, then the degree of A in $\mathcal{F}_n$ equals one. In the next theorem we show that the minimum degree of matrices in $\mathcal{B}_n$ is exponential in n. First we prove the following lemmas.

Lemma 9.1.7 *The degree of each matrix in $\mathcal{B}_3$ is at least 9.*

Proof Up to row and column permutations and multiplication of rows and columns by -1, there are three $(1, -1)$-matrices of order 3, namely,

$$G_1 = \begin{bmatrix} 1 & 1 & 1 \\ 1 & -1 & 1 \\ 1 & 1 & -1 \end{bmatrix}, \quad G_2 = \begin{bmatrix} 1 & 1 & 1 \\ 1 & 1 & 1 \\ 1 & 1 & -1 \end{bmatrix}, \text{ and } J_3.$$

The sign pattern of the inverse of the matrix

$$\begin{bmatrix} -2 & -1 & 3 \\ 1 & -2 & 3 \\ 1 & 1 & -2 \end{bmatrix},$$

and hence each matrix obtained from it by permuting rows and columns, is J_3. Thus the degree of J_3 in $\mathcal{B}_3$ is at least 36. The sign pattern of the inverse of

$$U = \begin{bmatrix} 0 & 1 & 1 \\ 1 & -1 & 0 \\ 1 & 0 & -1 \end{bmatrix}$$

equals G_1. By Lemma 9.1.2, each of the eight $(1, -1)$-matrices obtained from

U by replacing the 0's by 1 or -1 is joined in $\mathcal{B}_3$ to G_1. Since G_1 is also the sign pattern of the inverse of

$$\begin{bmatrix} 2 & -1 & -1 \\ -1 & 1 & 2 \\ -1 & 2 & 1 \end{bmatrix},$$

the degree of G_1 in $\mathcal{B}_3$ is at least 9. Consider the matrices of the form

$$\begin{bmatrix} 2 & -1 & \delta_1 \\ -1 & 1 & \delta_2 \\ \epsilon_1 & \epsilon_2 & -1 \end{bmatrix}.$$

Provided that the ϵ's and δ's are sufficiently small, such a matrix has a negative determinant and the sign pattern of its inverse is G_2 if and only if

$$\epsilon_1 + \epsilon_2 > 0, \ \ \epsilon_1 + 2\epsilon_2 > 0, \ \ \delta_1 + \delta_2 > 0, \ \text{and} \ \delta_1 + 2\delta_2 > 0.$$

It follows that G_2 is joined in $\mathcal{B}_3$ to each $(1, -1)$-matrix of the form

$$\begin{bmatrix} 1 & -1 & b_1 \\ -1 & 1 & b_2 \\ a_1 & a_2 & -1 \end{bmatrix}$$

such that not both a_1 and a_2 are negative and not both b_1 and b_2 are negative. Hence the degree of G_2 in $\mathcal{B}_3$ is at least 9.[1] □

Lemma 9.1.8 *Let*

$$A = \begin{bmatrix} A_{11} & A_{12} \\ A_{21} & A_{22} \end{bmatrix}$$

be a $(1, -1)$*-matrix of order n where* A_{11} *is a square matrix of order k. Then the degree of A in* $\mathcal{B}_n$ *is at least the product of the degrees of* A_{11} *and* A_{22} *in* $\mathcal{B}_k$ *and* $\mathcal{B}_{n-k}$, *respectively.*

Proof Let $\tilde{A}_{11}$ and $\tilde{A}_{22}$ be nonsingular matrices in $\mathcal{Q}(A_{11})$ and $\mathcal{Q}(A_{22})$, respectively, whose inverses have no zero entries. Then for a sufficiently small $\epsilon > 0$, the matrix

$$\begin{bmatrix} \tilde{A}_{11} & \epsilon A_{12} \\ \epsilon A_{21} & \tilde{A}_{22} \end{bmatrix}$$

---

[1] In fact, in $\mathcal{B}_3$ the degree of J_3 equals 102, the degree of G_1 equals 9, and the degree of G_2 equals 22 [1].

is nonsingular, and the sign pattern of its inverse is the same as the sign pattern of a matrix of the form

$$
\begin{bmatrix}
\widetilde{A}_{11}^{-1} & * \\
* & \widetilde{A}_{22}^{-1}
\end{bmatrix}.
\tag{9.2}
$$

It follows from Lemma 9.1.1 that there exists a matrix $\widetilde{A}$ in $\mathcal{Q}(A)$ the sign pattern of whose inverse has no zero entries and has the form (9.2). The lemma now follows.   □

Let δ_n be the smallest degree of a matrix in $\mathcal{B}_n$. The following lower bounds for δ_n improve the lower bound $\delta_n \geq 2^{\lfloor n/2 \rfloor}$ given in [1].

Theorem 9.1.9 *For each positive integer n we have*

$$
3^{2\lfloor n/3 \rfloor}.
$$

Indeed, if c_1 and c_2 are any positive constants, then $\delta_n > c_1 2^{c_2 n}$ for n sufficiently large.

Proof The first assertion follows by induction on n using Lemmas 9.1.7 and 9.1.8. Let k be a positive integer and let A be a $(1, -1)$-matrix of order $n \geq (k - 1)2^k + 1$. Then the matrix $A[\{1, 2, \ldots, k\}, :]$ has k identical columns. Without loss of generality we may assume that $A[\{1, 2, \ldots, k\}] = J_k$. Since the degree of J_k in $\mathcal{B}_k$ is at least $2^{k^2 - 2k}$, it follows from Lemma 9.1.8 that

$$
\delta_n \geq 2^{k^2 - 2k} \delta_{n-k}
$$

and hence

$$
\delta_n \geq (2^{k^2 - 2k})^{\left\lfloor \frac{n - (k-1)2^k - 1}{k} \right\rfloor}.
$$

The second assertion now follows.   □

Recall that the diameter of a connected graph is the maximum distance between a pair of its vertices.

Theorem 9.1.10 *Let n be an integer with $n \geq 4$. Then the diameter of $\mathcal{B}_n$ is at most 4, and the diameter of $\mathcal{F}_n$ is at most 6.*

Proof Let $W = I_n + C_n$. Since $n \geq 4$, both W and $J_n - W$ are fully indecomposable. Let $A = [a_{ij}]$ and $B = [b_{ij}]$ be two $(1, -1)$-matrices of order n. Let $M = [m_{ij}]$ be the $(1, -1)$-matrix such that $m_{ij} = b_{ij}$ if $(i, j) \in \{(1, 1), \ldots, (n, n), (1, 2), \ldots, (n - 1, n), (n, 1)\}$, and $m_{ij} = a_{ij}$ otherwise.

Then B and M agree on the nonzero positions of W, and A and M agree on the nonzero positions of $J_n - W$. It now follows from Lemma 9.1.5 that the distance between A and B in $\mathcal{B}_n$ is at most 4. Hence the diameter of $\mathcal{B}_n$ is at most 4, and using Lemma 9.1.1 we see that the diameter of $\mathcal{F}_n$ is at most 6.
$\square$

9.2 Inverse-positive sign patterns

Let $A = [a_{ij}]$ be a $(0, 1, -1)$-matrix of order n. The *bipartite digraph* of A is the digraph of order $2n$ with vertex bipartition $\{V, V'\}$ where

$$V = \{1, 2, \ldots, n\} \text{ and } V' = \{1', 2', \ldots, n'\},$$

and where the arcs are the ordered pairs (i, j') for which $a_{ij} = 1$ and the ordered pairs (j', i) for which $a_{ij} = -1$. Note that the bipartite digraph of A can be obtained from the bipartite graph of A by directing each edge. It is easy to verify that the bipartite digraph of A is strongly connected if and only if there do not exist permutation matrices P and Q such that

$$PAQ = \begin{bmatrix} A_{11} & A_{12} \\ A_{21} & A_{22} \end{bmatrix} \quad (A_{11} \text{ is } k \text{ by } \ell, 0 < k + \ell < 2n) \qquad (9.3)$$

where $A_{12} \geq O$ and $A_{21} \leq O$.

Let $\mathcal{G}$ be a graph and let $\mathcal{D}$ be a digraph obtained from $\mathcal{G}$ by replacing each edge $\{u, v\}$ by one of the arcs (u, v) and (v, u). Then $\mathcal{D}$ is strongly connected if and only if $\mathcal{G}$ is connected and each arc of $\mathcal{D}$ is contained in a directed cycle. Thus, if A is fully indecomposable, the bipartite digraph of A is strongly connected if and only if each of its arcs is contained in a directed cycle.

In this section we characterize the sign patterns of inverses of nonnegative matrices. We first characterize the $(0, 1, -1)$-matrices A such that A is the sign pattern of the inverse of a positive matrix, that is, $\{A, J_n\}$ is an edge of $\mathcal{F}_n$ [3] ([5] for the special case in which A is a $(1, -1)$-matrix). Note that a matrix of order $n \geq 2$ whose inverse is a positive matrix is fully indecomposable and has both a positive and negative entry in each row and column.

Theorem 9.2.1 *Let A be a fully indecomposable $(0, 1, -1)$-matrix of order $n \geq 2$. Then the following are equivalent:*

(i) *There exists a matrix $\widetilde{A}$ in $\mathcal{Q}(A)$ such that $\widetilde{A}^{-1}$ is positive.*

(ii) *There do not exist permutation matrices P and Q such that (9.3) holds where $A_{12} \geq O$ and $A_{21} \leq O$.*

(iii) *The bipartite digraph of A is strongly connected.*

(iv) *There exists a matrix $\tilde{A}$ in $\mathcal{Q}(A)$ each of whose row and column sums equals zero.*

(v) *There exists a matrix $\tilde{A}$ in $\mathcal{Q}(A)$ of rank $n - 1$ each of whose row and column sums equals zero.*

Proof First assume (i) but not (ii) holds. Let P and Q be permutation matrices such that (9.3) holds. Let

$$P\tilde{A}Q = \begin{bmatrix} \tilde{A}_{11} & \tilde{A}_{12} \\ \tilde{A}_{21} & \tilde{A}_{22} \end{bmatrix} \quad \text{and} \quad Q^T \tilde{A}^{-1} P^T = \begin{bmatrix} B_{11} & B_{12} \\ B_{21} & B_{22} \end{bmatrix},$$

where $\tilde{A}_{11}$ is k by ℓ and B_{11} is ℓ by k. Since $n \geq 2$, we have $0 < k < n$ and $0 < \ell < n$. The equations $\tilde{A}B = I_n$ and $B\tilde{A} = I_n$ imply

$$\tilde{A}_{11} B_{12} + \tilde{A}_{12} B_{22} = O \quad \text{and} \quad B_{21} \tilde{A}_{11} + B_{22} \tilde{A}_{21} = O.$$

Multiplying the first equation by B_{21} on the left and the second equation by B_{12} on the right and then subtracting, we obtain

$$B_{21} \tilde{A}_{12} B_{22} - B_{22} \tilde{A}_{21} B_{12} = O.$$

It now follows that $A_{12} = O$ and $A_{21} = O$, contradicting the full indecomposability of A. Hence (i) implies (ii).

We have already noted that (ii) implies (iii). Now assume that (iii) holds. Each arc α of the bipartite digraph is contained in a directed cycle γ_α. Let M_α be the $(0, 1, -1)$-matrix of order n whose bipartite digraph is the directed cycle γ_α. Then M_α and A agree on the nonzero positions of M_α, and each row and column sum of M_α equals zero. The matrix

$$\tilde{A} = \sum_\alpha M_{\gamma_\alpha},$$

where the summation is over all arcs α, satisfies (iv). Hence (iii) implies (iv).

Now assume that (iv) holds. Without loss of generality we assume that each of the main diagonal entries of A is nonzero. Since A is fully indecomposable, the digraph $\mathcal{D}(A)$ is strongly connected, and hence each arc β is contained in a directed cycle π_β. Let U_β be the $(0, 1)$-matrix of order n such that each entry on the main diagonal equals zero and $\mathcal{D}(U)$ is the directed cycle π_β. Let

$$U = \sum_\beta U_\beta$$

where the summation is over all arcs β. The sum of the entries in row i of U equals the sum of the entries in column i for each $i = 1, 2, \ldots, n$. Hence there exists a diagonal matrix D such that each entry on the main diagonal is positive and each row and column sum of $D - U$ equals zero. The matrix $D - U$ has the me zero pattern as A and hence is irreducible. Let r be the largest diagonal entry of D. Then $rI_n - D + U$ is an irreducible nonnegative matrix and each row and column sum equals r. By the Perron–Frobenius theory of nonnegative matrices, r is a simple eigenvalue of $rI_n - D + U$. It follows that the rank of $D - U$ is $n - 1$. Let E be a nonsingular submatrix of order $n - 1$ of $D - U$, and let E' be the corresponding submatrix of $\widetilde{A}$. We have $\det(E' + \epsilon E) = 0$ for at most $n - 1$ values of ϵ. Hence we may choose ϵ so that the rank of $E' + \epsilon E$ is $n - 1$ and $\widetilde{A} = \widetilde{A} + \epsilon(D - U)$ belongs to $\mathcal{Q}(A)$. Since the row and column sums of $\widetilde{A}$ equal zero, the rank of $\widetilde{A}$ equals $n - 1$. Hence (iv) implies (v).

Finally, assume that (v) holds. Since the rank of $\widetilde{A}$ is $n - 1$, the rank of its adjugate adj $\widetilde{A}$ is 1. Since the row and column sums of $\widetilde{A}$ equal zero, there exists a nonzero number c such that adj $\widetilde{A} = cJ_n$. Without loss of generality we assume that the $(1,1)$-entry of $\widetilde{A}$ is nonzero. Let E_{11} be the $(0,1)$-matrix of order n whose only 1 is in position $(1,1)$. Then $\det(\widetilde{A} + \epsilon E_{11}) = \epsilon c$ and for $\epsilon \neq 0$ we have

$$(\widetilde{A} + \epsilon E_{11})^{-1} = \frac{1}{\epsilon c} \operatorname{adj}(\widetilde{A} + \epsilon E_{11}).$$

Thus, for ϵ a sufficiently small positive number, the matrix $\widetilde{A} = \widetilde{A} + \epsilon E_{11}$ satisfies (i). Hence (v) implies (i). $\qquad\qquad\square$

The implication (iv) implies (v) in Theorem 9.2.1 is contained in [2]. Theorem 9.2.1 implies that $\{A, J_n\}$ is an edge of $\mathcal{F}_n$ if and only if $\{A, -J_n\}$ is.

Theorem 9.2.1 and the next theorem [4] imply that a fully indecomposable $(0, 1, -1)$-matrix is the sign pattern of the inverse of a nonnegative matrix if and only if it is the sign pattern of the inverse of a positive matrix.

Lemma 9.2.2 *Let*

$$\widetilde{A} = \begin{bmatrix} \widetilde{A}_{11} & \widetilde{A}_{12} \\ \widetilde{A}_{21} & \widetilde{A}_{22} \end{bmatrix}$$

be a nonsingular matrix of order $n \geq 2$ with inverse

$$B = \begin{bmatrix} B_{11} & B_{12} \\ B_{21} & B_{22} \end{bmatrix},$$

where $\widetilde{A}_{11}$ is k by ℓ, B_{11} is ℓ by k, and $0 < k + \ell < 2n$. Assume that $\widetilde{A}_{12} \leq O$, $\widetilde{A}_{21} \geq O$, and $B \geq O$. Then $B_{21} = O$.

Proof Since $\widetilde{A}B = I_n$, we have

$$\widetilde{A}_{11}B_{11} = I_k - \widetilde{A}_{12}B_{22} \quad \text{and} \quad -\widetilde{A}_{22}B_{21} = \widetilde{A}_{21}B_{11}.$$

It follows that

$$\widetilde{A}\begin{bmatrix} B_{11} \\ -B_{21} \end{bmatrix} = \begin{bmatrix} I_k - 2\widetilde{A}_{12}B_{21} \\ 2\widetilde{A}_{21}B_{11} \end{bmatrix} \geq O.$$

Since $\widetilde{A}^{-1} = B \geq O$, we have

$$\begin{bmatrix} B_{11} \\ -B_{21} \end{bmatrix} = B\begin{bmatrix} I_k - 2\widetilde{A}_{12}B_{21} \\ 2\widetilde{A}_{21}B_{11} \end{bmatrix} \geq O.$$

Thus $-B_{21} \geq O$ and therefore $B_{21} = O$.   □

Theorem 9.2.3 *Let A be a fully indecomposable $(0, 1, -1)$-matrix of order $n \geq 2$. Then there exists a matrix $\widetilde{A}$ in $Q(A)$ such that $\widetilde{A}^{-1}$ is nonnegative if and only if there do not exist permutation matrices P and Q such that (9.3) holds where $A_{12} \geq O$ and $A_{21} \leq O$.*

Proof Assume that $\widetilde{A}$ is a matrix in $Q(A)$ such that $B = \widetilde{A}^{-1}$ is a nonnegative matrix, and also assume that there exist permutation matrices P and Q such that (9.3) holds where $A_{12} \geq O$ and $A_{21} \leq O$. Without loss of generality we assume that $P = Q = I_n$ and thus that $\widetilde{A}$ and B have the forms in Lemma 9.2.2 where $\widetilde{A}_{12} \leq O$ and $\widetilde{A}_{21} \geq O$. By Lemma 9.2.2, $B_{21} = O$. Since $\widetilde{A}B = I_n$, we now have $\widetilde{A}_{21}B_{11} = O$. Since $\widetilde{A}_{21} \geq O$ and $B_{11} \geq O$, we conclude that the number p of zero columns of $\widetilde{A}_{21}$ and the number q of zero rows of B_{11} satisfy $p + q \geq \ell$. Hence $\widetilde{A}$ has an $n - k$ by p zero submatrix and B has an $n - \ell + q$ by k zero submatrix. Since $(n - k + p) + (n - \ell + q + k) \geq 2n$, either $\widetilde{A}$ or B is partly decomposable. Since the inverse of a partly decomposable matrix is also partly decomposable, we contradict the assumption that A is a fully indecomposable matrix. The theorem now follows from Theorem 9.2.1.   □

Finally, we characterize sign patterns of inverses of partly decomposable nonnegative matrices.[2]

---

[2] The characterization of sign patterns of inverses of partly decomposable nonnegative matrices given in [4] is not correct. The reason is that the necessary and sufficient conditions given are not invariant under simultaneous block row and column permutations which preserve block triangular form, but the property of having a nonnegative inverse is. For example, let

$$A = \begin{bmatrix} 1 & 0 & 0 & 0 \\ -1 & 1 & 0 & 0 \\ 0 & 0 & 1 & 0 \\ 1 & 0 & -1 & 1 \end{bmatrix}.$$

Then A satisfies the conditions in [4], but the matrix

$$\begin{bmatrix} 1 & 0 & 0 & 0 \\ 0 & 1 & 0 & 0 \\ -1 & 1 & 1 & 0 \\ 0 & -1 & 0 & 1 \end{bmatrix}$$

Let $\mathcal{D}$ be an acyclic digraph of order n. Recall from Chapter 7 that the arc (u, v) is a transitive arc of $\mathcal{D}$ provided there is a path from u to v which does not contain the arc (u, v), and that the transitive closure of $\mathcal{D}$ is the acyclic digraph of order n such that (x, y) is an arc if and only if there is a path in $\mathcal{D}$ from x to y. If we define $x > y$ provided there is a path in $\mathcal{D}$ from x to y, we obtain a partial order on the vertices of $\mathcal{D}$ such that the nontransitive arcs of $\mathcal{D}$ are the edges of its Hasse diagram.

Lemma 9.2.4 *Let $\mathcal{D}$ be an acyclic digraph and let (u, v) be a nontransitive arc of $\mathcal{D}$. Then the vertices of $\mathcal{D}$ can be put in the order $x_1, x_2, \ldots, x_n$ so that (x_i, x_j) is an arc only if $i > j$, and u and v are consecutive vertices in this order.*

Proof Since $\mathcal{D}$ is acyclic, its vertices can be put in the order $y_1, y_2, \ldots, y_n$ so that (y_i, y_j) is an arc only if $i > j$. Let $u = y_p$ and $v = y_q$. Since (u, v) is an arc, we have $p > q$. Let $i_1 < \cdots < i_k$ be the integers ℓ with $q < \ell < p$ such that there does not exist a path from u to y_ℓ. Let $j_1 < \cdots < j_{p-q-1-k}$ be the other integers ℓ with $q < \ell < p$. Then

$$y_1, \ldots, y_{q-1}, y_{j_1}, \ldots, y_{j_{p-q-1-k}}, v, u, y_{i_1}, \ldots, y_{i_k}, y_{p+1}, \ldots, y_n$$

is an order of the vertices of $\mathcal{D}$ satisfying the conclusions of the lemma. $\square$

Lemma 9.2.5 *Let $\mathcal{D}$ be a signed acyclic digraph of order n. Then the following are equivalent:*

(i) *If there is a path from i to j, then there is a path from i to j each of whose arcs is negative $(1 \leq i, j \leq n, i \neq j)$.*

(ii) *If there is a path from i to j, then there is a path from i to j whose first arc is negative $(1 \leq i, j \leq n, i \neq j)$.*

(iii) *Every nontransitive arc of $\mathcal{D}$ is negative.*

Proof Clearly, (i) implies (ii). If (i, j) is a nontransitive arc, then the only path from i to j is the path $i \to j$ of length one, and hence (ii) implies (iii). If there

obtained from A by simultaneous row and column permutations does not. Indeed, A is an S^2NS-matrix, and the sign pattern of the inverse of every matrix in $\mathcal{Q}(A)$ equals

$$\begin{bmatrix} 1 & 0 & 0 & 0 \\ 1 & 1 & 0 & 0 \\ 0 & 0 & 1 & 0 \\ -1 & 0 & 1 & 1 \end{bmatrix}.$$

is a path from i to j, then there is a path from i to j each of whose arcs is a nontransitive arc, and hence (iii) implies (i). $\qquad\square$

A *Hasse digraph* is a signed acyclic digraph which satisfies any of the equivalent conditions (i), (ii), and (iii) of Lemma 9.2.5.

Let A be a $(0, 1, -1)$-matrix whose determinant is not identically zero. There exist permutation matrices P and Q such that PAQ has the form

$$
\begin{bmatrix}
A_{11} & O & \cdots & O \\
A_{21} & A_{22} & \cdots & O \\
\vdots & \vdots & \ddots & \vdots \\
A_{k1} & A_{k2} & \cdots & A_{kk}
\end{bmatrix}
\quad (A_{11}, A_{22}, \ldots, A_{kk} \text{ fully indecomposable}).
$$

(9.4)

The *signed block digraph* of A is the signed digraph $\Gamma(A)$ of order k with vertices $1, 2, \ldots, k$, with a negative arc (i, j) if and only if $i > j$ and A_{ij} contains a negative entry, and with a positive arc (i, j) if and only if $i > j$ and A_{ij} is nonnegative and contains a positive entry. Clearly, $\Gamma(A)$ is acyclic.

In the next theorem we assume without loss of generality that A has the form (9.4).

Theorem 9.2.6 *Let A be a $(0, 1, -1)$-matrix whose determinant is not identically zero, and assume that A has the form (9.4). Let T be the $(0, 1)$-matrix of order n obtained from A by replacing each A_{ii} by J $(i = 1, 2, \ldots, k)$ and replacing A_{ij} by J if and only if (i, j) is an arc of the transitive closure of $\Gamma(A)$. Then there exists a matrix in $\mathcal{Q}(A)$ whose inverse is nonnegative if and only if the qualitative class of each fully indecomposable component of A contains a matrix whose inverse is nonnegative and $\Gamma(A)$ is a Hasse digraph. Moreover, if these conditions are satisfied, then there is a matrix in $\mathcal{Q}(A)$ whose inverse is nonnegative and whose zero pattern equals T.*

Proof First assume that there is a matrix $\widetilde{A}$ in $\mathcal{Q}(A)$ whose inverse is nonnegative. Then clearly there is a matrix in each $\mathcal{Q}(A_{ii})$ whose inverse is nonnegative. We prove that $\Gamma(A)$ is a Hasse digraph by showing that each nontransitive arc (u, v) is negative. By Lemma 9.2.4 we may assume that $u = v + 1$. Since $\widetilde{A}^{-1}$ is nonnegative, it follows that there exists a matrix in the qualitative class of

$$
\begin{bmatrix}
A_{vv} & O \\
A_{uv} & A_{uu}
\end{bmatrix}
$$

whose inverse

$$
\begin{bmatrix}
B_{vv} & O \\
B_{vu} & B_{uu}
\end{bmatrix}
$$

is nonnegative. Since $A_{uv} \neq O$, we may assume that $B_{vu} \neq 0$. Applying Lemma 9.2.2 we conclude that A_{uv} has a negative entry. Hence (u, v) is a negative arc of $\Gamma(A)$.

Now assume that $\Gamma(A)$ is a Hasse digraph and that each $\mathcal{Q}(A_{ii})$ contains a matrix with a nonnegative inverse. We prove by induction on k that there exists a matrix in $\mathcal{Q}(A)$ whose inverse is nonnegative and whose zero pattern equals T. If $k = 1$, there is nothing to prove. Let $k \geq 2$. Since $\Gamma(A)$ is a Hasse digraph, so is the digraph obtained from $\Gamma(A)$ by deleting vertex k. By the induction hypothesis there exists a nonsingular matrix

$$
\tilde{A}' = \begin{bmatrix} \tilde{A}_{11} & O & \cdots & O \\ \tilde{A}_{21} & \tilde{A}_{22} & \cdots & O \\ \vdots & \vdots & \ddots & \vdots \\ \tilde{A}_{k-1,1} & \tilde{A}_{k-1,2} & \cdots & \tilde{A}_{k-1,k-1} \end{bmatrix}
$$

$$
(\tilde{A}_{ij} \in \mathcal{Q}(A_{ij}), 1 \leq j \leq i \leq k-1)
$$

with inverse

$$
B = \begin{bmatrix} B_{11} & O & \cdots & O \\ B_{21} & B_{22} & \cdots & O \\ \vdots & \vdots & \ddots & \vdots \\ B_{k-1,1} & B_{k-1,2} & \cdots & B_{k-1,k-1} \end{bmatrix},
$$

where each B_{ii} is a positive matrix and where B_{ij} is a positive matrix if there is a path from i to j in $\Gamma(A)$ and is a zero matrix otherwise. By Theorems 9.2.1 and 9.2.3, there exists a matrix $\tilde{A}_{kk}$ in $\mathcal{Q}(A_{kk})$ whose inverse B_{kk} is a positive matrix. Let

$$
\tilde{A} = \left[\begin{array}{c|c} \tilde{A}' & O \\ \hline \tilde{A}_{k1} \quad \cdots \quad \tilde{A}_{k,k-1} & \tilde{A}_{kk} \end{array} \right]
$$

where $\tilde{A}_{ki}$ is an arbitrary matrix in $\mathcal{Q}(A_{ki})$, $(i = 1, \ldots, k-1)$. Then

$$
\tilde{A}^{-1} = \left[\begin{array}{c|c} B & O \\ \hline B_{k1} \quad \cdots \quad B_{k,k-1} & B_{kk} \end{array} \right]
$$

where

$$
B_{kj} = -\left(\sum_{i=j}^{k-1} B_{kk} \tilde{A}_{ki} B_{ij} \right) \quad (j = 1, 2, \ldots, k-1).
$$

If there does not exist a path from k to j in $\Gamma(A)$, then for each $i = j, \ldots, k - 1$, either $\widetilde{A}_{ki} = O$ or $B_{ij} = O$, and hence $B_{kj} = O$ for any choice of $\widetilde{A}_{k1}, \ldots, \widetilde{A}_{k,k-1}$. If there exists a path from k to j in $\Gamma(A)$, then since $\Gamma(A)$ is a Hasse digraph, there exists an i such that A_{ki} contains a negative entry and B_{ij} is a positive matrix. By choosing the negative entries of each $\widetilde{A}_{ki}$ sufficiently large in comparison with the positive entries, we see that B_{kj} is a positive matrix if there is a path from k to j in $\Gamma(A)$ and a zero matrix otherwise. Hence $\widetilde{A}$ is a matrix in $\mathcal{Q}(A)$ whose inverse is nonnegative with zero pattern equal to T.

$\square$

Bibliography

[1] M.A. Berger and A. Felzenbaum. Sign patterns of matrices and their inverses, *Linear Alg. Appls.*, 86:161–77, 1987.

[2] A. Berman and B.D. Saunders. Matrices with zero line sums and maximal rank, *Linear Alg. Appls.*, 40:229–35, 1981.

[3] M. Fiedler and R. Grone. Characterizations of sign patterns of inverse-positive matrices, *Linear Alg. Appls.*, 40:237–45, 1981.

[4] C.R. Johnson. Sign patterns of inverse nonnegative matrices, *Linear Alg. Appls.*, 55:69–80, 1983.

[5] C.R. Johnson, F.T. Leighton, and H.A. Robinson. Sign patterns of inverse-positive matrices, *Linear Alg. Appls.*, 24:75–83, 1979.

10

Sign stability

10.1 A problem from ecology

Consider an ecosystem, similar to that in [5], consisting of three populations, say coyotes, roadrunners, and grasshoppers. Since roadrunners are the natural prey of coyotes, it is reasonable to assume that an increase in the population of coyotes will cause a short-term decrease in the population of roadrunners, and similarly that an increase in roadrunners will cause a short-term increase in coyotes. Since grasshoppers are the staple food of roadrunners, it is also reasonable to assume that an increase in grasshoppers will result in an increase in roadrunners and an increase in roadrunners will result in a decrease in grasshoppers. Finally since coyotes, unlike roadrunners or grasshoppers, are a highly competive population, we assume that when the population of coyotes exceeds a certain level, the competition will cause the population to decrease. With these assumptions, the ecosystem can be modeled by the linear system of differential equations

$$
\begin{aligned}
\dot{x}_1 &= r_1 - ax_1 + bx_2 \\
\dot{x}_2 &= r_2 - dx_1 \quad\;\; + cx_3 \\
\dot{x}_3 &= r_3 \quad\;\; - fx_2
\end{aligned}
\tag{10.1}
$$

where $x_1 = x_1(t)$, $x_2 = x_2(t)$, and $x_3 = x_3(t)$ are, respectively, the population functions of coyotes, roadrunners, and grasshoppers, a, b, c, d, f are positive real numbers, and r_1, r_2, r_3 are the natural growth rates of the respective populations. Let $x = x(t) = (x_1, x_2, x_3)^T$ be the population vector. Then we can write (10.1) as

$$
\dot{x} = A(x - x_E)
\tag{10.2}
$$

239

where

$$A = \begin{bmatrix} -a & b & 0 \\ -d & 0 & c \\ 0 & -f & 0 \end{bmatrix} \tag{10.3}$$

and x_E satisfies $-Ax_E = (r_1, r_2, r_3)^T$. The solution to (10.2) with initial state $x = x_0$ at time $t = 0$ is

$$x = e^{At}(x_0 - x_E) + x_E. \tag{10.4}$$

If $x_0 = x_E$, then $x(t) = x_E$ for all t and hence x_E is a stable population for the ecosystem. There exists a diagonal matrix D with positive diagonal entries such that $D^{-1}AD$ has the form

$$B = \begin{bmatrix} -u & v & 0 \\ -v & 0 & w \\ 0 & -w & 0 \end{bmatrix}$$

where u, v, and w are positive real numbers. The characteristic polynomial of B is

$$p(x) = x^3 + ux^2 + (v^2 + w^2)x + uw^2.$$

Since each of the coefficients of $p(x)$ is a positive real number, either $p(x)$ has three negative real eigenvalues, or $p(x)$ has one negative real eigenvalue and two complex eigenvalues which are conjugates of each other. Since $p(0) > 0$ and $p(-u) = -uv^2 < 0$, $p(x)$ has a real root λ where $-u < \lambda < 0$. If λ is the only real root of $p(x)$, and μ and $\bar{\mu}$ are the other roots, then since

$$-u = \text{tr}(B) = \lambda + 2\text{Re}(\mu),$$

it follows that $\text{Re}(\mu) < 0$. Therefore, the real part of each eigenvalue of B is negative. It follows that $\lim_{t \to \infty} e^{Bt} = 0$ and hence that $\lim_{t \to \infty} e^{At} = 0$. Thus, for any initial state, the population vector x tends to the stable population x_E. Ecologists say that the vector x_E is an asymptotically stable equilibrium of the population. Note that in deriving the asymptotic stability of (10.2), we used only the sign pattern of the matrix A and not the magnitudes of its entries.

A real matrix A of order n is *stable* (respectively, *semi-stable*) provided each of its eigenvalues has negative (respectively, nonnegative) real part. A matrix A is *quasi-stable* provided it is semi-stable and each of its purely imaginary eigenvalues has equal algebraic and geometric multiplicity. We now discuss the fundamental role of stable and semi-stable matrices in the study of dynamical systems.

Consider a more general ecosystem consisting of n different populations which is modeled by the system

$$
\begin{aligned}
\dot{x}_1 &= f_1(x, t) \\
\dot{x}_2 &= f_2(x, t) \\
&\vdots \\
\dot{x}_n &= f_n(x, t)
\end{aligned}
\tag{10.5}
$$

where $x_i = x_i(t)$ is the size of the ith population at time t and $x = x(t) = (x_1, x_2, \ldots, x_n)^T$. Assume that the functions $f_1, f_2, \ldots, f_n$ are continuous and have continuous derivatives. A solution to (10.5) is called a *trajectory*. These assumptions on the f_j insure that for each initial state there is a unique trajectory. An initial state x_E is an *equilibrium* provided the corresponding trajectory $x(t)$ satisfies $x(t) = x_E$ for all $t \geq 0$. An equilibrium x_E is *asymptotically stable* provided the trajectories with initial state sufficiently close to x_E converge to x_E as t tends to infinity. Thus, if x_E is an asymptotically stable equilibrium, then a small perturbation in the populations will not have long-term effects. An equilibrium x_E is *stable* provided that for each $\epsilon > 0$ there is a $\delta > 0$ such that any trajectory whose initial state is within δ of x_E is also within ϵ of x_E for sufficiently large t. Thus, if x_E is a stable equilibrium, then a small perturbation in the populations may have long-term effects, but the resulting trajectory is near the equilibrium in the long term. A trajectory that is asymptotically stable is stable, but the converse does not hold.

Assume that (10.5) is a homogeneous system of linear differential equations with constant coefficients. Then (10.5) can be written as

$$
\dot{x} = Ax
\tag{10.6}
$$

for some matrix $A = [a_{ij}]$ of order n. The trajectories (solutions) of (10.6) are of the form

$$
x = e^{At} x_0.
$$

Thus, if A is a stable matrix, then $\lim_{t \to \infty} e^{At} = 0$, and hence $x_E = 0$ is an asymptotically stable equilibrium. If A has an eigenvalue λ with positive real part, then $\lim_{t \to \infty} ||e^{At} v|| = \infty$ for any eigenvector corresponding to λ. It follows that $x_E = 0$ is an asymptotically stable equilibrium if and only if A is a stable matrix. It is also known (see [5]) that $x_E = 0$ is a stable equilibrium if and only A is quasi-stable.

Asymptotic stability for the general system (10.5) is determined by considering its linearization. The linearization of (10.5) near x_E is the system of linear differential equations

$$\dot{x} = A(x - x_E)$$

where $A = [a_{ij}]$ is the matrix of order n with $a_{ij} = (\partial f_i/\partial x_j)(x_E, t)$. We assume, as is the case for most simple ecosystem models, that the system (10.5) is an *autonomous system*, that is, the entries of A do not depend on t. It can be shown that x_E is an asymptotically stable equilibrium of (10.5) if and only if it is an asymptotically stable equilibrium of the linearization, and hence if and only if A is stable. Similarly, the stability of x_E can be determined from A. Thus both the asymptotic stability and stability of an equilibrium state of (10.5) are determined by the stability and semi-stability of the matrix A.

The matrix A is known as the *community matrix*, and its entries reflect the effects on the ecosystem due to a small perturbation. More precisely, if $a_{ij} = \partial f_i/\partial x_j > 0$, then a small increase in population j causes an increase in population i. If $a_{ij} < 0$, then a small increase in population j causes a decrease in population i. If $a_{ij} = 0$, a small change in population j causes no change in population i. As in our introductory example, the signs of the entries of the community matrix can often be determined from general principles of ecology. However, the actual magnitudes of the entries are difficult to determine and can only be approximated. Thus it is natural to ask if it is possible to determine whether x_E is an asymptotically stable (or stable) equilibrium for (10.5) solely from the sign pattern of the community matrix. A square matrix A is *sign stable* (respectively, *sign semi-stable*) provided that every matrix in $\mathcal{Q}(A)$ is stable (respectively, semi-stable). A sign-stable matrix is clearly an SNS-matrix. A diagonal matrix is sign semi-stable if and only if each of its main diagonal entries is nonpositive and is sign stable if and only if each of its main diagonal entries is negative. The matrix (10.3) is a sign-stable matrix. For another example, let a, b, c, and d be positive numbers. Then the eigenvalues of

$$\begin{bmatrix} -a & b \\ -c & -d \end{bmatrix}$$

are

$$\frac{-(a+d) \pm \sqrt{(a+d)^2 - 4(ad+bc)}}{2},$$

and it follows that

$$\begin{bmatrix} -1 & 1 \\ -1 & -1 \end{bmatrix}$$

is a sign-stable matrix. Every matrix of the form

$$\begin{bmatrix} 0 & f \\ -g & 0 \end{bmatrix}$$

where f and g are positive has purely imaginary eigenvalues, and hence

$$\begin{bmatrix} 0 & 1 \\ -1 & 0 \end{bmatrix}$$

is an example of a sign semi-stable but not a sign-stable matrix.

We refer the reader to [6, 9, 12, 13] for a further discussion of the qualitative properties of ecosystem models and to [2, 15] for a qualitative analysis of the stability of certain nonlinear systems of differential equations. In [11] the theory of sign stability is described as "probably the most elegant and impressive of all mathematical contributions to theoretical population biology."

In the next section we characterize zero patterns of sign-stable and sign semi-stable matrices, in the third section we consider questions related to sign-solvable linear systems, and in the last section we study SNS-matrices with other restrictions on their eigenvalues.

10.2 Sign-stable matrices

Let $A = [a_{ij}]$ be a $(0, 1, -1)$-matrix of order n. Since A is sign stable (respectively, sign semi-stable) if and only if each of its irreducible components is sign stable (respectively, sign semi-stable), in investigating sign stability and sign semi-stability we assume that our matrices are irreducible.

If A is a sign semi-stable matrix of order n and B is a matrix of order n, at least one of whose eigenvalues has positive real part, then clearly B is not a limit of matrices in $Q(A)$. In particular, the diagonal entries of a sign semi-stable matrix are nonpositive. We begin by characterizing sign semi-stable matrices [14]. The characterization implies that the number of nonzero entries of an irreducible sign semi-stable matrix of order n is between $2n - 2$ and $3n - 2$. Recall that a doubly directed tree of order n is a digraph $\overset{\leftrightarrow}{T}$ obtained from a tree T with n vertices by replacing each edge $\{i, j\}$ with the oppositely directed arcs (i, j) and (j, i).

Theorem 10.2.1 *Let $A = [a_{ij}]$ be an irreducible $(0, 1, -1)$-matrix of order n. Then A is sign semi-stable if and only if each of the following properties holds:*

(i) *Each entry on the main diagonal of A is nonpositive.*

(ii) *If $i \neq j$, then $a_{ij}a_{ji} \leq 0$.*

(iii) *The digraph $\mathcal{D}(A)$ of A is a doubly directed tree.*

Proof First assume that A is sign semi-stable. We have already observed that (i) holds. Let γ be a directed cycle of $\mathcal{D}(A)$ of length k, and let M_γ be the $(0, 1, -1)$-matrix of order n obtained from A by replacing all entries not corresponding to arcs of γ by 0. The characteristic polynomial of M_γ equals $x^{n-k}(x^k - \text{sign } \gamma)$ where k is the length of γ. Hence, if $k \geq 3$ or if $k = 2$ and sign $\gamma = 1$, then M_γ has an eigenvalue with positive real part. Since M_γ is a limit of matrices in $\mathcal{Q}(A)$, we conclude that (ii) holds and that $\mathcal{D}(A)$ has no directed cycles of length 3 or more. Since A is irreducible, (iii) now holds.

Now assume that (i), (ii), and (iii) hold. Since (ii) and (iii) hold, there exists a diagonal matrix E of order n with positive diagonal entries such that $E^{-1}AE = D + S$ where D is a diagonal matrix with nonpositive diagonal entries and S is a skew-symmetric matrix. Let λ be an eigenvalue of A with corresponding eigenvector x. Then $y = E^{-1}x$ is an eigenvector of $D + S$ and

$$\lambda y^* y = y^* Dy + y^* Sy.$$

Since the real part of $y^* Sy$ is zero, and $y^* Dy$ is a nonpositive real number, the real part of λ is nonpositive. It follows that A is a sign semi-stable matrix. □

It is easy to decide whether or not conditions (i), (ii), and (iii) in Theorem 10.2.1 hold, and hence it is easy to decide whether or not a matrix is sign semi-stable. Note that the theorem implies that it is very rare for a matrix to be sign semi-stable. It is even rarer for a matrix to be sign stable. In characterizing sign-stable matrices one needs to find additional conditions on A that guarantee that no matrix in $\mathcal{Q}(A)$ has an eigenvalue with zero real part. The additional conditions are provided by (iv) and (v) in the following characterization [7].

Theorem 10.2.2 *Let $A = [a_{ij}]$ be an irreducible $(0, 1, -1)$-matrix of order n. Then A is sign stable if and only if each of the following properties holds:*

(i) *Each entry on the main diagonal of A is nonpositive.*
(ii) *If $i \neq j$, then $a_{ij}a_{ji} \leq 0$.*
(iii) *The digraph $\mathcal{D}(A)$ of A is a doubly directed tree.*
(iv) *A does not have an identically zero determinant.*
(v) *There does not exist a nonempty subset β of $\{1, 2, \ldots, n\}$ such that each diagonal element of $A[\beta]$ is zero, each row of $A[\beta]$ contains at least one nonzero entry, and no row of $A[\overline{\beta}, \beta]$ contains exactly one nonzero entry.*

Proof First assume that statements (i)–(v) hold, and suppose to the contrary that A is not sign stable. Since (i)–(iv) hold, and since a matrix with negative main diagonal whose signed digraph is a doubly directed tree is an SNS-matrix, A

is an SNS-matrix. In addition by Theorem 10.2.1, A is sign semi-stable. Since A is not sign stable, there exists a matrix $\widetilde{A}$ in $Q(A)$ which has a nonzero eigenvalue whose real part is 0. By premultiplying by a scalar matrix, we may further assume that $\pm i$ is an eigenvalue of $\widetilde{A}$. Since $\widetilde{A}$ is a real matrix, both i and $-i$ are eigenvalues of $\widetilde{A}$. Since $\mathcal{D}(A)$ is a tree, there exists a diagonal matrix E with positive diagonal entries such that $E^{-1}\widetilde{A}E = D + S$ where D is a diagonal matrix with nonpositive diagonal entries and S is a skew-symmetric matrix. Let $\widehat{A} = E^{-1}\widetilde{A}E$. Then $\widehat{A} \in Q(A)$ and i is an eigenvalue of $\widehat{A}$. Let $x = (x_1, x_2, \ldots, x_n)^T$ be an eigenvector of $\widehat{A}$ corresponding to i and let $\beta = \{\ell : x_\ell \neq 0\}$. Since $\widehat{A}x = ix$, each row of $\widehat{A}[\beta]$ has at least one nonzero entry, and no row of $\widehat{A}[\bar{\beta}, \beta]$ has exactly one nonzero entry. Since

$$ix^*x = x^*\widehat{A}x = x^*Dx + x^*Sx,$$

and since x^*Dx is real and x^*Sx is imaginary, we have

$$0 = x^*Dx = \sum_{k=1}^{n} a_{kk}|x_k|^2.$$

It now follows that $a_{kk} = 0$ for all $k \in \beta$ and that β is a counterexample to (v). Hence properties (i)–(v) imply that A is sign semi-stable.

Now assume that A is sign stable. Clearly (iv) holds, and since A is also sign semi-stable (i), (ii), and (iii) hold. Assume to the contrary that (v) does not hold. Let β be a counterexample to (v), and without loss of generality assume that $1 \in \beta$. We construct $\widetilde{A} = [\tilde{a}_{kl}]$ in $Q(A)$ and a nonzero vector $x = (x_1, x_2, \ldots, x_n)^T$ such that $\widetilde{A}x = ix$ as follows. Let $\mathcal{D}(A) = \overset{\leftrightarrow}{T}$ where T is a tree of order n which we consider rooted at vertex 1. For each vertex k there is a unique path in T from 1 to k. If $k \neq 1$, we let $p(k)$ denote the vertex on this path from 1 to k that immediately precedes k. Without loss of generality, we may assume that the vertices are labeled so that $p(k) < k$ for all $k \neq 1$. To facilitate the construction we make the following definitions. Let $N(1)$ denote the set of vertices in β which are adjacent to 1, and for $k \neq 1$ let $N(k)$ denote the set of vertices in $\beta \setminus \{p(k)\}$ which are adjacent to k. Let $\tau_1 = 0$, and for $k \neq 1$ let

$$\tau_k = \begin{cases} 0 & \text{if } p(k) \notin \beta \\ \frac{1}{2} & \text{if } p(k) \in \beta \text{ and } N(k) \neq \emptyset \\ 1 & \text{if } p(k) \in \beta \text{ and } N(k) = \emptyset. \end{cases}$$

Set

$$\begin{aligned}
x_1 &= & 1, & \\
\tilde{a}_{p(k)k} &= & a_{p(k)k} & (k \neq 1), \\
\tilde{a}_{kp(k)} &= & a_{kp(k)} & (k \notin \beta), \\
x_k &= & 0 & (k \notin \beta).
\end{aligned}$$

The values of $\tilde{a}_{sp(s)}$ and x_s for $s \in \beta$ and $s \neq 1$ are defined inductively as follows. Let k be an element of β with $k \neq 1$ and suppose that the numbers $\tilde{a}_{sp(s)}$ and x_s have been defined for all $s < k$. Let $j = p(k)$. Note that since no row of $A[\overline{\beta}, \beta]$ has exactly one nonzero entry, if $k \notin \beta$ then $|N(k)| \neq 0$. Also note that if $j \notin \beta$, then $k \in N(j)$ and hence $|N(j)| \neq 0$ and $\tau_j \neq 1$. These observations imply that the following definitions can be made. If $j \notin \beta$, then we define

$$x_k = -\frac{\tilde{a}_{jp(j)}x_{p(j)}}{a_{jk}|N(j)|} \quad \text{and} \quad \tilde{a}_{kj} = a_{kj}.$$

If $j \in \beta$, then we define

$$x_k = \frac{i(1 - \tau_j)x_j}{a_{jk}|N(j)|} \quad \text{and} \quad \tilde{a}_{kj} = \frac{-\tau_k(1 - \tau_j)}{a_{jk}|N(j)|}.$$

Since (ii) holds and $\tau_j \neq 1$, sign $\tilde{a}_{kj} = a_{kj}$. Thus the matrix $\tilde{A}$ constructed in this manner belongs to $\mathcal{Q}(A)$. It is easy to verify that

$$i\tau_k x_k = \tilde{a}_{kp(k)}x_{p(k)} \tag{10.7}$$

for all k in β with $k \neq 1$. We now show that $\tilde{A}x = ix$, by verifying that

$$\sum_{\ell=1}^{n} \tilde{a}_{k\ell}x_\ell = ix_k \quad (k = 1, 2, \ldots, n).$$

First assume that $k = 1$. Then

$$\begin{aligned}
\sum_{\ell=1}^{n} \tilde{a}_{1\ell}x_\ell &= \sum_{\ell \in N(1)} \tilde{a}_{1\ell}x_\ell \\
&= \sum_{\ell \in N(1)} \tilde{a}_{1\ell}\frac{(1 - \tau_1)ix_1}{a_{1\ell}|N(1)|} \\
&= (1 - \tau_1)ix_1 \\
&= ix_1.
\end{aligned}$$

Now assume that $k \in \beta$, $k \neq 1$, and $N(k) = \emptyset$. Since row k of $A[\beta]$ has a nonzero entry, $p(k) \in \beta$ and hence $\tau_k = 1$. It now follows from (10.7) that $\sum_{\ell=1}^{n} \tilde{a}_{k\ell} x_\ell = \tilde{a}_{kp(k)} x_{p(k)} = i x_k$. Now assume that $k \in \beta$, $k \neq 1$, and $N(k) \neq \emptyset$. Then by (10.7) we have that

$$
\begin{aligned}
\sum_{\ell=1}^{n} \tilde{a}_{k\ell} x_\ell &= \tilde{a}_{kp(k)} x_{p(k)} + \sum_{\ell \in N(k)} \tilde{a}_{k\ell} x_\ell \\
&= \tilde{a}_{kp(k)} x_{p(k)} + \sum_{\ell \in N(k)} \tilde{a}_{k\ell} \frac{(1-\tau_k)ix_k}{a_{k\ell}|N(k)|} \\
&= \tilde{a}_{kp(k)} x_{p(k)} + (1-\tau_k)ix_k \\
&= i x_k.
\end{aligned}
$$

Finally, assume that $k \notin \beta$. Then $N(k)$ is nonempty and (10.7) implies that

$$
\begin{aligned}
\sum_{\ell=1}^{n} \tilde{a}_{k\ell} x_\ell &= \tilde{a}_{kp(k)} x_{p(k)} + \sum_{\ell \in N(k)} \tilde{a}_{k\ell} x_\ell \\
&= \tilde{a}_{kp(k)} x_{p(k)} + \sum_{\ell \in N(k)} \tilde{a}_{k\ell} \left(\frac{-\tilde{a}_{kp(k)} x_{p(k)}}{a_{k\ell}|N(k)|} \right) \\
&= \tilde{a}_{kp(k)} x_{p(k)} - \tilde{a}_{kp(k)} x_{p(k)} \\
&= 0 \\
&= i x_k.
\end{aligned}
$$

Hence i is an eigenvalue of $\tilde{A}$ contradicting the assumption that A is sign stable. Therefore we conclude that (v) holds, and the proof is complete. $\qquad\square$

We now present an algorithm [10] for determining whether a matrix is sign stable.

Algorithm for recognizing sign-stable matrices

Let $A = [a_{ij}]$ be an irreducible $(0, 1, -1)$-matrix of order n.

(0) If A has an identically zero determinant, then A is not a sign-stable matrix. Otherwise,

(1) if there exist k and ℓ such that $k \neq \ell$ and $a_{k\ell} = a_{\ell k} \neq 0$ or if there exists k such that $a_{k,k} > 0$, then A is not a sign-stable matrix. Otherwise,

(2) if the digraph of A is not a doubly directed tree, then A is not a sign-stable matrix. Otherwise,

(3) let $\mathcal{D}(A) = \overset{\leftrightarrow}{T}$ where T is a tree of order n and let $\alpha = \{k : a_{kk} \neq 0\}$.

(4) If there exists a vertex $v \notin \alpha$ such that every vertex adjacent to v is in α, then replace α by $\alpha \cup \{v\}$ and repeat (4) with this new α. Otherwise,

(5) if there exists a vertex $v \notin \alpha$, and a vertex $w \in \alpha$ such that v is the only vertex not in α which is adjacent to w, then replace α by $\alpha \cup \{v\}$ and go back to (4). Otherwise,

(6) If $\alpha \neq \{1, 2, \ldots, n\}$, then A is not a sign-stable matrix. Otherwise, A is a sign-stable matrix.

The validity of the algorithm is implied by the following observations. If A satisfies (i)–(iv) of Theorem 10.2.2, but not (v), and β is a counterexample to (v), then $\beta \cap \alpha = \emptyset$ throughout the algorithm. Hence, if A satisfies (i)–(iv) but not (v), the algorithm halts with $\alpha \neq \{1, 2, \ldots, n\}$ and correctly concludes that A is not a sign-stable matrix. Conversely, if the algorithm halts with $\alpha \neq \{1, 2, \ldots, n\}$, then $\beta = \{1, 2, \ldots, n\} \setminus \alpha$ is a counterexample to (v), and A is not a sign-stable matrix.

The following corollary is a consequence of the algorithm and implies that each doubly directed tree is the digraph of some sign-stable matrix [10].

Corollary 10.2.3 *Let $A = [a_{ij}]$ be a $(0, 1, -1)$-matrix of order n which satisfies (i), (ii), (iii), and (iv) of Theorem 10.2.2. Let $\mathcal{D}(A) = \overset{\leftrightarrow}{T}$ where T is a tree of order n. If $a_{kk} = -1$ for each pendant vertex of A, then A is sign stable. If T is a path, say the path $1, 2, \ldots, n$, then A is sign stable if and only if $a_{11} = -1$, or $a_{nn} = -1$, or there exist integers p and q such that $a_{pp} = -1$, $a_{qq} = -1$, and $|p - q| \leq 2$.*

The matrix A is *sign quasi-stable* provided every matrix in $\mathcal{Q}(A)$ is quasi-stable. Sign quasi-stable matrices have been studied in [8].

10.3 Stably sign-solvable linear systems

As Samuelson [16] pointed out, in studying equilibriums of dynamical systems it is natural to assume that the coefficient matrices are stable. This assumption leads to the study of stably sign-solvable linear systems.

Let A be a $(0, 1, -1)$-matrix of order n. The *stable qualitative class* of A is the set $\mathcal{Q}^-(A)$ of all stable matrices in $\mathcal{Q}(A)$. Thus A is sign stable if and only if $\mathcal{Q}^-(A) = \mathcal{Q}(A)$. If A is a nonnegative matrix, then $\mathcal{Q}^-(A)$ is empty. If A has a negative main diagonal, then the matrix $A - nI_n$ is a stable matrix in $\mathcal{Q}(A)$ and hence $\mathcal{Q}^-(A)$ is nonempty. Since the real eigenvalues of a stable matrix are negative and the nonreal eigenvalues occur in conjugate pairs, the sign of the determinant of a stable matrix of order n is $(-1)^n$. The following lemma gives in particular a sufficient condition for a qualitative class to contain a stable matrix [4].

Lemma 10.3.1 *Let $B = [b_{k\ell}]$ be a matrix of order n such that*

$$\mathrm{sign}(\det B[\{1, 2, \ldots, m\}]) = (-1)^m \quad (m = 1, \ldots, n).$$

Then there exists a diagonal matrix D with positive diagonal entries such that DB is stable.

Proof The proof is by induction on n. If $n = 1$, then B is stable. Assume that $n > 1$. The matrix $B' = B[\{1, 2, \ldots, n-1\}]$ satisfies the inductive hypothesis, and hence there exists a diagonal matrix D' with positive diagonal entries such that $D'B'$ is stable. Let $D_\epsilon = D \oplus [\epsilon]$. The eigenvalues of $(D' \oplus [0])B$ are 0 together with the eigenvalues of $D'B'$. Hence, for a sufficiently small positive number ϵ, the real parts of $n - 1$ of the eigenvalues of $D_\epsilon B$ are negative. Since the number of eigenvalues with negative real parts of a real matrix of order n whose determinant has sign equal to $(-1)^n$ has the same parity as n, the real parts of all eigenvalues of $D_\epsilon B$ are negative, and hence $D_\epsilon B$ is stable. $\qquad\square$

Let A be a $(0, 1, -1)$-matrix of order n such that $\mathcal{Q}^-(A)$ is nonempty and let b be a nonzero n by 1 vector. Then the system $Ax = b$ is *stably sign-solvable* provided

$$\{\widetilde{A}^{-1}\widetilde{b} : \widetilde{A} \in \mathcal{Q}^-(A), \widetilde{b} \in \mathcal{Q}(b)\}$$

is a subset of a single qualitative class. Clearly, if $Ax = b$ is sign-solvable, then $Ax = b$ is stably sign-solvable. Since the sign of the determinant of each matrix in $\mathcal{Q}^-(A)$ equals $(-1)^n$, it follows that $Ax = b$ is stably sign-solvable if and only if the numbers

$$\det \widetilde{A}(j \leftarrow b) \quad (\widetilde{A} \in \mathcal{Q}^-(A))$$

have the same sign for each $j = 1, 2, \ldots, n$. Thus, if b has a unique nonzero entry, say in position r, then $Ax = b$ is stably sign-solvable if and only if the (r, j)-entries of the inverses of matrices in $\mathcal{Q}^-(A)$ have the same sign for each $j = 1, \ldots, n$.

The theory of stably sign-solvable linear systems is not yet as fully developed as that of sign-solvable linear systems. The remainder of this section discusses the known results in the area.

We now describe two families of stably sign-solvable linear systems. First, let A be a fully indecomposable $(0, 1, -1)$-matrix of order n with a negative main diagonal, and assume that each of the entries off the main diagonal of A is nonnegative. Let $\widetilde{A}$ be a matrix in $\mathcal{Q}^-(A)$. Then there exists a nonnegative number λ such that $\widetilde{A} = -\lambda I_n + \widetilde{B}$ where $\widetilde{B}$ is a nonnegative matrix each of whose eigenvalues has modulus less than λ. Thus

$$\left(\frac{-1}{\lambda}\widetilde{A}\right)^{-1} = \left(I_n - \frac{1}{\lambda}\widetilde{B}\right)^{-1} = \sum_{j=0}^{\infty} \frac{1}{\lambda^j}\widetilde{B}^j,$$

and we conclude that each entry of $\widetilde{A}^{-1}$ is negative. Hence the system $Ax = b$ is stably sign-solvable for any unsigned vector b. Note that this example shows that, unlike sign-solvable linear systems with a fully indecomposable coefficient matrix, there are stably sign-solvable linear systems $Ax = b$ with A fully indecomposable where b has more than one nonzero entry.

Next let A be an SNS*-matrix with a negative main diagonal such that each of the entries in its first row is negative. Let B be a matrix obtained from A by negating some of the entries in its first row and off of its main diagonal. Since $B(\{1\}, \{j\})$ is an SNS-matrix for $j = 1, 2, \ldots, n$, it follows that $Bx = e_1$ is stably sign-solvable where e_1 is the n by 1 column vector whose only nonzero entry is a 1 in its first position.

It follows from Corollary 3.2.2 that if A is an SNS-matrix with a negative main diagonal and $k \neq \ell$, then the (k, ℓ)-entries of the inverses of matrices in $\mathcal{Q}(A)$ all have the same sign if and only if all the paths in $\mathcal{D}(A)$ from k to ℓ have the same sign. The following lemma shows that if all of the (k, ℓ)-entries of the inverses of matrices in $\mathcal{Q}^-(A)$ have the same sign, then all the paths from k to ℓ have the same sign.

Lemma 10.3.2 *Let $A = [a_{k\ell}]$ be a matrix of order n with a negative main diagonal and let r and s be integers with $1 \leq r, s \leq n$. If there exists a path α from r to s in $\mathcal{D}(A)$, then there exists a matrix $\widetilde{A} \in \mathcal{Q}^-(A)$ such that the sign of the (r, s)-entry of $\widetilde{A}^{-1}$ equals the sign of α. Moreover,*

   (i) *if the sign of the (r, s)-entries of the inverses of matrices in $\mathcal{Q}^-(A)$ are all positive (respectively, negative), then every path from r to s is negative (respectively, positive);*

   (ii) *the sign of the (r, s)-entries of the inverses of matrices in $\mathcal{Q}^-(A)$ are all zero if and only if there does not exist a path in $\mathcal{D}(A)$ from r to s.*

Proof Suppose that α is a path from r to s in $\mathcal{D}(A)$. Without loss of generality we may assume that $r = 1$ and that α is the path

$$1 \rightarrow 2 \rightarrow \cdots \rightarrow s.$$

Let $\widetilde{A} = [a_{k\ell}]$ be the matrix of order n with $\tilde{a}_{kk} = -2$ if $k = 1, 2, \ldots, n$, $\tilde{a}_{k,k+1} = a_{k,k+1}$ if $k = 1, 2, \ldots, s - 1$, and $\tilde{a}_{k\ell} = \epsilon a_{k\ell}$ otherwise. If ϵ is positive, then $\widetilde{A}$ belongs to $\mathcal{Q}(A)$. $\widetilde{A}$ is stable for ϵ sufficiently small, and

$$\text{sign det } \widetilde{A}(s, 1) = (-1)^{n-s} \text{sign } \alpha.$$

Assertion (i) now follows. Assertion (ii) is a consequence of (i) and the fact that if there is no path from r to s, then $A(\{r\}, \{s\})$ has an identically zero determinant. $\square$

The converse of Lemma 10.3.2 is not true. For example, let

$$A = \begin{bmatrix} -1 & 1 & 0 & 0 \\ 0 & -1 & 1 & 0 \\ 0 & 0 & -1 & 1 \\ -1 & 0 & 1 & -1 \end{bmatrix}.$$

Then the only path from 1 to 2 in $\mathcal{D}(A)$ is the path of length 1, and the matrices

$$\begin{bmatrix} -1 & 1 & 0 & 0 \\ 0 & -1 & 1 & 0 \\ 0 & 0 & -1 & 1 \\ -\frac{1}{8} & 0 & 1 & -1 \end{bmatrix} \text{ and } \begin{bmatrix} -1 & .1 & 0 & 0 \\ 0 & -1 & .1 & 0 \\ 0 & 0 & -1 & .1 \\ -.1 & 0 & .1 & -1 \end{bmatrix}$$

are stable matrices in $\mathcal{Q}(A)$ such that the $(1,2)$-entry of the inverse of the first matrix is 0 and of the second matrix is negative. We do not know what conditions on the digraph of a matrix are necessary and sufficient to determine the sign of a given entry of the inverses of stable matrices in its qualitative class.

If A is an SNS-matrix with a negative main diagonal, then the inverse of each matrix in $\mathcal{Q}(A)$ has a negative main diagonal. A diagonal matrix each of whose diagonal entries is negative is a stable matrix whose inverse has a negative main diagonal. It follows that if A is a square matrix with a negative main diagonal, then there is a matrix in $\mathcal{Q}^-(A)$ whose inverse has a negative main diagonal. Let

$$A = \begin{bmatrix} -1 & 1 & 0 \\ 1 & -1 & 1 \\ 0 & -1 & -1 \end{bmatrix}.$$

Then the matrix

$$\begin{bmatrix} -1 & 1 & 0 \\ 1 & -\frac{1}{2} & 1 \\ 0 & -1 & -1 \end{bmatrix}$$

belongs to $\mathcal{Q}^-(A)$, and its inverse does not have a negative main diagonal. The next lemma gives necessary conditions for a diagonal entry of the inverse of each matrix in a stable qualitative class to be negative.

Lemma 10.3.3 *Let A be a matrix of order n with a negative main diagonal. Suppose that there exist directed cycles α and β such that α is negative and contains the vertex v, β is positive and does not contain v, and the intersection of α and β is a common subpath γ. Then there exists a matrix in $\mathcal{Q}^-(A)$ such that the (v,v)-entry of its inverse is positive.*

Proof Without loss of generality we assume that $v = 1$, α is the directed cycle $1 \to 2 \to \cdots \to t \to 1$, γ is the subpath $r \to \cdots \to s$ of α, and β is the directed cycle $r \to r+1 \to \cdots \to s \to t+1 \to t+2 \to \cdots \to u \to r$ where $1 \le r \le s \le t \le u \le n$. Let $B = [b_{k\ell}]$ be the matrix of order n with $b_{kk} = -1/2$ if $k = s$, and $b_{kk} = -1$ otherwise, and $b_{k\ell} = a_{k\ell}$ if $\{k, \ell\}$ is an arc of α or of β and $b_{k\ell} = 0$ otherwise. Then

$$\det B[\{1, 2, \ldots, m\}] = \begin{cases} (-1)^m & \text{if } m \in \{1, 2, \ldots, s-1\} \\ \frac{1}{2}(-1)^m & \text{if } m \in \{s, s+1, \ldots, t-1\} \\ \frac{3}{2}(-1)^m & \text{if } m \in \{t, t+1, t+2, \ldots, u\} \\ \frac{1}{2}(-1)^m & \text{otherwise,} \end{cases}$$

and $\det B(\{1\}) = (-1)^{n-1}(-1/2)$. By Lemma 10.3.1, there exists a nonnegative strict signing D such that DB is a stable matrix. The entry in the $(1, 1)$ position of the inverse of B and hence of DB is positive. Since B is a limit of matrices in $\mathcal{Q}(A)$, it follows that there exists a matrix in $\mathcal{Q}^-(A)$ whose inverse has a positive entry in position $(1, 1)$. $\qquad\square$

The converse of Lemma 10.3.3 is not true. For example, let A be a matrix of order n with a negative main diagonal whose digraph consists of the arcs of three directed cycles α, β, and γ such that α is negative and contains vertex 1, β is negative and contains exactly one vertex $v \ne 1$ of α, and γ is positive, contains no vertices of α, and contains exactly one vertex of β. Then A does not satisfy the hypothesis of Lemma 10.3.3, but by an argument similar to that in the proof of Lemma 10.3.3 there exists a matrix in $\mathcal{Q}^-(A)$ such that the $(1, 1)$-entry of its inverse is positive.

Theorem 10.3.4 *Let $Ax = b$ be a stably sign-solvable linear system where A is a fully indecomposable matrix with a negative main diagonal and where the only nonzero entry of b is in its first position. Then there exists a strict signing D such that every path in $\mathcal{D}(DAD)$ terminating at vertex 1 is positive, and no negative cycle containing 1 intersects a positive cycle which does not contain 1.*

Proof Since $Ax = b$ is stably sign-solvable, the sign patterns of the first columns of the inverses of matrices in $\mathcal{Q}^-(A)$ are all the same. Hence there exists a strict signing D such that the sign pattern of the first column of the inverse of each matrix in $\mathcal{Q}^-(DAD)$ equals $(-1, -1, \ldots, -1)^T$. It now follows from Lemma 10.3.2 that every path in $\mathcal{D}(DAD)$ terminating at 1 is positive. This implies that if there exists a negative directed cycle α containing 1 and a positive directed cycle β not containing 1 which intersect, then there exists a positive directed cycle γ not containing 1 which intersects α in a common subpath. The theorem now follows from Lemma 10.3.3. □

We do not know whether or not the conditions in Theorem 10.3.4 are sufficient for $Ax = b$ to be stably sign-solvable. In [1], it is claimed that another necessary condition is that every negative cycle which intersects a positive cycle contains vertex 1. The matrix obtained from the SNS*-matrix

$$\begin{bmatrix} -1 & -1 & -1 \\ 1 & -1 & -1 \\ 0 & 1 & -1 \end{bmatrix}$$

by replacing the $(1, 3)$-entry by a 1 is a counterexample to the claim.

Let $A = [a_{ij}]$ be a $(0, 1, -1)$-matrix of order n such that $Q^-(A) \neq \emptyset$. In general the inverses of matrices in $Q^-(A)$ need not have the same sign pattern. If there exists a matrix B such that

$$\{\widetilde{A}^{-1} : \widetilde{A} \in Q^-(A)\} \subseteq Q(B),$$

then A is an S^2NS-*matrix under stability*. Analogous to the situation for S^2NS-matrices, a matrix A is an S^2NS-matrix under stability if and only if $Ax = b$ is stably sign-solvable for each vector b with exactly one nonzero entry. Also, if A is an S^2NS-matrix under stability, then so is $DPAP^T D$ for any permutation matrix P and any strict signing D. Any S^2NS-matrix A for which $Q^-(A)$ is nonempty is an S^2NS-matrix under stability. As we have already noted, if A has a negative main diagonal and its off-diagonal entries are nonnegative, then the inverse of every matrix in $Q^-(A)$ is nonnegative, and hence A is an S^2NS-matrix under stability. We now give a characterization of fully indecomposable, S^2NS-matrices under stability with negative main diagonals [1].

Theorem 10.3.5 *Let A be a fully indecomposable, $(0, 1, -1)$-matrix with a negative main diagonal. Then A is an S^2NS-matrix under stability if and only if A is an S^2NS-matrix or there exists a strict signing D such that each entry off the main diagonal of DAD is nonnegative.*

Proof Assume that A is an S^2NS-matrix under stability. Lemma 10.3.2 implies that two paths in $\mathcal{D}(A)$ with the same initial vertex and the same terminal vertex have the same sign. First suppose that $\mathcal{D}(A)$ has has a positive directed cycle α. Let $A_1, A_2, \ldots, A_m$ be a recursive fully indecomposable construction of $\mathcal{D}(A)$ such that $\mathcal{D}(A_2)$ is the directed cycle α. We show by induction on k that there exists a strict signing D_k such that each entry off the main diagonal of $D_k A_k D_k$ is nonnegative. This is clear if $k = 1$ or $k = 2$. Assume that $k > 2$, and without loss of generality assume that each entry off the main diagonal of A_{k-1} is positive. Let α_k be the path consisting of those arcs in $\mathcal{D}(A_k)$ which are not in $\mathcal{D}(A_{k-1})$. It follows from Lemma 10.3.2 that α_k is positive. Hence, if α_k has length 1, then each entry off the main diagonal of A_k is nonnegative, and if α_k has length greater than 1, then there is a strict signing D_k such that each entry off the main diagonal of $D_k A_k D_k$ is nonnegative. Thus, by induction, there exists a strict signing D such that each entry off the main diagonal of DAD is nonnegative. This implies that every directed cycle of $\mathcal{D}(A)$ is positive. Hence we have shown that if $\mathcal{D}(A)$ has a positive directed cycle, then every directed cycle of $\mathcal{D}(A)$ is positive and there exists a strict signing D such that each entry off the main diagonal of DAD is nonnegative. Now suppose that $\mathcal{D}(A)$ does not have a positive directed cycle. Then A is an SNS-matrix and hence, by Corollary 3.2.3, A is an S^2NS-matrix.

The converse follows from the facts that an S^2NS-matrix is an S^2NS-matrix under stability and that a fully indecomposable matrix with a negative main diagonal whose off-diagonal entries are nonnegative is an S^2NS-matrix under stability. $\qquad\square$

10.4 SNS-matrices with other eigenvalue restrictions

Let A be a $(0, 1, -1)$-matrix of order n. Then A is an SNS-matrix if and only if 0 is not an eigenvalue of any matrix in $\mathcal{Q}(A)$. Theorems 10.2.1 and 10.2.2 classify the sign semi-stable, respectively sign-stable, matrices, that is, the matrices A for which the real parts of the eigenvalues of matrices in $\mathcal{Q}(A)$ are nonpositive, respectively negative. In this section, which is based on [3], we investigate other matrices for which the eigenvalues of matrices in $\mathcal{Q}(A)$ have a common property. Throughout this section we assume without loss of generality that A is irreducible.

We first classify the matrices A for which no matrix in $\mathcal{Q}(A)$ has a real eigenvalue.

Lemma 10.4.1 *Let B be a real matrix of order n such that λ is a real eigenvalue whose algebraic multiplicity equals 1. Then if B' is a real matrix sufficiently close to B, B' has a real eigenvalue.*

Proof The lemma follows from the facts that the eigenvalues of a matrix vary continuously with its entries and that the nonreal eigenvalues of a real matrix occur in complex conjugate pairs. $\qquad\square$

Theorem 10.4.2 *Let A be an irreducible matrix of order n. Then the following are equivalent:*

(i) *No matrix in $\mathcal{Q}(A)$ has a real eigenvalue.*

(ii) *A is an SNS-matrix each of whose diagonal entries is zero and every directed cycle of $\mathcal{D}(A)$ has even length and is negative.*

(iii) $-I_n + A$ *is an SNS-matrix and there exists a permutation matrix P such that*

$$P^T A P = \begin{bmatrix} O & B \\ C & O \end{bmatrix} \tag{10.8}$$

where B and C are SNS-matrices.

Proof First assume that (i) holds. Then clearly A is an SNS-matrix. A diagonal matrix with exactly one nonzero entry has a nonzero real eigenvalue with algebraic multiplicity equal to 1. In addition, a $(0, 1, -1)$-matrix each of whose diagonal entries equals 0 and whose digraph is a directed cycle of length k has a real eigenvalue with algebraic multiplicity equal to 1 if and only if k is odd or the directed cycle is positive. It now follows from Lemma 10.4.1 that (ii) holds. Now assume that (ii) holds. Theorem 3.2.1 implies that $-I_n + A$ is an SNS-matrix. Since $\mathcal{D}(A)$ is a strongly connected digraph and has no directed cycles of odd length, its vertices can be partitioned into two sets α and β so that every arc contains a vertex in α and a vertex in β. Hence there exists a permutation matrix P such that (10.8) holds. Since A is an SNS-matrix, B and C are also SNS-matrices. Hence (iii) holds. Finally, assume that (iii) holds. Without loss of generality, we may assume that $P = I_n$. Since $-I_n + A$ is an SNS-matrix, every directed cycle of $\mathcal{D}(A)$ is negative and has even length. This implies that $-I_n - A$ is also an SNS-matrix. Let $\widetilde{A}$ belong to $\mathcal{Q}(A)$ and let λ be a real number. Then the matrix $\lambda I_n - \widetilde{A}$ belongs to either $\mathcal{Q}(-I_n - A)$, $\mathcal{Q}(-A)$, or $\mathcal{Q}(I_n - A)$. Since each of these three classes is the qualitative class of an SNS-matrix, the matrix $\lambda I_n - \widetilde{A}$ is nonsingular and (i) holds. □

Theorem 10.4.2 implies that there exists a matrix in $\mathcal{Q}(A)$ which has a real eigenvalue if and only if A has a nonzero diagonal entry, $\mathcal{D}(A)$ has a directed cycle of odd length or $\mathcal{D}(A)$ has a positive directed cycle.

Let A be an irreducible matrix such that no matrix in $\mathcal{Q}(A)$ has a real eigenvalue. By Theorem 10.4.2 we may assume that

$$A = \begin{bmatrix} O & B \\ C & O \end{bmatrix}$$

where B and C are both SNS-matrices of order m. Then

$$\begin{bmatrix} -I_m & O \\ O & I_m \end{bmatrix} A \begin{bmatrix} -I_m & O \\ O & I_m \end{bmatrix} = -A.$$

Hence A and $-A$ are similar. It follows that if a complex number $\lambda = a + bi$ is an eigenvalue of A of multiplicity t, then $-\lambda$, $\bar\lambda$, and $-\bar\lambda$ are also eigenvalues of A of multiplicity t. Thus, for each matrix in $\mathcal{Q}(A)$, the eigenvalues which are not purely imaginary occur in quadruplets. In particular, if m is odd, then A has at least two purely imaginary eigenvalues.

Let A be a skew-symmetric $(0, 1, -1)$-matrix whose digraph is a doubly directed tree. If $\tilde A \in \mathcal{Q}(A)$, then there exists a diagonal matrix with a positive diagonal such that $D\tilde A D$ is skew-symmetric, and hence all of the eigenvalues of each matrix in $\mathcal{Q}(A)$ are purely imaginary. The converse is contained in the following theorem.

Theorem 10.4.3 *Let A be an irreducible $(0, 1, -1)$-matrix. Then the following are equivalent:*

   (i) *The real part of each eigenvalue of each matrix in $\mathcal{Q}(A)$ is zero.*

   (ii) *Each diagonal entry of A equals 0, and each directed cycle of $\mathcal{D}(A)$ is negative and has length 2.*

  (iii) *A is a skew-symmetric matrix whose digraph is a doubly directed tree.*

Proof We have already shown that (iii) implies (i). A real diagonal matrix with exactly one nonzero entry has an eigenvalue whose real part is nonzero. A $(0, 1, -1)$-matrix B each of whose main diagonal entries equals 0 and whose digraph is a directed cycle γ of length k has minimal polynomial $x^k - \text{sign}\,\gamma$. Thus B has an eigenvalue with a nonzero real part if and only if $k \neq 2$ and γ is positive. That (i) implies (ii) is now a consequence of the continuity of eigenvalues. Since A is irreducible, $\mathcal{D}(A)$ is strongly connected, and hence (ii) implies (iii).   □

Corollary 10.4.4 *Let A be an irreducible $(0, 1, -1)$-matrix. Then the following are equivalent:*

   (i) *Each eigenvalue of each matrix in $\mathcal{Q}(A)$ is nonzero and has zero real part.*

   (ii) *Each diagonal entry of A equals 0, each directed cycle of $\mathcal{D}(A)$ has length 2, and both A and $-I_n + A$ are SNS-matrices.*

  (iii) *A is a skew-symmetric SNS-matrix whose digraph is a doubly directed tree.*

Proof Clearly (i) implies that A is an SNS-matrix. As in the proof of Theorem 10.4.3, (i) also implies that each diagonal entry of A is zero, each directed cycle of $\mathcal{D}(A)$ is negative and has length 2. Hence, by Theorem 3.2.1, (i) implies (ii). The remainder of the proof follows as in Theorem 10.4.3. □

We conclude by classifying the sign patterns A for which the eigenvalues of each matrix in $\mathcal{Q}(A)$ are real.

Theorem 10.4.5 *Let A be an irreducible $(0, 1, -1)$-matrix. Then the following are equivalent:*

(i) *Each eigenvalue of each matrix in $\mathcal{Q}(A)$ is real.*

(ii) *Each diagonal entry of A equals 0, and each directed cycle of $D(A)$ is positive and has length 2.*

(iii) *A is a symmetric matrix whose digraph is a doubly directed tree.*

Proof If A is a symmetric matrix whose digraph is a tree and $\widetilde{A} \in \mathcal{Q}(A)$, then there exists a diagonal matrix D with a positive diagonal such that $D\widetilde{A}D$ is symmetric. Since all of the eigenvalues of a real symmetric matrix are real, (iii) implies (i). Arguments similar to those in the proof of Theorem 10.4.1 show the equivalence of (i) and (ii) and that (ii) implies (iii) . □

Bibliography

[1] L. Bassett, J. Maybee, and J. Quirk. Qualitative economics and the scope of the correspondence principle, *Econometrica*, 36:544–63, 1968.

[2] T. Bone, C. Jeffries, and V. Klee. A qualitative analysis of $\dot{x} = Ax + b$, *Discrete Appl. Math.*, 20:9–30, 1988.

[3] C.A. Eschenbach and C.R. Johnson. Sign patterns that require real, non-real, and purely imaginary eigenvalues, *Linear Multilin. Alg.*, 29:299–311, 1991.

[4] M.E. Fisher and A.T. Fuller. On the stabilization of matrices and the convergence of linear iterative processes, *Proc. Camb. Philos. Soc.*, 54:417–25, 1958.

[5] C. Jeffries. Qualitative stability and digraphs in model ecosystems, *Ecology*, 55:1415–19, 1974.

[6] C. Jeffries. *Mathematical Modelling in Ecology: A Workbook for Students*, Birkhäuser, Boston, 1989.

[7] C. Jeffries, V. Klee, and P. van den Driessche. When is a matrix sign stable? *Canad. J. Math.*, 29:315–26, 1977.

[8] C. Jeffries, V. Klee, and P. van den Driessche. Qualitative stability of linear systems, *Linear Alg. Appls.*, 87:1–48, 1987.

[9] V. Klee. Sign-patterns and stability, in *Applications of Combinatorics and Graph Theory to the Biological and Social Sciences* (F. Roberts, ed.), IMA Volumes in Mathematics and Its Applications, Springer, New York, 17:203–19, 1989.

[10] V. Klee and P. van den Driessche. Linear algorithms for testing the sign stability of a matrix and for finding Z-maximum matchings in acyclic graphs, *Numer. Math.,* 28:273–85, 1977.

[11] D. Logofet. *Matrices and Graphs: Stability Problems in Mathematical Ecology,* CRC Press, Boca Raton, 1992.

[12] R.M. May. Qualitative stability in model ecosystems, *Ecology,* 54:638–44, 1973.

[13] R.M. May. *Stability and Complexity in Model Ecosystems,* Princeton Univ. Press, Princeton, N.J., 1973.

[14] J. Quirk and R. Ruppert. Qualitative economics and the stability of equilibrium, *Rev. Economic Studies,* 32:311–26, 1965.

[15] R. Redheffer and Z. Zhou. Sign semistability and global asymptotic stability, *Ann. Diff. Eq.,* 5:145–53, 1989.

[16] P.A. Samuelson. *Foundations of Economic Analysis,* Harvard University Press, Cambridge, 1947, Atheneum, New York, 1971.

11

Related topics

11.1 Conditional sign-solvability

Recall from section 1.2 that a linear system $Ax = b$ is defined to be sign-solvable provided that (i) $\widetilde{A}x = \widetilde{b}$ is solvable for all $\widetilde{A}$ in $Q(A)$ and all $\widetilde{b}$ in $Q(b)$, and (ii) the solutions of the linear systems $\widetilde{A}x = \widetilde{b}$ have the same sign pattern. In this section, which is based on [3], we replace (i) by the condition that $\widetilde{A}x = \widetilde{b}$ is solvable for at least one $\widetilde{A}$ and $\widetilde{b}$.[1]

A linear system $Ax = b$ is *conditionally sign-solvable* provided

$$\{\tilde{x} : \text{ there exists } \widetilde{A} \in Q(A) \text{ and } \widetilde{b} \in Q(b) \text{ with } \widetilde{A}\tilde{x} = \widetilde{b}\} \qquad (11.1)$$

is a nonempty set which is contained in a single qualitative class. If $Ax = b$ is conditionally sign-solvable, then (11.1) is the *qualitative solution class* of $Ax = b$ and, as for sign-solvable linear systems, is denoted by $Q(Ax = b)$. It follows easily that $Q(Ax = b)$ is an entire qualitative class. The linear system $Ax = b$ is sign-solvable provided it is conditionally sign-solvable and $\widetilde{A}x = \widetilde{b}$ is solvable for each $\widetilde{A}$ in $Q(A)$ and each $\widetilde{b}$ in $Q(b)$.

Consider a linear system $Ax = b$ where

$$A = \begin{bmatrix} -1 & 1 & 1 \\ 1 & -1 & 1 \\ 1 & 1 & -1 \\ 1 & 1 & 1 \end{bmatrix} \text{ and } b = \begin{bmatrix} 0 \\ 0 \\ 0 \\ 1 \end{bmatrix}. \qquad (11.2)$$

Suppose $\widetilde{A}x = \widetilde{b}$ has a solution $u = (u_1, u_2, u_3)^T$. In order that the first three equations are satisfied, $u_1, u_2,$ and u_3 must have the same sign. In order that

---

[1] Sign-solvability where in (i) and (ii) the column vector $\widetilde{b}$ and the columns of $\widetilde{A}$ belong to cones more general than qualitative classes is investigated in [11, 12].

the last equation is also satisfied,

$$\text{sign } u = \begin{bmatrix} 1 \\ 1 \\ 1 \end{bmatrix}.$$

Thus, if $\widetilde{A}x = \widetilde{b}$ has a solution, we have determined its sign pattern knowing only the sign patterns of $\widetilde{A}$ and $\widetilde{b}$. By replacing each -1 of A by -2 we obtain a matrix $\widetilde{A}$ such that $\widetilde{A}x = b$ is solvable, and hence $Ax = b$ is conditionally sign-solvable with $Q(Ax = b) = Q((1, 1, 1)^T)$. Since the matrix obtained from A by deleting the last row is nonsingular, $Ax = b$ is not solvable, and hence $Ax = b$ is not sign-solvable.

By Theorem 1.2.12, sign-solvable linear systems are characterized in terms of L-matrices and S^*-matrices. Conditionally sign-solvable linear systems are characterized in terms of L-matrices and a generalization of S^*-matrices which we now define.

Let A be an m by n matrix. Then A is a *conditionally S-matrix*, abbreviated a CS-matrix, provided that

    (i) A has no zero rows,

   (ii) for each $\widetilde{A}$ in $Q(A)$, the right null space of $\widetilde{A}$ either is trivial or is spanned by a vector with only positive entries, and

  (iii) at least one $\widetilde{A}$ in $Q(A)$ has a nontrivial right nullspace.

It follows that A is a CS-matrix if and only if there is a matrix $\widetilde{A}$ in $Q(A)$ with linearly dependent columns, and for each such matrix $\widetilde{A}$, its column vectors are the vertices of an $(n-1)$-simplex whose relative interior contains the origin. Clearly, if a matrix A is a conditionally S-matrix, then A^T is not an L-matrix, but every matrix obtained from A^T by deleting a row is an L-matrix. If w is a column vector with no zero entries, then the matrix $[w \ -w]$ with two columns is a CS-matrix. The matrix A is a *conditionally S^*-matrix*, abbreviated a CS*-matrix, provided there exists a strict signing D such that AD is a conditionally S-matrix.

An S-matrix is a CS-matrix, and an S^*-matrix is a CS*-matrix. Clearly, if A is an m by n CS-matrix, then $n \leq m+1$, and if $n = m+1$, then A is an S-matrix. It follows from example (11.2) that

$$[A \ -b] = \begin{bmatrix} -1 & 1 & 1 & 0 \\ 1 & -1 & 1 & 0 \\ 1 & 1 & -1 & 0 \\ 1 & 1 & 1 & -1 \end{bmatrix} \tag{11.3}$$

is a CS-matrix.

The characterizations of CS-matrices, CS*-matrices, and conditionally sign-solvable linear systems in the following lemma are consequences of definitions and Lemma 2.1.1.

Lemma 11.1.1 *Let A be an m by n matrix and let b be an m by 1 column vector. Then*

(i) *A is a CS*-matrix if and only if A has no zero rows and there exists a strict signing D such that AD and $A(-D)$ are the only column signings of A each of whose rows is balanced.*

(ii) *A is a CS-matrix if and only if A has no zero rows and $AI_n = A$ and $A(-I_n) = -A$ are the only column signings of A each of whose rows is balanced.*

(iii) *$Ax = b$ is conditionally sign-solvable if and only if there exists a unique signing of the form $D = \mathrm{diag}(d_1, \ldots, d_n, 1)$ such that each row of $[A \; -b]D$ is balanced.*

In [3] a subclass of the CS*-matrices is shown to be essentially equivalent to the L-indecomposable, barely L-matrices (see section 2.2) having a column with no zero entries.

By Corollary 1.2.2, the transpose of the coefficient matrix of a sign-solvable linear system is an L-matrix. Suppose that $Ax = b$ is conditionally sign-solvable. If $\widetilde{A}$ is in $\mathcal{Q}(A)$, then it can happen that there does not exist a vector $\widetilde{b}$ in $\mathcal{Q}(b)$ such that $\widetilde{A}x = \widetilde{b}$ is solvable. For instance, suppose A and b are as in (11.2) and let $\widetilde{A} = A$. Then $\widetilde{A}x = \widetilde{b}$ does not have a solution for any $\widetilde{b}$ in $\mathcal{Q}(b)$. Thus, if $Ax = b$ is only conditionally sign-solvable, an argument different from that given in the proof of Corollary 1.2.2 is needed to show A^T is an L-matrix.

Theorem 11.1.2 *If $Ax = b$ is a conditionally sign-solvable linear system, then A^T is an L-matrix.*

Proof Let A be an m by n matrix. Suppose that A^T is not an L-matrix. We prove that $Ax = b$ is not conditionally sign-solvable by showing that there does not exist a unique signing of the form $D = \mathrm{diag}(d_1, \ldots, d_n, 1)$ such that each row of $[A \; -b]D$ is balanced, and then applying Lemma 11.1.1. If no such signing D exists, then the argument is complete. We now assume that such a signing D exists and show its non-uniqueness. Since A^T is not an L-matrix, there exists a signing $E = \mathrm{diag}(e_1, \ldots, e_n)$ such that each row of AE is balanced.

After permutations of the columns of A, we may assume that $E = E_1 \oplus O$

where E_1 is a strict signing of order ℓ where $1 \leq \ell \leq n$. After permutations of rows we may assume that $[A \; -b]$ has the form

$$\left[\begin{array}{c|c|c} A_1 & A_3 & -b^1 \\ \hline O & A_2 & -b^2 \end{array}\right],$$

where A_1 is a k by ℓ matrix with no zero rows. Each row of the matrices

$$[A \; -b]\mathrm{diag}(e_1, \ldots, e_\ell, d_{\ell+1}, \ldots, d_n, 1)$$

and

$$[A \; -b]\mathrm{diag}(-e_1, \ldots, -e_\ell, d_{\ell+1}, \ldots, d_n, 1)$$

is balanced and the theorem now follows. □

Corollary 11.1.3 *Let $Ax = b$ be a linear system where A is a square matrix. Then $Ax = b$ is conditionally sign-solvable if and only if it is sign-solvable.*

Proof Suppose that $Ax = b$ is conditionally sign-solvable. By Theorem 11.1.2, A is an L-matrix and hence each matrix $\widetilde{A}$ in $\mathcal{Q}(A)$ is invertible. Thus $\widetilde{A}x = \widetilde{b}$ is always solvable and hence $Ax = b$ is sign-solvable. The converse is obvious. □

A matrix A of order n is a *conditionally S^2NS-matrix* provided A does not have an identically zero determinant and the inverses of the nonsingular matrices in $\mathcal{Q}(A)$ belong to a single qualitative class. The conditionally S^2NS-matrices of order n are precisely the vertices of the inverse sign pattern graph $\mathcal{I}_n$ which belong to exactly one edge.

Theorem 11.1.4 *A matrix is a conditionally S^2NS-matrix if and only if it is an S^2NS-matrix.*

Proof Let A be a conditionally S^2NS-matrix of order n. We prove that A is an S^2NS-matrix by showing that the assumption that A is not an SNS-matrix leads to a contradiction. Assume that $\mathcal{Q}(A)$ contains a singular matrix. Then there exist matrices $\widetilde{A}$ and $\widehat{A}$ in $\mathcal{Q}(A)$ such that $\widetilde{A}$ is nonsingular and $\widehat{A}$ is singular. The rank of a matrix obtained from $\widetilde{A}$ by changing a single entry is at least $n - 1$. This observation allows us to assume that the rank of $\widehat{A}$ equals $n - 1$ and that $\widehat{A}$ differs from $\widetilde{A}$ in exactly one position, say the entry in position $(1,1)$ of $\widetilde{A}$ equals c and of $\widehat{A}$ equals d where $c \neq d$. Since $\widetilde{A}$ is nonsingular and $\widehat{A}$ is singular, it follows that $\det \widehat{A}(\{1\}) \neq 0$. Let $\widehat{A}_\epsilon$ be the matrix obtained from $\widehat{A}$ by adding ϵ to its entry in position $(1,1)$. Then $\det \widehat{A}_\epsilon = \epsilon \det \widehat{A}(\{1\})$. Hence, if $0 < |\epsilon| < |d|$, $\widehat{A}_\epsilon$ is a nonsingular matrix in $\mathcal{Q}(A)$ and the entry in position

(1,1) of $\widehat{A}_\epsilon^{-1}$ equals ϵ. This contradicts the fact that A is a conditionally S^2NS-matrix. Since an S^2NS-matrix is a conditionally S^2NS-matrix, the theorem now follows. $\qquad\square$

The following theorem characterizes conditionally sign-solvable linear systems.

Theorem 11.1.5 *Let $A = [a_{ij}]$ be an m by n matrix and let b be an m by 1 column vector. Assume that $z = (z_1, z_2, \ldots, z_n)^T$ is a solution of the linear system $Ax = b$. Let*

$$\beta = \{j : z_j \neq 0\} \quad and \quad \alpha = \{i : a_{ij} \neq 0 \, for \, some \, j \in \beta\}.$$

Then $Ax = b$ is conditionally sign-solvable if and only if the matrix

$$\begin{bmatrix} A[\alpha, \beta] & -b[\alpha] \end{bmatrix}$$

is a CS*-matrix and the matrix $A(\alpha, \beta)^T$ is an L-matrix.

Proof Without loss of generality we may assume that $\beta = \{1, 2, \ldots, \ell\}$ and $\alpha = \{1, 2, \ldots, k\}$ for some nonnegative integers k and ℓ. Thus

$$A = \begin{bmatrix} A_1 & A_3 \\ O & A_2 \end{bmatrix}$$

where A_1 is a k by ℓ matrix with no zero rows. Let $b' = b[\{1, 2, \ldots, k\}]$. Then the linear system $Ax = b$ can be written as

$$A_1 x^{(1)} + A_3 x^{(2)} = b' \tag{11.4}$$

$$A_2 x^{(2)} = 0. \tag{11.5}$$

First assume that $[A_1 \ -b']$ is a CS*-matrix and A_2^T is an L-matrix. We use Lemma 11.1.1 to show that $Ax = b$ is conditionally sign-solvable. Let D be a signing of order $n + 1$ of the form

$$\text{diag}(d_1, \ldots, d_n, 1). \tag{11.6}$$

Since A_2^T is an L-matrix, each row of $[A \ -b]D$ is balanced if and only if $d_{\ell+1} = \cdots = d_n = 0$ and each row of $[A_1 \ -b']\text{diag}(d_1, \ldots, d_\ell, 1)$ is balanced. Since $[A_1 \ -b']$ is a CS*-matrix, it follows from (i) of Lemma 11.1.1 that there is a unique signing of the form (11.6) which balances all the rows of $[A \ -b]$. By (iii) of Lemma 11.1.1, $Ax = b$ is conditionally sign-solvable.

Now assume that $Ax = b$ is conditionally sign-solvable. By Theorem 11.1.2, A^T is an L-matrix. By (iii) of Lemma 11.1.1, there is a unique signing D of the form diag$(d_1, \ldots, d_n, 1)$ such that each row of $[A \ -b]D$ is balanced. Since $Az = b$, we have $d_i = \text{sign } z_i$ for $i = 1, 2, \ldots, n$. If $F = \text{diag}(f_{\ell+1}, \ldots, f_n)$ is a signing such that each row of $A_2 F$ is balanced, then each row of

$$[A \ -b]\text{diag}(d_1, \ldots, d_\ell, f_{\ell+1}, \ldots, f_n, 1)$$

is balanced, contradicting the uniqueness of D. Hence A_2^T is an L-matrix. We now show that $[A_1 \ -b']$ is a CS*-matrix. Since $d_{\ell+1} = \cdots = d_n = 0$, each row of $[A_1 \ - b']\text{diag}(d_1, \ldots, d_\ell, 1)$ is balanced. Let $E = \text{diag}(e_1, \ldots, e_\ell, e_{\ell+1})$ be any signing such that each row of $[A_1 \ - b']E$ is balanced. Then each row of

$$[A \ -b]\text{diag}(e_1, \ldots, e_\ell, 0, \ldots, 0, e_{\ell+1})$$

is balanced. Since A^T is an L-matrix, $e_{\ell+1} \neq 0$. The uniqueness of D now implies that $E = \pm\text{diag}(d_1, \ldots, d_\ell, 1)$. By (i) of Lemma 11.1.1, $[A_1 \ - b']$ is a CS*-matrix.   □

Since a CS*-matrix with the property that every matrix in its qualitative class has a nontrivial right nullspace is an S*-matrix, Theorem 11.1.5 extends Theorem 1.2.12. However, there is a significant computational difference in applying Theorem 1.2.12 to test for sign-solvability and in applying Theorem 11.1.5 to test for conditional sign-solvability. Let $Ax = b$ be a linear system. To apply Theorem 1.2.12 to test whether $Ax = b$ is sign-solvable, we first solve $Ax = b$. If there is no solution, the linear system is not sign-solvable. Otherwise, we test whether a certain matrix is an S*-matrix and another is an L-matrix. To apply Theorem 11.1.5 to test whether $Ax = b$ is conditionally sign-solvable, it is not enough to first solve $Ax = b$. This is because if $Ax = b$ has no solution, we cannot conclude that $Ax = b$ is not conditionally sign-solvable. We must first find an $\widetilde{A}$ in $Q(A)$ and a $\tilde{b}$ in $Q(b)$ such that $\widetilde{A}x = \tilde{b}$ has a solution. This is essentially equivalent to finding a signing D whose last diagonal entry equals 1 such that each row of $[A \ - b]D$ is balanced.[2] If no such D exists, then $Ax = b$ is not conditionally sign-solvable. Otherwise, we test whether a certain matrix is a CS*-matrix and another is an L-matrix.

_____

[2]Let B be an m by n matrix. The problem ($*$) of deciding whether there exists a signing D of order n with a specified diagonal entry equal to 1 such that each row of BD is balanced is clearly in the class NP. The L-matrix recognition problem is polynomially reducible to ($*$): Since the matrix B^T is not an L-matrix if and only if for some i with $1 \leq i \leq n$, there exists a signing D whose ith diagonal entry equals 1 such that each row of BD is balanced, n applications of an algorithm for problem ($*$) will determine whether B^T is not an L-matrix. Since the L-matrix recognition problem is NP-complete, so is problem ($*$).

11.2 Least squares sign-solvability

In this section we discuss a generalization of sign-solvability introduced in [15] and motivated by the following example. Let $p, q, \ldots, v$ be positive numbers and consider the linear system $Ax = b$ where

$$A = \begin{bmatrix} p & r & t \\ q & 0 & 0 \\ 0 & s & 0 \\ 0 & 0 & u \end{bmatrix} \quad \text{and } b = \begin{bmatrix} 0 \\ 0 \\ 0 \\ v \end{bmatrix}.$$

Since $Ax = b$ has no solution, $Ax = b$ is not sign-solvable. Let us now consider the least squares solution to $Ax = b$. Since A^T is an L-matrix, the least squares solution to $Ax = b$ is the solution to the normal equation $A^T Ax = A^T b$. Computing the normal equation we have

$$\begin{bmatrix} p^2 + q^2 & pr & pt \\ pr & r^2 + s^2 & rt \\ pt & rt & t^2 + u^2 \end{bmatrix} x = \begin{bmatrix} 0 \\ 0 \\ uv \end{bmatrix}. \tag{11.7}$$

The linear system (11.7) is not sign-solvable; however, its solution is

$$\frac{uv}{\det A^T A} \begin{bmatrix} -ps^2 t \\ -q^2 rt \\ p^2 s^2 + q^2 r^2 + q^2 s^2 \end{bmatrix}.$$

Since A^T is an L-matrix, $A^T A$ is a positive definite matrix and in particular $\det A^T A > 0$. It follows that the least squares solution to $Ax = b$ belongs to $\mathcal{Q}([\ -1 \quad -1 \quad 1 \ ]^T)$, regardless of the magnitudes of $p, q, \ldots, v$. Thus the signs of the entries of the least squares solution to $Ax = b$ depend only on the signs of the entries of A and of b.

This example leads to the following definitions. Let A be an m by n matrix and let b be an n by 1 column vector. The linear system $Ax = b$ is *least squares sign-solvable* provided the vectors in

$$\{u : \text{there exist } \widetilde{A} \in \mathcal{Q}(A) \text{ and } \widetilde{b} \in \mathcal{Q}(b) \text{ with } \|\widetilde{A}u - \widetilde{b}\| = \min_{x \in R^n} \|\widetilde{A}x - \widetilde{b}\|\}$$

are contained in a single qualitative class. Clearly, a sign-solvable linear system is also least squares sign-solvable. The example shows that

$$\begin{bmatrix} 1 & 1 & 1 \\ 1 & 0 & 0 \\ 0 & 1 & 0 \\ 0 & 0 & 1 \end{bmatrix} x = \begin{bmatrix} 0 \\ 0 \\ 0 \\ 1 \end{bmatrix}$$

is a least squares sign-solvable system which is not sign-solvable. Clearly, if P and Q are permutation matrices and D and E are invertible diagonal matrices, then $Ax = b$ is least squares sign-solvable if and only if $(PDAQE)x = PDb$ is least squares sign-solvable. We now show that, as for sign-solvable systems (see Corollary 1.2.2) and conditionally sign-solvable systems (see Theorem 11.1.2), the transpose of the coefficient matrix of a least squares sign-solvable linear system is an L-matrix.

Lemma 11.2.1 *If $Ax = b$ is a least squares sign-solvable linear system, then A^T is an L-matrix.*

Proof Suppose that A^T is not an L-matrix. Then there exists a matrix $\widetilde{A} \in \mathcal{Q}(A)$ and a nonzero vector y such that $\widetilde{A}y = 0$. Let z be a least squares solution to $\widetilde{A}x = b$. Then since $\widetilde{A}(z+\lambda y) = \widetilde{A}z$, the vector $z+\lambda y$ is also a least squares solution to $\widetilde{A}x = b$ for all real numbers λ. For some choice of λ, the vectors z and $z + \lambda y$ belong to different qualitative classes, and therefore $Ax = b$ is not least squares sign-solvable.   □

The following corollary relates least squares sign-solvability and conditionally sign-solvability.

Corollary 11.2.2 *If $Ax = b$ is a least squares sign-solvable linear system, then either $Ax = b$ is conditionally sign-solvable or $\begin{bmatrix} A & -b \end{bmatrix}^T$ is an L-matrix.*

Proof Suppose that $Ax = b$ is least squares sign-solvable and that $\begin{bmatrix} A & -b \end{bmatrix}^T$ is not an L-matrix. Then there exists a matrix $\widehat{A} \in \mathcal{Q}(A)$ and a vector $\widehat{b} \in \mathcal{Q}(A)$ such that the columns of $\begin{bmatrix} \widehat{A} & -\widehat{b} \end{bmatrix}$ are linearly dependent. By Lemma 11.2.1, A^T is an L-matrix, and hence the columns of $\widehat{A}$ are linearly independent. It follows that $\widehat{A}x = \widehat{b}$ has a solution. Since

$$\{u : \text{there exists } \widetilde{A} \in \mathcal{Q}(A) \text{ and } \tilde{b} \in \mathcal{Q}(b) \text{ such that } \widetilde{A}u = \tilde{b}\}$$

is a subset of

$$\{u : \text{there exists } \widetilde{A} \in \mathcal{Q}(A) \text{ and } \tilde{b} \in \mathcal{Q}(b) \text{ where } u \text{ is a}$$
$$\text{least squares solution to } \widetilde{A}u = \tilde{b}\},$$

$Ax = b$ is conditionally sign-solvable.   □

If the columns of an m by n matrix A are linearly independent, then the zero vector is the least squares solution to $Ax = b$ if and only if each column of

A is orthogonal to b. The next corollary characterizes the least squares sign-solvable linear systems whose solution is the zero vector. Two vectors $u = (u_1, u_2, \ldots, u_n)^T$ and $v = (v_1, v_2, \ldots, v_n)^T$ are *combinatorially orthogonal* if $\tilde{u}^T \tilde{v} = 0$ for all $\tilde{u} \in \mathcal{Q}(u)$ and all $\tilde{v} \in \mathcal{Q}(v)$. Clearly, u and v are combinatorially orthogonal if and only if $u_i v_i = 0$ for $i = 1, 2, \ldots, n$.

Corollary 11.2.3 *The linear system $Ax = b$ is least squares sign-solvable and its solution is the zero vector if and only if A^T is an L-matrix and each column of A is combinatorially orthogonal to b.*

Proof Suppose that A^T is an L-matrix and that each column of A is combinatorially orthogonal to b. Let $\tilde{A} \in \mathcal{Q}(A)$ and $\tilde{b} \in \mathcal{Q}(b)$. Then the least squares solution to $\tilde{A}x = \tilde{b}$ is the solution to $\tilde{A}^T \tilde{A}x = \tilde{A}^T \tilde{b} = 0$. Since A^T is an L-matrix, $\tilde{A}^T \tilde{A}$ is nonsingular and thus the only solution to $\tilde{A}^T \tilde{A}x = \tilde{A}^T \tilde{b}$ is the zero vector. Hence $Ax = b$ is least squares sign-solvable and its solution is the zero vector.

Conversely, suppose that $Ax = b$ is least squares sign-solvable and its solution is the zero vector. By Lemma 11.2.1, A^T is an L-matrix. If $\tilde{A} \in \mathcal{Q}(A)$ and $\tilde{b} \in \mathcal{Q}(b)$, then the zero vector is the least squares solution to $\tilde{A}x = \tilde{b}$, and hence $\tilde{A}^T \tilde{b} = \tilde{A}^T \tilde{A}(0) = 0$. It follows that each column of A is combinatorially orthogonal to column b. $\qquad\square$

The next corollary can be used to combine two least squares sign-solvable linear systems into a larger least squares sign-solvable system; it follows from the definitions and Lemma 11.2.1.

Corollary 11.2.4 *Let A_1 and A_2 be m_1 by n_1 and m_2 by n_2 matrices, respectively, and let b_1 and b_2 be m_1 by 1 and m_2 by 1 column vectors, respectively. Then*

$$\begin{bmatrix} A_1 & O \\ O & A_2 \end{bmatrix} x = \begin{bmatrix} b_1 \\ b_2 \end{bmatrix} \tag{11.8}$$

is least squares sign-solvable if and only if both $A_1 u = b_1$ and $A_2 v = b_2$ are least squares sign-solvable.

The next two results, whose proofs are given in [15], provide more examples of least squares sign-solvable linear systems.

Theorem 11.2.5 *Let $A = [a_{ij}]$ be an m by $m-1$ $(0, 1)$-matrix which is the vertex-edge incidence matrix of a tree with $m \geq 2$ vertices. Let b be an m by 1 column vector with exactly one nonzero entry. Then the linear system $Ax = b$ is least squares sign-solvable.*

It is easy to verify that if A and b satisfy the hypothesis of Theorem 11.2.5, then $\begin{bmatrix} A & b \end{bmatrix}^T$ is an L-matrix. Hence it follows that the least squares sign-solvable linear systems described in Theorem 11.2.5 are not conditionally sign-solvable. Let B be an m by $m - 1$ matrix such that the matrix A obtained from B by replacing each nonzero entry by a 1 is the vertex-edge incidence matrix of a tree. Then it is easy to show that there exist invertible diagonal matrices D and E such that $DBE = A$. It now follows from Theorem 11.2.5 that if b has exactly one nonzero entry, then $Bx = b$ is least squares sign-solvable.

Corollary 11.2.6 *Let A be an m by $m - 1$ matrix which is the vertex-edge incidence matrix of a tree, and let b be a column vector with exactly one nonzero entry. Then for $j = 1, 2, \ldots, m - 1$, the linear system*

$$A(j \leftarrow b)x = A_j$$

is least squares sign-solvable, where A_j is the jth column of A.

We now study the structure of least squares sign-solvable linear systems. The main result of this section is a generalization of the relationship between sign-solvable linear systems (see Theorem 1.2.12) and S- and L-matrices, and the relationship between conditionally sign-solvable linear systems and conditionally S-matrices and L-matrices (see Theorem 11.1.5).

An m by n matrix A is *balanceable* provided there exists a strict signing D of order m such that each column of DA is balanced. If A is balanceable, then there exists a matrix $\widetilde{A} \in \mathcal{Q}(A)$ such that each row of $\widetilde{A}$ is a linear combination of the other rows of $\widetilde{A}$. If $Ax = b$ is a least squares sign-solvable linear system, then the ith entry of the system is *exact* if for each $\widetilde{A} \in \mathcal{Q}(A)$ and each $\tilde{b} \in \mathcal{Q}(b)$ the ith entry of $\tilde{b} - \widetilde{A}u$ is zero where u is the least squares solution to $\widetilde{A}x = \tilde{b}$. For example, it can be verified that only the third entry of the least squares sign-solvable linear system

$$\begin{bmatrix} 1 & 0 \\ 1 & 0 \\ 1 & 1 \end{bmatrix} x = \begin{bmatrix} 1 \\ 0 \\ 0 \end{bmatrix}$$

is exact. Suppose that $Ax = b$ is a least squares sign-solvable linear system whose ith entry is exact. Let u be a least squares solution to $Ax = b$. Then $Au - b$ has a 0 in its ith entry and is orthogonal to the columns of A. Thus $A[\overline{\{i\}}, :]u - b[\overline{\{i\}}]$ is orthogonal to the columns of $A[\overline{\{i\}}, :]$. This implies that u is a least squares solution to $A[\overline{\{i\}}, :]x = b[\overline{\{i\}}]$.

Lemma 11.2.7 *Let $Ax = b$ be a least squares sign-solvable linear system such that the least squares solution to $Ax = b$ has no zero entries. If each row of A is nonzero and A is balanceable, then no entry of the system $Ax = b$ is exact.*

Proof Assume that A each row of A is nonzero and that A is balanceable. Suppose to the contrary that the last entry of $Ax = b$ is exact. Let B be the matrix obtained from A by removing its last row and let w^T be the last row of A. Since A is balanceable, there exist a matrix $\widetilde{B} \in \mathcal{Q}(B)$ and a vector $\tilde{w} \in \mathcal{Q}(w)$ such that $\tilde{w}$ belongs to the row space of $\widetilde{B}$. Let u be the least squares solution to

$$\begin{bmatrix} \widetilde{B} \\ \tilde{w}^T \end{bmatrix} x = b. \tag{11.9}$$

Some entry, say the first, of $\tilde{w}^T$ is nonzero and without loss of generality is positive. Let v be the least squares solution to

$$\begin{bmatrix} \widetilde{B} \\ \tilde{w}^T + e_1^T \end{bmatrix} x = b \tag{11.10}$$

where $e_1^T = (1, 0, 0, 0, \ldots, 0)$. Since the last entry of $Ax = b$ is exact,

$$\tilde{w}^T u = (\tilde{w}^T + e_1^T)v, \tag{11.11}$$

and both u and v are least squares solutions to $\widetilde{B}x = b'$ where b' is the vector obtained from b by deleting its last row. Thus $\widetilde{B}(u - v) = 0$. Because $\tilde{w}^T$ belongs to the row space of $\widetilde{B}$, we conclude $\tilde{w}^T(u - v) = 0$. It follows from (11.11) that $e_1^T v = 0$ and hence that v has a zero entry, contrary to assumption. Therefore, each entry of $Ax = b$ is exact. $\qquad\square$

The next two theorems generalize Theorems 1.2.12 and 11.1.5.

Theorem 11.2.8 *Let*

$$\begin{bmatrix} A & O & O \\ B & C & D \\ O & O & E \end{bmatrix} \begin{bmatrix} x_1 \\ x_2 \\ x_3 \end{bmatrix} = \begin{bmatrix} b_1 \\ b_2 \\ b_3 \end{bmatrix} \tag{11.12}$$

be a linear system in block matrix form such that the entries of b_1, b_2, and b_3 are nonnegative. Assume that

(i) *each row of A is nonzero and A is balanceable;*
(ii) *$Ax_1 = b_1$ is least squares sign-solvable and its least squares solution has only positive entries;*
(iii) *each row of $\begin{bmatrix} B & -b_2 \end{bmatrix}$ is nonnegative or nonpositive;*

(iv) *the matrix* $\begin{bmatrix} C & b_2' \end{bmatrix}$ *is an S-matrix where b_2' is the row sum vector of*
$\begin{bmatrix} B & -b_2 \end{bmatrix}$;
(v) *either E^T is an L-matrix or E^T has no rows and columns; and*
(vi) *the columns of E are combinatorially orthogonal to b_3.*

Then (11.12) is a least squares sign-solvable linear system.

Proof Let M be the coefficient matrix of (11.12) and let

$$b = \begin{bmatrix} b_1 \\ b_2 \\ b_3 \end{bmatrix}.$$

Consider a linear system

$$\begin{bmatrix} \widetilde{A} & O & O \\ \widetilde{B} & \widetilde{C} & \widetilde{D} \\ O & O & \widetilde{E} \end{bmatrix} \begin{bmatrix} x_1 \\ x_2 \\ x_3 \end{bmatrix} = \begin{bmatrix} \tilde{b}_1 \\ \tilde{b}_2 \\ \tilde{b}_3 \end{bmatrix} \qquad (11.13)$$

where the vectors and matrices belong to the appropriate qualitative classes.
Let $\widetilde{M}$ and $\tilde{b}$ be the coefficient matrix and the vector on the right-hand side of
(11.13), respectively. By (ii) and Lemma 11.2.1, A^T is an L-matrix. It follows
from (iv) that C is an SNS-matrix, $Cx_2 = -b_2'$ is sign-solvable, and each entry
of the solution to $Cx_2 = -b_2'$ is positive. It now follows from (v) that M^T is
an L-matrix and hence that there is a unique least squares solution u to (11.13).
Let u_1 be the least squares solution to $\widetilde{A}x = \tilde{b}_1$. Then the distance between
the column space of $\widetilde{A}$ and $\tilde{b}_1$ equals $\|\widetilde{A}u_1 - \tilde{b}_1\|$. It follows from (v), (vi),
and Corollary 11.2.3 that the distance between the column space of $\widetilde{E}$ and $\tilde{b}_3$
equals $\|\tilde{b}_3\|$. Thus the distance between the column space of $\widetilde{M}$ and $\tilde{b}$ is at least
$\sqrt{\|\widetilde{A}u_1 + \tilde{b}_1\|^2 + \|\tilde{b}_3\|^2}$. By (iii) and (iv), $Cx_2 = \tilde{b}_2 - \widetilde{B}u_1$ is sign-solvable.
Let u_2 be the solution to $\widetilde{C}x_2 = \tilde{b}_2 - \widetilde{B}u_1$, and let

$$u = \begin{bmatrix} u_1 \\ u_2 \\ 0 \end{bmatrix}.$$

Then the distance between $\widetilde{M}u$ and $\tilde{b}$ equals $\sqrt{\|\widetilde{A}u_1 + \tilde{b}_1\|^2 + \|\tilde{b}_3\|^2}$. There-
fore u is the least squares solution to $\widetilde{M}x = \tilde{b}$. Since the sign patterns of u_1
and of u_2 are determined, it follows that (11.12) is a least squares sign-solvable
linear system. $\qquad \square$

We now prove a converse to Theorem 11.2.8. Note that if $Ax = b$ is least
squares sign-solvable with least squares solution u, then there exist strict signings D and E such that Db and Eu are nonnegative, and $(DAE)x = Db$ is

least squares sign-solvable. Hence there is no loss in generality in assuming that both b and u are nonnegative. The existence of the sets γ and δ in the statement of the next theorem is shown in Theorem 3.1.4.

Theorem 11.2.9 *Let $M = [m_{ij}]$ be an m by n matrix and let b be an m by n vector such that $Mx = b$ is a least squares sign-solvable linear system, b is nonnegative, and the least squares solution $u = (u_1, u_2, \ldots, u_n)^T$ to $Mx = b$ is nonnegative. Let*

$$\beta = \{j : u_j \neq 0\} \quad and \quad \alpha = \{i : m_{ij} \neq 0 \text{ for some } j \in \beta\}.$$

Then $M[\alpha, \beta]x = b[\alpha]$ is least squares sign-solvable and in particular $M[\alpha, \beta]^T$ is an L-matrix. Let γ and δ be the unique subsets of α and β, respectively, such that $M[\gamma, \beta \setminus \delta] = O$, $M[\alpha \setminus \gamma, \beta \setminus \delta]$ is an SNS-matrix and $M[\gamma, \delta]$ is balanceable. Then $Mx = b$ has the form (11.12) and satisfies (i)–(vi) of Theorem 11.2.8 where $A = M[\gamma, \delta]$, $B = M[\alpha \setminus \gamma, \delta]$, $C = M[\alpha \setminus \gamma, \beta \setminus \delta]$, $D = M[\alpha \setminus \gamma, \overline{\beta}]$, $E = M[\overline{\alpha}, \overline{\beta}]$, $b_1 = b[\gamma]$, $b_2 = b[\alpha \setminus \gamma]$, and $b_3 = b[\overline{\alpha}]$.

Proof Since $Mx = b$ is least squares sign-solvable and $u[\overline{\beta}] = 0$, if $\widetilde{N} \in \mathcal{Q}(M[\alpha, \beta])$, $\tilde{c} \in \mathcal{Q}(b[\alpha])$, and v is a least squares solution to $\widetilde{N}x = \tilde{c}$, then

$$\begin{bmatrix} v \\ 0 \end{bmatrix}$$ is the least squares solution to $\widetilde{M}x = \tilde{b}$ for each $\widetilde{M} \in \mathcal{Q}(M)$ and

$\tilde{b} \in \mathcal{Q}(b)$ such that $\widetilde{M}[\alpha, \beta] = \widetilde{N}$ and $\tilde{b}[\alpha] = \tilde{c}$. Thus $M[\alpha, \beta]x = b[\alpha]$ is a least squares sign-solvable linear system. By Lemma 11.2.1, $M[\alpha, \beta]^T$ is an L-matrix. Let A, B, C, D, E, b_1, b_2, and b_3 be defined as in the statement of the theorem.

Statement (i) holds by definition, and since $M[\alpha, \beta]^T$ is an L-matrix, C is an SNS-matrix. Thus, if $\widetilde{A} \in \mathcal{Q}(A)$, $\widetilde{B} \in \mathcal{Q}(B)$, $\widetilde{C} \in \mathcal{Q}(C)$, and $\tilde{b} \in \mathcal{Q}(b)$ and if v_1 is a least squares solution to

$$\begin{bmatrix} \widetilde{A} \\ \widetilde{B} \end{bmatrix} x = \begin{bmatrix} \tilde{b}_1 \\ \tilde{b}_2 \end{bmatrix}$$

and v_2 is the solution to $\widetilde{C}x = \tilde{b}_2 - \widetilde{B}v_1$, then

$$\begin{bmatrix} v_1 \\ v_2 \\ 0 \end{bmatrix}$$

is the least squares solution to any system $\widetilde{M}x - \tilde{b}$ such that $\widetilde{M} \in \mathcal{Q}(M)$ and

$$\widetilde{M}[\alpha, \beta] = \begin{bmatrix} \widetilde{A} & O \\ \widetilde{B} & \widetilde{C} \end{bmatrix}.$$

It follows that (ii) holds. In addition, each linear system $Cx = \bar{b}_2 - \widetilde{B}v_1$ is sign-solvable and each entry of its solution is positive. It follows that each of the matrices $[\; \widetilde{C} \;\; \bar{b}_2 - \widetilde{B}v_1 \;]$ is an S-matrix. By Theorem 4.3.1, up to multiplication of rows by -1, there is at most one S-matrix with a given zero-pattern. It now follows that (iii) and (iv) hold.

Suppose that $M[\gamma, \bar{\beta}] \neq 0$. Let $i \in \gamma$, $j \in \bar{\beta}$ be integers such that $m_{ij} \neq 0$. By Lemma 11.2.7 there exists a matrix $\widetilde{A} \in \mathcal{Q}(A)$ and a vector $\bar{b}_1 \in \mathcal{Q}(b_1)$ such that the ith entry of $\widetilde{A}v_1 - \bar{b}_1$ is nonzero, where v is the least squares solution to $\widetilde{A}x = \bar{b}_1$. Let $\widetilde{M}$ be the matrix in $\mathcal{Q}(M)$ such that $\widetilde{M}[\gamma, \delta] = \widetilde{A}$, the (i, j)-entry of $\widetilde{M}$ equals 1, and all other entries of $\widetilde{M}$ are 0, ϵ, or $-\epsilon$. Let $\bar{b}$ be a vector in $\mathcal{Q}(b)$ such that $\bar{b}[\gamma] = \bar{b}_1$. Let v_2 be the solution to $\widetilde{A}x = \bar{b}_2 - \widetilde{B}v_1$.

Then $\begin{bmatrix} v_1 \\ v_2 \\ 0 \end{bmatrix}$ is the least squares solution to $\widetilde{M}x = \bar{b}$. But for ϵ sufficiently small the jth column of $\widetilde{M}$ is not orthogonal to $\widetilde{M}v - \bar{b}$, a contradiction. Thus $M[\gamma, \bar{\beta}] = O$.

Since M^T is an L-matrix and C is an SNS-matrix, (v) holds. If $b_3 = 0$, then clearly (vi) holds. Assume that $b_3 \neq 0$. By Corollary 11.2.3, to show that (vi) holds, it suffices to show that $Ex = b_3$ is least squares sign-solvable and its solution is the zero vector. Since E^T is an L-matrix, each $\widetilde{E}x = \bar{b}_3$ $(\widetilde{E} \in \mathcal{Q}(E), \bar{b}_3 \in \mathcal{Q}(b_3))$ has exactly one least squares solution. Suppose that the least squares solution v'_3 to $\widetilde{E}x = \bar{b}_3$ is nonzero. Let v_1 be the least squares solution to $Ax = b_1$ and let v'_2 be the solution to $Cx = b_2 - Bv_1 - Dv'_3$. Then

$$\|\widetilde{M} \begin{bmatrix} v_1 \\ v'_2 \\ v'_3 \end{bmatrix} - \bar{b}\| < \|\widetilde{M} \begin{bmatrix} v_1 \\ v_2 \\ 0 \end{bmatrix} - \bar{b}\|$$

where v_2 is the solution to $Cx = b_2 - Bv_1$,

$$\widetilde{M} = \begin{bmatrix} A & O & O \\ B & C & D \\ O & O & \widetilde{E} \end{bmatrix} \quad \text{and} \quad \bar{b} = \begin{bmatrix} b_1 \\ b_2 \\ \bar{b}_3 \end{bmatrix}.$$

This contradicts the fact that $\begin{bmatrix} v_1 \\ v_2 \\ 0 \end{bmatrix}$ is the least squares solution to $\widetilde{M}x = \bar{b}$.

Thus $Ex = b_3$ is least squares sign-solvable and has the zero vector as its least squares solution. Therefore (vi) holds, and the proof is complete. $\square$

Let $Mx = b$ be a least squares sign-solvable linear system, and suppose the sets α and γ have been defined as in the statement of Theorem 11.2.9. Then by

Lemma 11.2.7, if $i \in \gamma$, then the ith entry of $Mx = b$ is not exact. It is easy to verify that if $i \in \alpha \setminus \gamma$, then the ith entry of $Mx = b$ is exact, and if $i \in \bar{\alpha}$, then the ith entry of $Mx = b$ is exact if and only if the ith entry of b is zero. Thus, if $Mx = b$ is sign-solvable, the matrix A has no rows and no columns.

Theorems 11.2.8 and 11.2.9 show that the study of least squares sign-solvable linear systems reduces to the study of L-matrices and to least squares sign-solvable linear systems for which no entry is exact and for which the least squares solution has no zero entries. This reduction is quite similar to those for sign-solvable linear systems and conditionally sign-solvable linear systems described in sections 1.1 and 11.1, respectively.

We conclude this section by discussing a generalization of S^2NS-matrices. Let M be an m by n matrix whose columns are linearly independent. The *generalized inverse* of M, denoted by $M^\dagger$, is the matrix $(M^T M)^{-1} M^T$. Let $A = [a_{ij}]$ be an m by n matrix such that A^T is an L-matrix, and let p and q be integers such that $1 \leq p \leq n$ and $1 \leq q \leq m$. Then the (p, q)-entry of $A^\dagger$ is *signed* provided the (p, q)-entries of the matrices in

$$\{\widetilde{A}^\dagger : \widetilde{A} \in \mathcal{Q}(A)\}$$

all have the same sign. For example let

$$A = \begin{bmatrix} 1 & 1 \\ 1 & 1 \\ 1 & 0 \end{bmatrix}.$$

Then A^T is an L-matrix and for $\widetilde{A} \in \mathcal{Q}(A)$ there exist positive numbers a, b, c, d, e such that

$$\widetilde{A} = \begin{bmatrix} a & b \\ c & d \\ e & 0 \end{bmatrix}.$$

Then

$$\widetilde{A}^\dagger = \frac{1}{\det \widetilde{A}^T \widetilde{A}} \begin{bmatrix} ad^2 - cbd & cb^2 - dab & e(b^2 + d^2) \\ -acd + bc^2 + be^2 & -cab + da^2 + de^2 & -e(ab + cd) \end{bmatrix}.$$

Thus it follows that the $(1, 3)$ and $(2, 3)$ entries of $A^\dagger$ are signed, and no other entry of $A^\dagger$ is signed. More generally, each entry in the jth column of $A^\dagger$ is signed if and only if $Ax = e_j$ is least squares sign-solvable where e_j is the $(0, 1)$-vector whose only nonzero entry is a 1 in its jth row. The following results, whose proofs are in [15], describe the structure of matrices whose generalized inverse is signed.

Theorem 11.2.10 *Let A be an m by n matrix which has a signed generalized inverse. Then each submatrix of A of order n either has an identically zero determinant or is an S^2NS-matrix.*

Clearly, an S^2NS-matrix of order m has a signed generalized inverse and each of its submatrices of order m is an S^2NS-matrix. Let A be an m by $m - 1$ matrix such that the matrix obtained from A by replacing its nonzero entries by 1's is the vertex-edge incidence matrix of a tree. By Corollary 11.2.6, each system $Ax = b$ where b is a vector that has exactly one nonzero entry is least squares sign-solvable. Hence A has a signed generalized inverse. Since no submatrix of A of order $m - 1$ has an identically zero determinant, Theorem 11.2.10 implies that each submatrix of A of order $m - 1$ is an S^2NS-matrix.

Theorem 11.2.11 *Let A be an m by n matrix with $n \geq 2$ such that A has a signed generalized inverse and no submatrix of A of order n has an identically zero determinant. Then either $m = n$ and A is an S^2NS-matrix, or $m = n + 1$ and the matrix obtained from A by replacing its nonzero entries by 1's is the vertex-edge incidence matrix of a tree.*

There do exist m by n matrices with signed generalized inverses some of whose submatrices of order m have identically zero determinant. For example, the matrix

$$A = \begin{bmatrix} 1 & 0 \\ 1 & 0 \\ 0 & 1 \\ 0 & 1 \end{bmatrix}$$

has a signed generalized inverse and $A[\{1, 2\}, \{1, 2\}]$ has an identically zero determinant. The following theorem shows that matrices with a signed generalized inverse have a very specific structure.

Theorem 11.2.12 *Let A be an m by n matrix with no zero rows. If A has a signed generalized inverse, then there exist permutation matrices P and Q, diagonal matrices D and E, and an integer k such that PAQ has the form*

$$\begin{bmatrix} A_1 & O & \cdots & O \\ A_{21} & A_2 & \cdots & O \\ \vdots & \vdots & \ddots & \vdots \\ A_{k1} & A_{k2} & \cdots & A_k \end{bmatrix} \tag{11.14}$$

where each A_i is either a matrix of one column each of whose entries is 1, an S^2NS-matrix, or the vertex-edge incidence matrix of a tree.

Let A be an m by n matrix of the form (11.14) such that each A_i is a matrix of one column and each entry is 1, an S^2NS-matrix, or the vertex-edge incidence matrix of a tree. Necessary and sufficient conditions on the matrices A_{ij} in order that A have a signed generalized inverse are not known at present and are a topic of continuing research.

11.3 Variation of rank over a qualitative class

Let A be an m by n $(0, 1, -1)$-matrix. In this section we consider the ranks of matrices in the qualitative class $Q(A)$. Recall that the term rank of A equals the largest order of a square submatrix of A which does not have an identically zero determinant. Clearly, the *maximum rank* r^A of a matrix in $Q(A)$ equals the term rank of A. Let B be a submatrix of A of order r^A whose term rank equals r^A. Since the determinant of B is a nonzero polynomial in the entries of B, the set of all nonsingular matrices in $Q(B)$ is a dense subset of $Q(B)$. It follows that the set of all matrices in $Q(A)$ with rank r^A is a dense subset of $Q(A)$. Now assume that A is a CS*-matrix. If A is not an S*-matrix, then the rank of a matrix in $Q(A)$ is $n-1$ or n, and there exist matrices in $Q(A)$ of each rank. The matrices of rank n in $Q(A)$ are dense in $Q(A)$. This implies that the relationship of S*-matrices to S-matrices is different from that of CS*-matrices to CS-matrices for the following reason. There exists a strict signing D such that AD is a CS-matrix. If A is an S*-matrix, then $D = \mathrm{diag}(\mathrm{sign}\ u_1, \ldots, \mathrm{sign}\ u_n)$ where $u = (u_1, \ldots, u_{n+1})^T$ is any nonzero solution of $Ax = 0$. If A is not an S*-matrix, then to determine D we first need to find a matrix $\tilde{A}$ in $Q(A)$ of rank $n-1$, thus a matrix of $Q(A)$ in the complement of a dense subset, and then a nonzero solution u of $\tilde{A}x = 0$.

If A is an L-matrix with m rows, then the rank of each matrix in $Q(A)$ equals m. More generally, we have the following [9].

Theorem 11.3.1 *Let A be an m by n $(0, 1, -1)$-matrix. Then the rank of each matrix in $Q(A)$ equals r if and only if there exist nonnegative integers e and f with $e + f = r$ and permutation matrices P and Q such that PAQ has the form*

$$\begin{bmatrix} X & O \\ Z & Y \end{bmatrix} \tag{11.15}$$

where Z is an e by f matrix, and X^T and Y are L-matrices.

Proof First assume that the rank of each matrix in $Q(A)$ equals r. Then r is the term rank of A, and by König's theorem (see section 5.3) we may assume that A has the form (11.15) where X is an $m - e$ by f matrix with term rank f, Y is an e by $n - f$ matrix with term rank e, and $r = e + f$. Since Y is e by $n - f$, each matrix in $\mathcal{Q}(Y)$ has at most e linearly independent columns, and since each matrix in $\mathcal{Q}(A)$ has exactly r linearly independent columns, each matrix in $\mathcal{Q}(Y)$ has exactly e linearly independent columns. It follows that the rows of each matrix in $\mathcal{Q}(Y)$ are linearly independent and hence that Y is an L-matrix. In a similar way we conclude that X^T is an L-matrix.

Now assume that A has the form (11.15) where Z is an e by f matrix with $e + f = r$, and X^T and Y are L-matrices. Every submatrix of A of order $r + 1$ has a p by q zero submatrix for some integers p and q with $p + q = r + 2$ and hence has an identically zero determinant. Thus the rank of each matrix in $\mathcal{Q}(A)$ is at most r. Since every matrix in $\mathcal{Q}(X)$ contains a nonsingular submatrix of order f and every matrix in $\mathcal{Q}(Y)$ contains a nonsingular submatrix of order e, it follows that every matrix in $\mathcal{Q}(A)$ contains a nonsingular submatrix of order $e + f = r$.   □

It follows from the comment at the end of section 2.1 and an argument similar to that in the proof of Theorem 11.3.1, that each matrix with the same zero pattern as a matrix A has rank r if and only if r is the term rank of A and there exists an invertible triangular matrix B of order r such that PBQ is a submatrix of A for some permutation matrices P and Q [9].

Let A be an m by n $(0, 1, -1)$-matrix with $m \leq n$. Since r^A equals the term rank of A, the maximum rank depends only on the zero pattern of A. The *minimum rank* r_A of matrices in $\mathcal{Q}(A)$ is more difficult to compute and in general is not determined solely by the zero pattern of A. It is easy to verify that $r_A = 1$ if and only if there exist nonzero $(0, 1, -1)$-vectors x and y such that $A = xy^T$. Theorem 11.3.1 characterizes those sign patterns for which the rank of each matrix in $\mathcal{Q}(A)$ equals r_A. Although there always exists an m by n $(0, 1, -1)$-matrix A such that $r_A = m$, there need not exist a $(1, -1)$-matrix A with this property. For example, since every SNS-matrix of order 3 has a zero entry, there does not exist a 3 by 3 $(1, -1)$-matrix A such that $r_A = 3$. Let

$$r_{m,n} = \max\{r_A : A \text{ an } m \text{ by } n \ (1, -1)\text{-matrix}\}$$

be the smallest integer t such that for every m by n $(1, -1)$-matrix A there exists a matrix in $\mathcal{Q}(A)$ whose rank is at most t. If A contains a submatrix equal to the k by 2^{k-1} L-matrix Λ_k defined in section 2.1, then $r_A \geq k$. It follows that

$$r_{m,n} \geq \min\{m, \log_2 n + 1\}. \tag{11.16}$$

It seems likely that equality holds in (11.16).[3]

---

[3] In the case that $m = n$, this question has been raised by C.R. Johnson.

A matrix is *proper* provided no column is a multiple of another. Let m and k be positive integers with $k \le m$. Let $d(m, k)$ be the smallest integer n such that $r_A \ge k$ for every m by n proper $(1, -1)$-matrix. Let $d'(m, k)$ be the smallest integer n such that every m by n proper $(1, -1)$-matrix A contains a submatrix equal to $P \Lambda_k Q D$ for some permutation matrices P and Q and strict signing D. Clearly, $d(m, k) \le d'(m, k)$. Theorem 6 of [4] implies that $d(m, k)$ equals

$$1 + \sum_{i=1}^{k-1} \binom{m-1}{i-1}. \tag{11.17}$$

Adapting the technique of the proof of Theorem 1.2 in [1], one can show that $d'(m, k)$ also equals (11.17). Thus $d(m, k) = d'(m, k)$. In [4] an m by $\sum_{i=1}^{k-1} \binom{m-1}{i-1}$ proper $(1, -1)$-matrix B with $r_B \le k - 1$ is constructed. This construction and the fact that

$$d'(m, k) = 1 + \sum_{i=1}^{k-1} \binom{m-1}{i-1}$$

gives an alternative proof that

$$d(m, k) = 1 + \sum_{i=1}^{k-1} \binom{m-1}{i-1}.$$

Let $\widetilde{A}$ and $\widehat{A}$ be matrices in $\mathcal{Q}(A)$ with ranks r_A and r^A, respectively. Since changing a single entry of a matrix can change the rank by at most 1, and since $\widehat{A}$ can be obtained from $\widetilde{A}$ by changing one entry at a time, for each integer p with $r_A \le p \le r^A$, there exists a matrix in $\mathcal{Q}(A)$ with rank p.[4]

11.4 SkewSNS-matrices and pfaffians

Let $B = [b_{ij}]$ be a skew-symmetric matrix of order n. Let E_n denote the set of permutations of $\{1, 2, \ldots, n\}$ each of whose cycles has even length. The skew-symmetry of B implies that

$$\det B = \sum_{\sigma \in E_n} \text{sgn}(\sigma) b_{1i_1} b_{2i_2} \cdots b_{ni_n} \tag{11.18}$$

where the summation extends over all permutations $\sigma = (i_1, i_2, \ldots, i_n)$ in E_n. If n is odd, then $E_n = \emptyset$, and hence $\det B = 0$. We call (11.18) the *skew-symmetric determinant expansion* of the skew-symmetric matrix B.

---

[4] As observed by C.R. Johnson.

Let

$$B = \begin{bmatrix} 0 & 1 & -1 & 0 & 0 & 0 \\ -1 & 0 & -1 & 1 & 0 & 0 \\ 1 & 1 & 0 & 0 & -1 & 0 \\ 0 & -1 & 0 & 0 & -1 & -1 \\ 0 & 0 & 1 & 1 & 0 & 1 \\ 0 & 0 & 0 & 1 & -1 & 0 \end{bmatrix}.$$

(11.19)

There are four nonzero terms in the skew-symmetric determinant expansion (11.18), and each of these terms is positive. Therefore $\det \widetilde{B} > 0$ for every skew-symmetric matrix $\widetilde{B}$ in $\mathcal{Q}(B)$. Hence every skew-symmetric matrix in $\mathcal{Q}(B)$ is nonsingular, although not every matrix in $\mathcal{Q}(B)$ is nonsingular since B is not an SNS-matrix. A skew-symmetric matrix B is a *skewSNS-matrix* provided every skew-symmetric matrix in $\mathcal{Q}(B)$ is nonsingular. In particular, a skewSNS-matrix has even order. Every SNS-matrix which is skew-symmetric is a skewSNS-matrix, but as the example shows, not every skewSNS-matrix is a skew-symmetric, SNS-matrix. If A is a square matrix, then A is an SNS-matrix if and only if

$$\begin{bmatrix} O & A \\ -A^T & O \end{bmatrix}$$

is a skew-symmetric, SNS-matrix.

Let $B = [b_{ij}]$ be a skew-symmetric matrix of even order n. If there is a nonzero term in the skew-symmetric determinant expansion (11.18) and all nonzero terms are positive, then clearly B is a skewSNS-matrix. We use the pfaffian function in order to prove the converse also holds. Let F_n be the set of all permutations $\sigma = (i_1, i_2, \ldots, i_n)$ of $\{1, 2, \ldots, n\}$ such that

$$i_1 < i_2, i_3 < i_4, \ldots, i_{n-1} < i_n$$

and

$$i_1 < i_3 < \cdots < i_{n-1}.$$

The *pfaffian* of the skew-symmetric matrix $B = [b_{ij}]$ of order n is defined by

$$\text{pf } B = \sum_{\sigma \in F_n} \text{sgn}(\sigma) b_{i_1 i_2} b_{i_3 i_4} \cdots b_{i_{n-1} i_n}.$$

(11.20)

There is a one-to-one correspondence between the nonzero terms in the pfaffian (11.20) and the perfect matchings of the graph of the symmetric matrix obtained from B by replacing each entry with its absolute value. It is a classical result that

$$\det B = (\text{pf } B)^2,$$

(11.21)

that is,

$$\det B = \sum_{\sigma \in E_n} \text{sgn}(\sigma) b_{1i_1} b_{2i_2} \cdots b_{ni_n} = (\text{pf } B)^2$$

$$= \left(\sum_{\sigma \in F_n} \text{sgn}(\sigma) b_{i_1 i_2} b_{i_3 i_4} \cdots b_{i_{n-1} i_n} \right)^2$$

is a polynomial identity on the set of all skew-symmetric matrices B of order n.

Theorem 11.4.1 *Let $B = [b_{ij}]$ be a skew-symmetric matrix of even order n. Then the following are equivalent:*

  (i) *B is a skewSNS-matrix.*
 (ii) *There is a nonzero term in the pfaffian expansion (11.20) and every nonzero term has the same sign.*
(iii) *There is a nonzero term in the skew-symmetric determinant expansion (11.18) and every nonzero term is positive.*

Proof Clearly, (iii) implies (i), and by (11.21) (ii) implies (iii). For each nonzero term in the pfaffian expansion (11.20) there is a skew-symmetric matrix in $\mathcal{Q}(B)$ whose pfaffian has the same sign. Since the pfaffian is a continuous function and the set of skew-symmetric matrices in $\mathcal{Q}(B)$ is a connected set, it follows that (i) implies (ii). $\qquad\square$

It follows from Theorem 11.4.1 that if B is a $(0, 1, -1)$ skewSNS-matrix, then the number of perfect matchings of the graph of the symmetric matrix obtained from B by replacing each -1 with 1 is

$$|\text{pf } B| = \sqrt{\det B}.$$

This fact accounts for the combinatorial interest in skewSNS-matrices.

The following theorem characterizes the zero patterns of skewSNS-matrices [13].[5]

Theorem 11.4.2 *Let A be a symmetric $(0, 1)$-matrix of even order n such that each main diagonal element of A equals zero and the graph $\mathcal{G}(A)$ has a perfect matching. Then A is not the zero pattern of a skewSNS-matrix if and only if*

---

[5]The graphs of skewSNS-matrices are called *pfaffian graphs* in the literature. Theorem 11.4.2 and Corollary 11.4.3 are formulations in terms of skewSNS-matrices of theorems originally formulated in terms of pfaffian graphs.

there is a spanning subgraph of the $\mathcal{G}(A)$ which is the vertex disjoint union of a subdivision of $K_{3,3}$ and a matching.

The characterization of the zero patterns of SNS-matrices given in Corollary 6.5.6 is a special case of Theorem 11.4.2. The proof of Theorem 11.4.2 given in [13] contains a nonbacktracking algorithm which either finds a skewSNS-matrix B with zero pattern A or concludes that no such B exists. This algorithm applied to matrices of the form

$$\begin{bmatrix} O & C \\ C^T & O \end{bmatrix}$$

reduces to the algorithm given in section 6.1 for the construction of an SNS-matrix with a presribed zero pattern C.

The following corollary is contained in [10] (see also [14]).

Corollary 11.4.3 *Let $A \neq O$ be a symmetric $(0, 1)$-matrix of even order n each of whose main diagonal elements equals zero. Assume that $\mathcal{G}(A)$ is planar and that each edge is contained in a perfect matching. Then A is the zero pattern of a skewSNS-matrix.*

Proof By Kuratowski's theorem [8], $\mathcal{G}(A)$ does not contain a subdivision of $K_{3,3}$, and the corollary follows from Theorem 11.4.2. If $\mathcal{G}(A)$ is bipartite, this corollary is Theorem 6.4.3. The proof of Theorem 6.4.3 can be modified to give a different proof of Corollary 11.4.3 as follows. The matrix $B = [b_{ij}]$ is defined so that $b_{ij} = 1$ or -1 according to whether the arc joining i and j is oriented from i to j or from j to i. The fact that every nonzero term in the pfaffian of B has the same sign is established in a manner similar to that used in the proof of Theorem 6.4.3 to establish that every nonzero term in the determinant of B has the same sign. $\qquad\square$

Let B be a skew-symmetric $(0, 1, -1)$-matrix of order n such that $\mathcal{G}(B)$ is connected. Let $D = \text{diag}(d_1, \ldots, d_n)$ and $E = \text{diag}(e_1, \ldots, e_n)$ be strict signings. If $E = \pm D$, then DBE is skew-symmetric. Let $\alpha = \{i : d_i = e_i\}$ and let $\beta = \{1, 2, \ldots, n\} \setminus \alpha$. If DBE is skew-symmetric, then $B[\alpha, \beta] = O$ and $B[\beta, \alpha] = O$, and the connectivity of $\mathcal{G}(B)$ implies that $E = \pm D$. Now assume that B is a skewSNS-matrix. Then for each strict signing D, $\pm DBD$ is a skewSNS-matrix with the same zero pattern as B. However, as the following example shows, there may be other skewSNS-matrices with the same zero pattern as B.[6] Let

$$
B = \begin{bmatrix}
0 & -1 & 1 & 0 & -1 & 1 \\
1 & 0 & -1 & 1 & 0 & 0 \\
-1 & 1 & 0 & 1 & 0 & 0 \\
0 & -1 & -1 & 0 & 1 & 1 \\
1 & 0 & 0 & -1 & 0 & 1 \\
-1 & 0 & 0 & -1 & -1 & 0
\end{bmatrix}.
$$

Then B is a skewSNS-matrix and so is the matrix B' obtained by negating the entries in the submatrix $B[\{1, 2, 3\}]$ of B. But there does not exist a strict signing D such that $B' = \pm DBD$.

There has been some interest in determining the maximum number of nonzero entries in a skewSNS-matrix of even order n. In particular, it has been conjectured that this maximum is $(n^2+2n)/2$ [6, 7]. By Theorem 8.1.1, the maximum number of nonzero entries in an SNS-matrix of order n is $(n^2 + 3n - 2)/2$.

There exist a skew-symmetric matrix A of order n and a column vector b such that $Ax = b$ is not sign-solvable but the solutions of the linear systems $\widetilde{A}x = \widetilde{b}$, where $\widetilde{A}$ is a skew-symmetric matrix in $\mathcal{Q}(a)$ and $\widetilde{b}$ is in $\mathcal{Q}(B)$, belong to a single qualitative class. It is possible to characterize such a linear system $Ax = b$ in the spirit of Theorem 1.2.12.

Finally, we remark that there isn't an analogous theory for symmetric matrices. This is because it is not difficult to show that if A is a symmetric matrix, then every symmetric matrix in $\mathcal{Q}(A)$ is nonsingular if and only if A is an SNS-matrix.

11.5 SNS-matrix pairs

Let B be a matrix of order n. Then B is an SNS-matrix if and only if the Hadamard product $B * X$ is a nonsingular matrix for all positive matrices X of order n. If B is a skew-symmetric matrix, then B is a skewSNS-matrix if and only if $B * X$ is a nonsingular matrix for all symmetric positive matrices X.

Let C be another matrix of order n. Then (B, C) is an *SNS-matrix pair* provided $B * X + C$ is a nonsingular matrix for all positive matrices X of order n. Thus (B, C) is an SNS-matrix pair if and only if $\widetilde{B} + C$ is a nonsingular matrix for each matrix $\widetilde{B}$ in $\mathcal{Q}(B)$. Note that (O, C) is an SNS-matrix pair if and only if C is a nonsingular matrix and that (B, O) is an SNS-matrix pair if and only if B is an SNS-matrix. Clearly, the property of (B, C) being an SNS-matrix pair does not depend on the magnitude of the entries of B, but in

[6]This observation should be contrasted with Theorem 6.1.3.

general it does depend on the magnitude of the entries of C. The pair (B, C) where

$$B = \begin{bmatrix} 1 & 0 \\ 0 & 0 \end{bmatrix} \quad \text{and} \quad C = \begin{bmatrix} 3 & 6 \\ 2 & 4 \end{bmatrix}$$

is an SNS-matrix pair. For each matrix C there exists a matrix B such that (B, C) is an SNS-matrix pair. In fact, if C is nonsingular, then we may take B to be a zero matrix. Otherwise, we may assume that $C[\{1, 2, \ldots, k\}]$ is a nonsingular matrix where k is the rank of C and then take $B = O \oplus I_{n-k}$. Note also that if D and E are invertible diagonal matrices, then (B, C) is an SNS-matrix pair if and only if (DBE, DCE) is an SNS-matrix pair. If D and E have positive main diagonals, then (B, C) is an SNS-matrix pair if and only if (B, DCE) is an SNS-matrix pair.

The following lemma is a generalization of Theorem 1.2.5 [2].

Lemma 11.5.1 *Let B and C be matrices of order n. Then (B, C) is an SNS-matrix pair if and only if*

$$\det(B * X + C), \tag{11.22}$$

considered as a polynomial in the entries of X, is not identically zero and all of its nonzero coefficients have the same sign.

Proof Assume that (B, C) is an SNS-matrix pair. Then (11.22) is not identically zero. For each nonzero coefficient α of (11.22) there exists a positive matrix Y such that the sign of $\det(B * Y + C)$ equals the sign of α. Since (11.22) is a continuous function and the positive matrices form a connected set, all nonzero coefficients of (11.22) have the same sign. The converse is immediate. □

The following theorem generalizes inequality (8.1) of Theorem 8.1.1 and is contained in [2].

Theorem 11.5.2 *Let B and C be matrices of order n such that (B, C) is an SNS-matrix pair. Let k be the degree of the polynomial (11.22). Then the maximum number of nonzero entries of S equals*

$$\frac{n^2 - n + 4k - 2}{2} \quad \text{if } k \neq 0$$
$$\frac{n^2 - n}{2} \quad \text{if } k = 0.$$

The zero patterns of the matrices B for which equality holds in Theorem 11.5.2 are characterized in [2].

We now discuss SNS-matrix pairs (B, C) for certain classes of matrices C.

Theorem 11.5.3 *Let $B = [b_{ij}]$ and $C = [c_{ij}]$ be matrices of order n such that the bipartite graph of C is a forest. Then (B, C) is an SNS-matrix pair if and only if $\widetilde{B} + \widetilde{C}$ is a nonsingular matrix for all $\widetilde{B}$ in $\mathcal{Q}(B)$ and all $\widetilde{C}$ in $\mathcal{Q}(C)$.*

Proof Let $\widetilde{B}$ and $\widetilde{C}$ be any matrices in $\mathcal{Q}(B)$ and $\mathcal{Q}(C)$, respectively. Since the bipartite graph of C is a forest, an easy induction shows that there exist diagonal matrices D and E with positive main diagonals such that $\widetilde{C} = DCE$. Then

$$\widetilde{B} + \widetilde{C} = D(D^{-1}\widetilde{B}E^{-1} + C)E.$$

Since $D^{-1}\widetilde{B}E^{-1}$ is in $\mathcal{Q}(B)$, it follows that if (B, C) is an SNS-matrix pair, then $\widetilde{B} + \widetilde{C}$ is nonsingular. The converse clearly holds. $\quad\square$

The set of matrices of the form $\widetilde{B} + \widetilde{C}$ in Theorem 11.5.3 is a set $\mathcal{R}$ consisting of all matrices in which each element is prescribed to be either positive, negative, or zero, or is completely arbitrary. If $b_{ij}c_{ij} \geq 0$ for all i and j, then $\mathcal{R} = \mathcal{Q}(B + C)$.

Corollary 11.5.4 *Let $B = [b_{ij}]$ and $C = [c_{ij}]$ be matrices of order n such that B is fully indecomposable. If (B, C) is an SNS-matrix pair, then $b_{ij}c_{ij} \geq 0$ for all i and j.*

Proof The full indecomposability assumption implies that for each i and j there exists a matrix $\widetilde{B}$ in $\mathcal{Q}(B)$ such that $\det(\widetilde{B} + C)(\{i\}, \{j\}) \neq 0$. If $b_{ij}c_{ij} < 0$ for some i and j, then there exist positive matrices X_1 and X_2 such that $\det(B * X_1 + C)$ and $\det(B * X_2 + C)$ are nonzero and of opposite sign. The corollary now follows by continuity. $\quad\square$

Corollary 11.5.5 *Let B and C be matrices of order n such that C is a nonsingular diagonal matrix. Then (B, C) is an SNS-matrix pair if and only if the diagonal vectors of B and C are conformal and $B + C$ is an SNS-matrix.*

Proof Assume that (B, C) is an SNS-matrix pair. By Theorem 11.5.3, $\widetilde{B} + \widetilde{C}$ is nonsingular for all $\widetilde{B}$ in $\mathcal{Q}(B)$ and all $\widetilde{C}$ in $\mathcal{Q}(C)$. If the diagonal vectors of B and C are not conformal, then it is easy to see that there are matrices of the form $\widetilde{B} + \widetilde{C}$ with nonzero determinants of opposite sign and hence, by continuity, matrices of this form with zero determinant. Thus the diagonal vectors of B and C are conformal, and hence the set of all matrices of the form $\widetilde{B} + \widetilde{C}$ equals

$Q(B + C)$. Therefore $B + C$ is an SNS-matrix. The converse clearly holds.

□

11.6 Nonsingular strength of a matrix

Let A be a matrix of order n. Let D and E be diagonal matrices with positive diagonal vectors u and v, respectively. Then

$$DAE = A * (uv^T)$$

is in the qualitative class $Q(A)$ and is nonsingular if and only if A is. Let u' and v' also be positive vectors. Then

$$A * (uv^T) + A * (u'v'^T) = A * R \tag{11.23}$$

is in $Q(A)$ where $R = uv^T + u'v'^T$ is a positive matrix of rank at most 2. It may happen that A is nonsingular and (11.23) is singular. For example, let

$$A = \begin{bmatrix} 1 & 5 \\ 1 & 6 \end{bmatrix} \tag{11.24}$$

and let

$$uv^T = \begin{bmatrix} 1 & 1 \\ 2 & 2 \end{bmatrix} \quad \text{and} \quad u'v'^T = \begin{bmatrix} 1 & 3 \\ 1 & 3 \end{bmatrix}.$$

Then A is nonsingular but $A * R$ is singular. We define the *nonsingular strength* of a square nonsingular matrix A to be the largest integer $s(A)$ such that $A * R$ is nonsingular for all positive matrices R of rank at most $s(A)$. It follows from these remarks that the nonsingular strength of a nonsingular matrix is at least one. The nonsingular strength of a singular matrix is defined to be zero. A matrix A is an SNS-matrix if and only if $s(A) = n$. If A is the matrix in (11.24), then $s(A) = 1$. The results in this section are reformulations in terms of nonsingular strength of results contained in [5].

Lemma 11.6.1 *Let m, n, and k be positive integers with $1 \leq k \leq \min\{m, n\}$. Then the set $R_{m,n}^k$ of all positive m by n matrices of rank at most k is connected.*

Proof Let $A = [a_{ij}]$ be a matrix in $R_{m,n}^k$. We prove the lemma by showing that there exists a continuous path in $R_{m,n}^k$ joining A and the matrix $J_{m,n}$. Without loss of generality we assume that the row space of A is spanned by the first k rows. Every other row is a linear combination of the first k rows where at least one of the coefficients is positive. We continuously change the negative coefficients in these linear combinations to zero to obtain a positive matrix A' each of whose last $m - k$ rows is a nonnegative linear combination of its first k rows. We now continuously change the entries of A' in its first k rows to one while at the same time changing the entries in the last $m - k$ rows so as not to disturb these nonnegative linear combinations and thereby obtain a positive matrix A'' each of whose rows is of the form $(c, c, \ldots, c)$. Continuously changing the c's to 1 completes the proof. ☐

The following corollary is an immediate consequence of Lemma 11.6.1 and the continuity of the determinant.

Corollary 11.6.2 *Let A be a nonsingular matrix of order n and let k be a positive integer. Then $s(A) \geq k$ if and only if all of the matrices $A * R$ with R in $R_{m,n}^k$ have determinants of the same sign.*

The main result of this section is that if the strength of a matrix of order n is sufficiently large, then the strength equals n.

Lemma 11.6.3 *Let A be a nonsingular matrix of order n such that $s(A) \geq k$. Let $A[\alpha, \beta]$ be a square submatrix of order $p < k$ such that $A(\alpha, \beta)$ is nonsingular. Then $A[\alpha, \beta]$ either is an SNS-matrix or has an identically zero determinant.*

Proof Without loss of generality we assume that $\alpha = \beta = \{1, 2, \ldots, p\}$. Suppose that $A[\alpha, \beta]$ does not have an identically zero determinant. Let Y be any positive matrix of order p, and for each positive number ϵ let

$$R_\epsilon = \begin{bmatrix} Y & \epsilon J_{p,n-p} \\ J_{n-p,p} & J_{n-p,n-p} \end{bmatrix}.$$

Then R_ϵ is in $R_{n,n}^k$ and, for ϵ sufficiently small, $\det A * R_\epsilon$ and $(\det A[\alpha, \beta] * Y)$ $\det A(\alpha, \beta)$ have the same sign. Using Corollary 11.6.2, we see that $\det A[\alpha, \beta] * Y$ has the same sign for all positive matrices Y, and hence $A[\alpha, \beta]$ is an SNS-matrix. ☐

Lemma 11.6.4 *Let $n \geq 4$ and $k \geq 2$ be integers with $n \geq 2k$, and let G be a graph of order n. If G contains at least*

$$(k - 2)(2n - 2k + 1) + n$$

edges, then G has a matching of k edges.

Proof The lemma is easily verified for $k = 2$. An edge of G is incident to at most $2n - 4$ other edges. We now argue by induction on the graph obtained from G by deleting the two vertices of an edge and complete the proof. $\square$

Theorem 11.6.5 *Let k be a positive integer. Then for n sufficiently large, a matrix A of order n such that $s(A) \geq n - k$ is an SNS-matrix.*

Proof It suffices to prove the theorem for $k \geq 2$. Let $A = [a_{ij}]$ be a nonsingular matrix of order n such that $s(A) \geq n - k$, but A is not an SNS-matrix. Without loss of generality we may assume that A has a positive main diagonal. Since A is not an SNS-matrix, we may furthermore assume that $\det A < 0$. Let G be the graph of order n with vertices $1, 2, \ldots, n$ in which there is an edge joining vertices i and j if and only if $i \neq j$ and at least one of a_{ij} and a_{ji} is zero. Suppose that G has a matching of size k. The edges of this matching correspond to zeros of A, and without loss of generality we assume that $a_{12} = a_{34} = \cdots = a_{2k-1,2k} = 0$. For $\epsilon > 0$ let $R_\epsilon = [r_{ij}]$ be the matrix in $R_{n,n}^{n-k}$ such that $r_{ij} = 1$ if $i = j$ or if either (i, j) or (j, i) is one of $(1, 2), (3, 4), \ldots, (2k - 1, 2k)$, and $r_{ij} = \epsilon$ otherwise. Since A has a positive main diagonal, $\det A * R_\epsilon > 0$ for ϵ sufficiently small. Since $s(A) \geq n - k$ and $\det A < 0$, we contradict Corollary 11.6.2. Thus G does not have a matching of size k. By Lemma 11.6.4, either $n < 2k$ or A has fewer than $2((k - 2)(2n - 2k + 1) + n)$ zeros. Assume that $n \geq 2k$. Then A has fewer than $2((k - 2)(2n - k + 1) + n)$ zeros. Since A is nonsingular, there exists a nonsingular submatrix $A[\alpha, \beta]$ of A of order $n - k - 1$ such that $A(\alpha, \beta)$ is nonsingular. By Lemma 11.6.3, both $A[\alpha, \beta]$ and $A(\alpha, \beta)$ are SNS-matrices. Applying Theorem 8.1.1 to $A[\alpha, \beta]$, we conclude that A has at least

$$\binom{n - k - 1}{2}$$

zeros. Hence

$$\binom{n - k - 1}{2} < 2((k - 2)(2n - 2k + 1) + n),$$

implying that

$$n < \frac{10k - 9 + \sqrt{64k^2 - 120k + 51}}{2}. \tag{11.25}$$

If $n < 2k$, then n and k satisfy (11.25). Therefore, if

$$n \geq \frac{10k - 9 + \sqrt{64k^2 - 120k + 51}}{2},$$

every matrix of order n whose nonsingular strength is at least $n - k$ is an SNS-matrix. □

Bibliography

[1] R.P. Anstee. A forbidden configuration theorem of Alon, *J. Combin. Theory, Ser. A*, 47:16–27, 1988.

[2] R.A. Brualdi and K.L. Chavey. Sign-nonsingular matrix pairs, *SIAM J. Matrix Anal. Appl.*, 13:36–40, 1992.

[3] R.A. Brualdi, K.L. Chavey, and B.L. Shader. Conditional sign-solvability, *Math. Comp. Model.*, 17:141–8, 1993.

[4] P. Delsarte and Y. Kamp. Low rank matrices with a given sign pattern, *SIAM J. Disc. Math.*, 2:51–63, 1989.

[5] J. Drew, C.R. Johnson, and P. van den Driessche. Strong forms of nonsingularity, *Linear Alg. Appls.*, 162–4:187–204, 1992.

[6] P.M. Gibson. The pfaffian and 1-factors of graphs, *Trans. New York Acad. Sci.*, 34:52–7, 1972.

[7] P.M. Gibson. The pfaffian and 1-factors of graphs II, in *Graph Theory and Applications: Lecture Notes in Mathematics* vol. 303 (Y. Alavi, D.R. Lick, and A.T. White, eds.) Springer-Verlag, New York, 89–98, 1972.

[8] F. Harary. *Graph Theory*, Addison-Wesley, Reading, 1969.

[9] D. Hershkowitz and H. Schneider. Ranks of zero patterns and sign patterns, *Linear Multilin. Alg.*, 34:3–21, 1993.

[10] P.W. Kasteleyn. Dimer statistics and phase transitions, *J. Math. Phys.*, 287–93, 1963.

[11] V. Klee, B. Von Hohenbalken, and T. Lewis. On the recognition of S-systems. *Linear Alg. Appls.*, 192:187–204, 1993.

[12] V. Klee, B. Von Hohenbalken, and T. Lewis. S-systems, L-systems and an extension of sign-solvability, in preparation.

[13] C.H.C. Little. A characterization of convertible $(0, 1)$-matrices, *J. Combin. Theory, Ser. B*, 18:187–208, 1975.

[14] E.W. Montroll. Lattice statistics, in *Applied Combinatorial Mathematics* (E.F. Beckenbach, ed.), Wiley, New York, 96–143, 1964.

[15] B.L. Shader. Least squares sign-solvability, *SIAM J. Matrix Anal. Appl.*, to appear.

Master bibliography

R. Aharoni, R. Manber, and B. Wajnryb. Special parity of perfect matchings in bipartite graphs, *Discrete Math.*, 79:221–8, 1990.

M. Allingham and M. Morishima. Qualitative economics and comparative statics, in *Theory of Demand—Real and Monetary* (M. Morishima et al., eds.), Clarendon Press, Oxford, 3–69, 1973.

N. Alon, R.A. Brualdi, and B.L. Shader. Multicolored trees in bipartite decompositions, *J. Combin. Theory, Ser. B*, 53:143–8, 1991.

N. Alon and N. Lineal. Cycles of length 0 modulo k in directed graphs, *J. Combin. Theory, Ser. B*, 47:114–19, 1989.

N. Alon and J.H. Spencer. *The Probabilistic Method*, Wiley, New York, 1992.

T. Ando and R.A. Brualdi. Sign-central matrices, *Linear Alg. Appls.*, 208/209:283–95, 1994.

R.P. Anstee. A forbidden configuration theorem of Alon, *J. Combin. Theory, Ser. A*, 47:16–27, 1988.

L. Bassett, J. Maybee, and J. Quirk. Qualitative economics and the scope of the correspondence principle, *Econometrica*, 36:544–63, 1968.

M.A. Berger and A. Felzenbaum. Sign patterns of matrices and their inverses, *Linear Alg. Appls.*, 86:161–77, 1987.

A. Berman and B.D. Saunders. Matrices with zero line sums and maximal rank, *Linear Alg. Appls.*, 40:229–35, 1981.

J.C. Bermond and C. Thomassen. Cycles in digraphs–a survey, *J. Graph Theory*, 5:1–43, 1981.

T. Bone, C. Jeffries, and V. Klee. A qualitative analysis of $\dot{x} = Ax + b$, *Discrete Appl. Math.*, 20:9–30, 1988.

P. Botta. On the conversion of the determinant into the permanent, *Canad. Math. Bull.*, 11:31–4, 1968.

F. Brouwer, J. Maybee, and P. Nijkamp. Sign-solvability in economic models through plausible restrictions, *Atlantic Econ. Journal*, 17:21–6, 1988.

R.A. Brualdi. Counting permutations with restricted positions: Permanents of (0, 1)–matrices. A tale in four parts., in The 1987 Utah State University Department of Mathematics Conference Report by L. Beasley and E.E. Underwood, *Linear Alg. Appls.*, 104:173–83, 1988.

R.A. Brualdi. The many facets of combinatorial matrix theory, in *Matrix Theory and Applications Proceeding Symposia Applied Mathematics*, vol. 40 (C.R. Johnson, ed.), Amer. Math. Soc., Providence, 1–35, 1990.

R.A. Brualdi. Graphs and matrices and graphs, *Congressus Numerantium*, 83:129–47, 1991.

R.A. Brualdi. The symbiotic relationship of combinatorics and matrix theory, *Linear Alg. Appls.*, 162–4:65–105, 1992.

R.A. Brualdi and K.L. Chavey. Sign-nonsingular matrix pairs, *SIAM J. Matrix Anal. Appl.*, 13:36–40, 1992.

290   *Master bibliography*

R.A. Brualdi, K.L. Chavey, and B.L. Shader. Conditional sign-solvability, *Math. Comp. Model.*, 17:141–8, 1993.

R.A. Brualdi, K.L. Chavey, and B.L. Shader. Rectangular *L*-matrices, *Linear Alg. Appls.*, 196:37–61, 1994.

R. A. Brualdi, K. L. Chavey, and B. L. Shader. Bipartite graphs and inverse sign patterns of strong sign-nonsingular matrices, *J. Combin. Theory, Ser. B*, 62:133–52, 1994.

R.A. Brualdi and T. Foregger. Matrices with constant permanental minors, *Linear Multilin. Alg.*, 3:227–43, 1975.

R.A. Brualdi and H.J. Ryser. *Combinatorial Matrix Theory*, Cambridge University Press, New York, 1991.

R.A. Brualdi and B.L. Shader. Matrix factorizations of determinants and permanents, *J. Combin. Theory, Ser. A*, 54:132–4, 1990.

R.A. Brualdi and B.L. Shader. On sign-nonsingular matrices and the conversion of the permanent into the determinant, in *Applied Geometry and Discrete Mathematics* (P. Gritzmann and B. Sturmfels, eds.), Amer. Math. Soc., Providence, 117–34, 1991.

R.A. Brualdi and B.L. Shader. Cutsets in bipartite graphs, *Linear Multilin. Alg.*, 34:51–4, 1993.

D. Carlson. Nonsingularity criteria for matrices involving combinatorial considerations, *Linear Alg. Appls.*, 107:41–56, 1988.

L. Ching. The maximum determinant of an $n \times n$ lower Hessenberg (0, 1)-matrix, *Linear Alg. Appls.*, 183:147–53, 1993.

F.R.K. Chung, W. Goddard, and D.J. Kleitman. Even cycles in directed graphs, *SIAM J. Disc. Math.*, 7:474–83, 1994.

G.V. Davydov and I.M. Davydov. Solubility of the system $Ax = 0$, $x \geq 0$ with indefinite coefficients, *Soviet Math. (IZ.VUZ.)*, 34-9:108–12, 1990.

P. Delsarte and Y. Kamp. Low rank matrices with a given sign pattern, *SIAM J. Disc. Math.*, 2:51–63, 1989.

J.H. Drew, B.C.J. Green, C.R. Johnson, D.D. Olesky, and P. van den Driessche. Bipartite characterization of sign-nonsingularity, preprint.

J. Drew, C.R. Johnson, and P. van den Driessche. Strong forms of nonsingularity, *Linear Alg. Appls.*, 162–4:187–204, 1992.

G. Engel and H. Schneider. Cyclic and diagonal products on a matrix, *Linear Alg. Appls.*, 7:301–35, 1973.

C. Eschenbach, F. Hall, and C.R. Johnson. Self-inverse sign patterns, in *Combinatorial and Graph-Theoretical Problems in Linear Algebra* (R. A. Brualdi, S. Friedland, and V. Klee, eds.), Springer-Verlag, 245–57, 1993.

C.A. Eschenbach and C. R. Johnson. Sign patterns that require real, nonreal and purely imaginary eigenvalues, *Linear Multilin. Alg.*, 29:299–311, 1991.

M. Fiedler. Combinatorial properties of sign-patterns in some classes of matrices, in *Lecture Notes in Mathematics*, vol. 1018 (M. Borowiecki, J.W. Kennedy, and M.M. Syslo, eds.), Springer-Verlag, Berlin, 28–32, 1983.

M. Fiedler and R. Grone. Characterizations of sign patterns of inverse-positive matrices, *Linear Alg. Appls.*, 40:237–45, 1981.

M.E. Fisher and A.T. Fuller. On the stabilization of matrices and the convergence of linear iterative processes, *Proc. Camb. Philos. Soc.*, 54:417–25, 1958.

P. Fontaine, M. Garbely, and M. Gilli. Qualitative solvability in economic models, *Computer Science in Economics and Management*, 4:285–301, 1991.

S. Friedland. Every 7-regular digraph contains an even cycle, *J. Combin. Theory, Ser. B*, 46:249–52, 1989.

M.R. Garey and D.S. Johnson. *Computers and Intractability, A guide to the theory of NP-completeness*, W.H. Freeman and Company, San Francisco, 1979.

P.M. Gibson. Combinatorial matrix functions and 1-factors of graphs, *SIAM J. Applied Math.*, 19:330–3, 1970.

P.M. Gibson. Conversion of the permanent into the determinant, *Proc. Amer. Math. Soc.*, 27:471–6, 1971.

P.M. Gibson. The pfaffian and 1-factors of graphs, *Trans. New York Acad. Sci.*, 34:52–7, 1972.

P.M. Gibson. The pfaffian and 1-factors of graphs II, in *Graph Theory and Applications: Lecture Notes in Mathematics*, vol. 303 (Y. Alavi, D.R. Lick, and A.T. White, eds.), Springer-Verlag, New York, 89–98, 1972.

W.M. Gorman. A wider scope for qualitative economics, *Rev. Econ. Studies*, 31:65–8, 1964.

R.L. Graham and H.O. Pollak. On addressing the problem for loop switching, *Bell System Tech. J.*, 50:2495–519, 1971.

P. Hansen. Recognizing sign-solvable graphs, *Discrete Appl. Math.*, 6:237–41, 1983.

F. Harary. *Graph Theory*, Addison-Wesley, Reading, 1969.

F. Harary, J.R. Lundgren, and J.S. Maybee. On signed digraphs with all cycles negative, *Discrete Appl. Math.*, 12:155–64, 1985.

D. Hershkowitz and H. Schneider. Ranks of zero patterns and sign patterns, *Linear Multilin. Alg.*, 34:3–21, 1993.

C. Jeffries. Qualitative stability and digraphs in model ecosystems, *Ecology*, 55:1415–19, 1974.

C. Jeffries. *Mathematical Modelling in Ecology: A Workbook for Students*, Birkhäuser, Boston, 1989.

C. Jeffries, V. Klee, and P. van den Driessche. When is a matrix sign stable? *Canad. J. Math.*, 29:315–26, 1977.

C. Jeffries, V. Klee, and P. van den Driessche. Qualitative stability of linear systems, *Linear Alg. Appls.*, 87:1–48, 1987.

C.R. Johnson. Sign patterns of inverse nonnegative matrices, *Linear Alg. Appls.*, 55:69–80, 1983.

C.R. Johnson, F.T. Leighton, and H.A. Robinson. Sign patterns of inverse-positive matrices, *Linear Alg. Appls.*, 24:75–83, 1979.

C.R. Johnson and J.S. Maybee. *Qualitative analysis of Schur complements*, in *Applied Geometry and Discrete Mathematics* (P. Gritzmann and B. Sturmfels, eds.), Amer. Math. Soc., Providence, 359–65, 1991.

C.R. Johnson, D.D. Olesky, and P. van den Driessche. Sign determinancy in LU factorization of *P*-matrices, preprint.

P.W. Kasteleyn. The statistics of dimers on a lattice, *Physica*, 27:1209–25, 1961.

P.W. Kasteleyn. Dimer statistics and phase transitions, *J. Math. Phys.*, 287–93, 1963.

P. W. Kasteleyn. Graph theory and crystal physics, in *Graph Theory and Theoretical Physics* (Frank Harary, ed.), Academic Press, New York, 44–110, 1967.

V. Klee. Recursive structure of *S*-matrices and an $O(m^2)$ algorithm for recognizing strong sign solvability, *Linear Alg. Appls.*, 96:233–47, 1987.

V. Klee. Sign-patterns and stability, in *Applications of Combinatorics and Graph Theory to the Biological and Social Sciences* (F. Roberts, ed.), IMA Volumes in Mathematics and Its Applications, Springer, New York, 17:203–19, 1989.

V. Klee, B. Von Hohenbalken, and T. Lewis. On the recognition of *S*-systems, *Linear Alg. Appls.*, 192:187–204, 1993.

V. Klee, B. Von Hohenbalken, and T. Lewis. *S*-systems, *L*-systems and an extension of sign-solvability, in preparation.

V. Klee and R. Ladner. Qualitative matrices: Strong sign-solvability and weak satisfiability, in *Computer-assisted Analysis and Model Simplification* (H. Greenberg and J. Maybee, eds.), Academic Press, New York, 293–320, 1981.

V. Klee, R. Ladner, and R. Manber. Signsolvability revisited, *Linear Alg. Appls.*, 59:131–57, 1984.

V. Klee and P. van den Driessche. Linear algorithms for testing the sign stability of a matrix and for finding *Z*-maximum matchings in acyclic graphs, *Numer. Math.*, 28:273–85, 1977.

K. Koh. Even circuits in directed graphs and Lovász' conjecture, *Bull. Malayasian Math. Soc.*, 7:47–52, 1976.

A. Kräuter and N. Seifter. On convertible (0, 1)-matrices, *Linear Multilin. Alg.*, 13:311–22, 1983.

G.M. Lady. The structure of qualitatively determinate relationships, *Econometrica*, 51:197–218, 1983.

G.M. Lady, T.J. Lundy, and J. Maybee. Nearly sign-nonsingular matrices, preprint.

G.M. Lady and J. Maybee. Qualitatively invertible matrices, *Math. Social Sciences*, 6:397–407, 1983.

K. Lancaster. The scope of qualitative economics, *Rev. Econ. Studies*, 29:99–132, 1962.

K. Lancaster. The theory of qualitative linear systems, *Econometrica*, 33:395–408, 1964.

K. Lancaster. Partitionable systems and qualitative economics, *Rev. Econ. Studies*, 31:69–72, 1964.

K. Lancaster. The solution of qualitative comparative static problems, *Quart. J. Econ.*, 8:279–95, 1966.

C.C. Lim. Nonsingular sign patterns and the orthogonal group, *Linear Alg. Appls.*, 184:1–12, 1993.

C.C. Lim and D.A. Schmidt. Full sign-invertibility and symplectic matrices, *Linear Alg. Appls.*, to appear.

C.H.C. Little. An extension of Kåsteleyn's method of enumerating the 1-factors of planar graphs, in *Combinatorial Mathematics, Proceedings 2nd Australian Conference: Lecture Notes in Mathematics*, vol. 403 (D. Holton, ed.), Springer-Verlag, Berlin, 63–72, 1974.

C.H.C. Little. A characterization of convertible (0, 1)-matrices, *J. Combin. Theory, Ser. B*, 18:187–208, 1975.

C.H.C. Little. On the number of parity sets in a graph, *Canad. J. Math*, 28:1167–71, 1976.

C.H.C. Little. Another characterisation of pfaffian bipartite graphs, *J. Austral. Math. Soc. Ser. A*, 39:132–42, 1985.

C.H.C. Little and F. Rendl. Operations preserving the pfaffian property of a graph, *J. Austral. Math. Soc. Ser. A*, 50:248–57, 1991.

C.H.C. Little and J.M. Pla. Sur l'utilisation d'un pfaffien dans l'étude des couplages parfaits d'un graphe, *C. R. Acad. Sci.*, 274:447, 1972.

D. Logofet. *Matrices and Graphs: Stability Problems in Mathematical Ecology*, CRC Press, Boca Raton, 1992.

L. Lovász. Problem 2 in *Advances in Graph Theory, Proc. Symp. Prague* (M. Fiedler, ed.), Academia Praha, Prague, 1974.

L. Lovász. On determinants, matchings and random algorithms, in *Proceedings of Fundamentals of Computation Theory*, Akademie Verlag, Berlin, 567–74, 1979.

L. Lovász and M.D. Plummer. *Matching Theory*, Elsevier, Amsterdam, 1986.

R. Lundgren and J.S. Maybee. A class of maximal L-matrices, *Congressus Numerantium*, 44:239–50, 1984.

T.J. Lundy and J.S. Maybee. Inverses of sign nonsingular matrices, preprint.

T.J. Lundy and J.S. Maybee. Zero submatrices and digraph connectivity, preprint.

T. Lundy, J. Maybee, and J. van Buskirk. On maximal sign-nonsingular matrices, preprint.

R. Manber. Graph-theoretical approach to qualitative solvability of linear systems, *Linear Alg. Appls.*, 48:457–70, 1982.

R. Manber and J. Shao. On digraphs with the odd-cycle property, *J. Graph Theory*, 10:155–65, 1986.

M. Marcus and H. Minc. On the relation between the determinant and the permanent, *Illinois J. Math.*, 5:376–81, 1961.

R.M. May. Qualitative stability in model ecosystems, *Ecology*, 54:638–44, 1973.

R.M. May. *Stability and Complexity in Model Ecosystems*, Princeton Univ. Press, Princeton, N.J., 1973.

J.S. Maybee. Combinatorially symmetric matrices, *Linear Alg. Appls.*, 8:529–37, 1974.

J. S. Maybee. Some possible new directions for combinatorial matrix analysis, *Linear Alg. Appls.*, 107:23–40, 1988.

S. Maybee. A method for identifying sign-solvable systems, Master's thesis, Department of Computer Science, University of Colorado, Boulder, 1982.

J.S. Maybee and L. Quirk. Qualitative problems in matrix theory, *SIAM Review*, 11:30–51, 1969.

J.S. Maybee and G.M. Wiener. From qualitative matrices to quantitative restrictions, *Linear Multilin. Alg.*, 22:229–48, 1988.

W. McCuaig. Intercyclic digraphs, *Graph Structure Theory, Contemporary Mathematics* 147, Amer. Math. Soc., Providence, 203–45, 1993.

E.C. Milnor. A combinatorial theorem on systems of sets, *J. London Math. Soc.*, 43:204–206, 1968.

H. Minc. *Nonnegative matrices,* Wiley, New York, 1988.

E.W. Montroll. Lattice statistics, in *Applied Combinatorial Mathematics* (E.F. Beckenback, ed.), Wiley, New York, 96–143, 1964.

T. Muir. A relation between permanents and determinants, *Proc. Royal Society,* 22:134–6, 1897.

J.M. Pla. Sur l'utilisation d'un pfaffien dans l'étude des couplages parfait d'un graphe, *C. R. Acad. Sci. Paris,* 260:2967–70, 1965.

G. Pólya. Aufgabe 424, *Arch. Math. Phys.,* 20:271, 1913.

J. Quirk. Qualitative stability of matrices and economic theory: A survey article, in *Computer–assisted Analysis and Model Simplification* (H. Greenberg and J. Maybee, eds.), Academic Press, New York, 113–64, 1981.

J. Quirk and R. Ruppert. Qualitative economics and the stability of equilibrium, *Rev. Economic Studies,* 32:311–26, 1965.

R. Rado. A theorem on independence relations, *Quart. J. Math. Oxford,* 13:83–9, 1942.

T. Rader. Impossibility of qualitative economics: Excessively strong correspondence principles in production-exchange economies, *Zeitschrift für Nationalökonomie,* 32:397–416, 1972.

R. Redheffer and Z. Zhou. Sign semistability and global asymptotic stability, *Ann. Diff. Eq.,* 5:145–53, 1989.

S. Reich. Another solution of an old problem of Pólya, *Amer. Math. Monthly,* 78:649–50, 1971.

F.S. Roberts. *Discrete Mathematical Modelling: With Applications to Social, Biological, and Environmental Problems,* Prentice Hall, New Jersey, 1976.

H.J. Ryser. Indeterminates and incidence matrices, *Linear Multilin. Alg.,* 1:149–57, 1973.

P.A. Samuelson. *Foundations of Economic Analysis,* Harvard University Press, Cambridge, 1947, Atheneum, New York, 1971.

P.D. Seymour. On the two-colouring of hypergraphs, *Quart. J. Math. Oxford,* 25:303–12, 1974.

P. Seymour and C. Thomassen. Characterization of even directed graphs, *J. Combin. Theory, Ser. B,* 42:36–45, 1987.

B.L. Shader. Maximal convertible matrices, *Congressus Numerantium,* 81:161–72, 1991.

B.L. Shader. Convertible, nearly decomposable and nearly reducible matrices, *Linear Alg. Appls.,* 184:37–53, 1993.

B.L. Shader. Least squares sign-solvability, *SIAM J. Matrix Anal. Appl.,* to appear.

R. Sinkhorn and P. Knopp. Problems involving diagonal products in nonnegative matrices, *Trans. Amer. Math. Soc.,* 136:67–75, 1969.

J.L. Stuart. Determinants of Hessenberg *L*-matrices, *SIAM J. Matrix Anal. Appl.,* 12:7–15, 1991.

G. Szegö. Lösung zu Aufgabe 424, *Arch. Math. Phys.,* 21:291, 1913.

C. Thomassen. Even cycles in directed graphs, *European J. Combin.,* 6:85–90, 1985.

C. Thomassen. The 2-linkage problem for acyclic digraphs, *Discrete Math.,* 55:73–87, 1985.

C. Thomassen. Sign-nonsingular matrices and even cycles in directed graphs, *Linear Alg. Appls.,* 75:27–41, 1986.

C. Thomassen. On digraphs with no two disjoint cycles, *Combinatorics,* 7:145–50, 1987.

C. Thomassen. When the sign pattern of a square matrix determines uniquely the sign pattern of its inverse, *Linear Alg. Appls.,* 119:27–34, 1989.

C. Thomassen. Disjoint cycles in digraphs, *Combinatorics* 3:393–6, 1989.

C. Thomassen. The even cycle problem for directed graphs, *J. Amer. Math. Soc.* 5:217–30, 1992.

C. Thomassen. The even cycle problem for planar directed graphs, *J. Algorithms,* 15:61–75, 1993.

M.F. Tinsley. Permanents of cyclic matrices, *Pacific J. Math.,* 10:1067–82, 1960.

L.G. Valiant. The complexity of computing the permanent, *Theoretical Computer Science,* 8:189–201, 1979.

V. Vazirani and M. Yannakakis. Pfaffian orientations, 0–1 permanents and even cycles in directed graphs, *Discrete Appl. Math.,* 25:179–90, 1989.

Index

$\mathcal{A}_i$, 24–27, 32–33, 218–219, 222
acyclic orientation, 51
adjacency matrix, 50
algorithms for
  constructing an SNS-matrix with a
    prescribed zero pattern, 112
  constructing an S^2NS-matrix with a
    prescribed zero pattern, 179, 189
  recognizing combinatorially symmetric
    SNS-matrices, 162
  recognizing fully indecomposable
    S^2NS-matrices, 177, 182
  recognizing partly decomposable
    S^2NS-matrices, 186
  recognizing sign-stable matrices, 247
  recognizing S-matrices, 70
  recognizing S^*-matrices, 71
  recognizing SNS*-matrices, 70
  recognizing the inverse sign-pattern of an
    S^2NS-matrix, 201–202
antichain, 219
  intersecting, 221
anticonformal vectors, 71
autonomous system, 242

balanceable matrix, 268
balanced
  vector, 18
  signing, 18
barely L-matrix, 18–34, 218–223
  characterizations of, 19, 25–28, 33
  examples of, 20, 218–220
$\mathcal{B}_n$ digraph, 225–238
biclique, 82
  decomposition, 83
bipartite digraph of a matrix, 231–233

bipartite graph
  even subdivision of, 79
  of a matrix, 79, 153, 158, 178
  subdivision of, 79
block, 158
  end-block, 164–165
blocker, 100–102
bridge, 158

central matrix, 99
clockwise orientation, 142
clutter, 100–102
colored row multigraph, 82–85
combinatorially orthogonal vectors, 267
combinatorially symmetric SNS-matrix,
  157–165
  maximal, 163
community matrix, 242
condenser, 26–27
conditional
  sign-solvability, 259–264
  S-matrices, 260–261, 264, 275
  S^*-matrices, 260–261, 264, 275
  S^2NS-matrices, 262–263
conformal
  contraction, 66–70, 171
  copying of entries, 174–175, 198–199
  matrices, 200
  splitting, 72–73
  vectors, 66
contraction
  conformal, 66–70, 171
  elementary, 75, 147
  on an entry, 147
  of columns, 75
  principal, 147
  pseudo-elementary, 154